The Hyperuniverse Project and Maximality

 Birkhäuser

Carolin Antos • Sy-David Friedman •
Radek Honzik • Claudio Ternullo

Editors

The Hyperuniverse Project and Maximality

Editors

Carolin Antos
Zukunftskolleg
University of Konstanz
Konstanz, Germany

Sy-David Friedman
Kurt Gödel Research Center
University of Vienna
Vienna, Austria

Radek Honzik
Charles University
Department of Logic
Prague, Czech Republic

Claudio Ternullo
Kurt Gödel Research Center
University Wien
Wien, Austria

ISBN 978-3-319-87432-6   ISBN 978-3-319-62935-3   (eBook)
https://doi.org/10.1007/978-3-319-62935-3

This book is published under the trade name Birkhäuser, www.birkhauser-science.com
The registered company is Springer International Publishing AG
The registered company address is: Gewerbestrasse 11, 6330 Cham, Switzerland

Preface

Set theory provides an excellent foundation for the field of mathematics; however, it suffers from Gödel's incompleteness phenomenon: There are important statements, such as the continuum hypothesis, that remain undecidable using the standard axioms. It is therefore of great value to find well-justified approaches to the discovery of new axioms of set theory.

The Hyperuniverse Project, funded by the John Templeton Foundation (JTF) from January 2013 until September 2015, was the first concerning Friedman's Hyperuniverse Programme, a valuable such approach based on the intrinsic maximality features of the set-theoretic universe. In the course of this project, the participants Carolin Antos, Radek Honzik, Claudio Ternullo and Friedman discovered an optimal form of "height maximality" and generated numerous "width maximality" principles which are currently under intensive mathematical investigation. The project also featured prominently in the important Symposia on the Foundations of Mathematics held in Vienna (7–8 July 2014, 21–23 September 2015) and London (12–13 January 2015); see https://sotfom.wordpress.com/.

The project resulted in 12 chapters, collected in this volume, which together provide the necessary background to gain an understanding of maximality in set theory and related topics.

Konstanz, Germany   Carolin Antos
Vienna, Austria   Sy-David Friedman
Prague, Czech Republic   Radek Honzik
Wien, Austria   Claudio Ternullo

Introduction: On the Development of the Hyperuniverse Project

In brief, the Hyperuniverse Programme (HP) aims to generate mathematical principles expressing the maximality of the set-theoretic universe in height and in width, to analyse and synthesise these principles and ultimately to arrive at an optimal maximal principle whose first-order consequences can be regarded as intrinsically justified axioms of set theory.

The primary goal of the Templeton-funded project was to provide a robust and convincing philosophical justification for the Hyperuniverse Programme, which mainly consisted in providing intrinsic evidence for the acceptance of the maximality principles taken into consideration by the programme. And a secondary goal was to systematically formulate mathematical criteria of maximality for the set-theoretic universe and to develop the necessary mathematical tools for analysing them.

We achieved our primary goal, that of providing the HP with a firm foundation, and made significant progress with our secondary goal, the mathematical unfolding of the programme. However, it is now clear that the mathematical challenges for the advancement of the programme are even greater than we had imagined, although we are pleased with the very significant progress that we have made.

The Philosophical Grounding of the HP

At the start of the programme, we in fact considered a number of different features of the set-theoretic universe that might be regarded as "intrinsic".

However, we concluded that in fact the only feature for which there is a definitive case for intrinsicness is the maximality feature of V (= the universe of sets).

Maximality naturally breaks into two forms, height maximality and width maximality. Our initial approach was to treat them analogously, from both a height-potentialist and width-potentialist perspective. However, thanks to the input of several leading scholars in the philosophy of set theory, we later came to realise that the programme is most appropriately (although not exclusively) formulated as a height-potentialist and width-actualist programme.

Height potentialism was further analysed and developed in the Friedman-Honzik theory of sharp generation ("On Strong Reflection Principles in Set Theory", in this volume), what we feel to be the ultimate, strongest formulation of height maximality. However, width maximality presented a serious challenge, as, formally speaking, width actualism does not allow for the existence of thickenings (widenings) of V, blocking the easy formulation of width-maximality principles in which V is compared to wider universes. The resolution of this dilemma constituted a major new discovery of the project: the use of V-logic to internally express, consistently with width actualism, width-maximality principles which refer to possible thickenings of V. A further important point was to realise that the principles expressed in V-logic, although not first-order, are nevertheless first-order over a mild lengthening (heightening) of V called V+ (the least admissible universe past V) and of course such lengthenings are entirely permissible from a height-potentialist perspective. The reason that this point is important is that it then allows the use of the downward Löwenheim-Skolem theorem to reduce the study of width-maximality principles for V to their study over countable transitive models of set theory, quantifying solely over the collection of all countable transitive such models. The latter collection is what is termed the "Hyperuniverse", hence the name of the programme.

In this way, we feel that the HP is well-justified philosophically and its conceptual framework is sound. But of course there remains further work to be done from a philosophical perspective: How is one to justify the "synthesis" of initially conflicting maximality principles? How does one support the claim that the generation and analysis of further maximality principles will ultimately converge upon a single "optimal" maximality criterion? How can the programme be developed from a height-actualist perspective?

The Mathematical Development of the HP

As already mentioned, height maximality is nicely captured using the notion of sharp generation, which has a clean and convincing mathematical formulation. However, the most natural form of width maximality, the inner model hypothesis (IMH), is in conflict with sharp generation. Honzik and I succeeded in "synthesising" the two, arriving at a consistent combined maximality principle IMH-sharp.

However, we did not reach our goal of establishing the consistency of SIMH, the strong IMH. This will be a major achievement, as it will yield a well-motivated form of width maximality that resolves Cantor's continuum problem. Ideally, we aim to then further synthesise the SIMH with sharp generation, arriving at a consistent principle SIMH-sharp, which not only resolves the continuum problem but is also compatible with height maximality (and with large cardinal axioms).

A useful way of organising maximality principles is via the maximality protocol. According to this, maximality is developed by first maximising the ordinals (via sharp generation), then maximising the cardinals through the so-called CardMax principles and finally maximising width via the cardinal-preserving IMH with absolute parameters. This is a satisfying, systematic approach to maximality. However, we have not yet succeeded in finding the mathematical tools needed to establish the consistency of the principles generated in this way. That remains for the further development of the HP.

Two other appealing forms of maximality regard width indiscernibility and omniscience. The former is an analogue for width of sharp generation for height. It postulates that V occurs at stage Ord in a sequence of length Ord+Ord of increasing universes which form a chain under elementary embeddings and which are indiscernible in an appropriate sense. The consistency of this has not yet been established, yet this form of maximality is especially appealing as it helps to restore a symmetry between the notions of maximality in height and in width. Omniscience asserts that the satisfiability of sentences with parameters from V in outer models of V is V-definable. Here we have made definite progress: Honzik and I showed ("Definability of Satisfaction in Outer Models", in this volume) that one can obtain the consistency of the omniscience principle (together with a definable well-order of the universe) from just an inaccessible cardinal. What remains is to verify that it can be successfully synthesised with other forms of maximality, such as the IMH-sharp.

To summarise: The main success of the JTF-funded Hyperuniverse Project was to establish a conceptually sound approach to the discovery of new set-theoretic axioms based on the intrinsic maximality features of V. In addition, significant progress was made on the mathematical formulation of maximality principles, on their synthesis and on establishing their consistency. Thanks to this project, the HP is now well-positioned to make important discoveries regarding set-theoretic truth based on intrinsic evidence and through the use of as yet undiscovered mathematical techniques.

The Chapters in Brief

The 12 chapters of this volume document some of the major advances of the JTF Hyperuniverse Project.

A key technique in the mathematical development of the project is the method of class-forcing. Chapter "Class Forcing in Class Theory" provides the proper setting for class-forcing, which had formerly been done by reducing to versions of ZFC. A further technique is hyperclass-forcing, the foundations for which is provided in Chap. "Hyperclass Forcing in Morse-Kelley Class Theory". Chapter "Multiverse Conceptions in Set Theory" provides a broad analysis of multiverse conceptions in set theory, taking into account different views regarding actualism and potentialism in height and in width. Chapter "Evidence for Set-Theoretic Truth and the Hyperuniverse Programme" is currently the most up-to-date full presentation of the Hyperuniverse Programme. Chapter "On the Set-Generic Multiverse" provides a modern treatment of Bukovsky's characterisation of set-generic extensions, an important feature of the set-generic multiverse. Chapters "On Strong Forms of

Reflection in Set Theory" and "Definability of Satisfaction in Outer Models" are the already-mentioned chapters on height maximality and omniscience. Chapters "The Search for New Axioms in the Hyperuniverse Programme" and "Explaining Maximality Through the Hyperuniverse Programme" take a deeper look at how the HP analyses maximality in set theory. Finally, Chaps. "Large Cardinals and the Continuum Hypothesis", "Gödel's Cantorianism", and "Remarks on Buzaglo's Concept Expansion and Cantor's Transfinite" provide insights into related topics, such as the role of large cardinals, the Cantorian features of Gödel's philosophy of sets and Buzaglo's treatment of concept expansion.

Contents

Class Forcing in Class Theory

Carolin Antos

Abstract In this article we show that Morse-Kelley class theory (MK) provides us with an adequate framework for class forcing. We give a rigorous definition of class forcing in a model $(M, \mathcal{C})$ of MK, the main result being that the Definability Lemma (and the Truth Lemma) can be proven without restricting the notion of forcing. Furthermore we show under which conditions the axioms are preserved. We conclude by proving that Laver's Theorem does not hold for class forcings.

1 Introduction

The idea of considering a forcing notion with a (proper) class of conditions instead of with a set of conditions was introduced by W. Easton in 1970. He needed the forcing notion to be a class to prove the theorem that the continuum function 2^κ, for κ regular, can behave in any reasonable way and as changes in the size of 2^κ are bounded by the size of a set forcing notion, the forcing has to be a class. Two problems arise when considering a class sized forcing: the forcing relation might not be definable in the ground model and the extension might not preserve the axioms. This was addressed in a general way in S. Friedman's book (see [3]) where he presented class forcings which are definable (with parameters) over a model $\langle M, A \rangle$. This is called a model of ZF if M is a model of ZF and Replacement holds in M for formulas which mention A as a predicate. We will call such forcings *A-definable class forcings*, their generics G *A-definable class-generics* and the resulting new model *A-definable class-generic outer model*. Friedman showed that for such A-definable class forcing which satisfy an additional condition called tameness the Definability Lemma, the Truth Lemma and the preservation of the Axioms of ZFC hold.

Originally published in C. Antos, Class forcing in class theory (submitted). arXiv:1503.00116.

C. Antos (✉)
Zukunftskolleg/Department of Philosophy, University of Konstanz, Konstanz, Germany
e-mail: carolin.antos-kuby@uni-konstanz.de

© Springer International Publishing AG 2018
C. Antos et al. (eds.), *The Hyperuniverse Project and Maximality*,
https://doi.org/10.1007/978-3-319-62935-3_1

1

In this article we consider class forcing in the framework of Morse-Kelley class theory. In difference to the case of A-definable class forcings we are able to prove in MK that the Definability Lemma holds for all forcing notions (without having to restrict to tame forcings). For the preservation of the axioms however we still need the property of tameness.[1]

In the following we will introduce Morse-Kelley class theory and define the relevant notions like names, interpretations and the extension for class forcing in Morse-Kelley. Then we will show that the forcing relation is definable in the ground model, that the Truth Lemma holds and we characterize P-generic extensions which satisfy the axioms of MK. We will show that Laver's Theorem fails for class forcings.

2 Morse-Kelley Class Theory

In ZFC we can only talk about classes as abbreviations for formulas as our only objects are sets. In class theories like Morse-Kelley (MK) or Gödel-Bernays (GB) the language is two-sorted, i.e. the object are sets and classes and we have corresponding quantifiers for each type of object.[2] We denote the classes by upper case letters and sets by lower case letters, the same will hold for class-names and set-names and so on. Hence atomic formulas for the $\in$-relation are of the form "$x \in X$" where x is a set-variable and X is a set- or class-variable. The models $\mathcal{M}$ of MK are of the form $\langle M, \in, \mathcal{C} \rangle$, where M is a transitive model of ZFC, $\mathcal{C}$ the family of classes of $\mathcal{M}$ (i.e. every element of $\mathcal{C}$ is a subset of M) and $\in$ is the standard $\in$ relation (from now on we will omit mentioning this relation).

The axiomatizations of class theories which are often used and closely related to ZFC are MK and GBC. Their axioms which are purely about sets coincide with the corresponding ZFC axioms such as pairing and union and they share class axioms like the Global Choice Axiom. Their difference lies in the Comprehension Axiom in the sense that GB only allows quantification over sets whereas MK allows

---

[1]In [2] R. Chuaqui follows a similar approach and defines forcing for Morse-Kelley class theory. However there is a significant difference between our two approaches. To show that the extension preserves the axioms Chuaqui restricts the generic G for an arbitrary forcing notion P in the following way: A subclass G of a notion of forcing P is *strongly P-generic over* a model $(M, \mathcal{C})$ of MK iff G is P-generic over $(M, \mathcal{C})$ and for all ordinals $\beta \in M$ there is a set $P' \in M$ such that $P' \subseteq P$ and for all sequences of dense sections $\langle D_\alpha : \alpha \in \beta \rangle$, there is a $q \in G$ satisfying

$$\forall \alpha (\alpha \in \beta \to \exists p \, (p \in P' \cap G \wedge \text{the greatest lower bound of } p \text{ and } q \text{ exists}$$

$$\text{and is an element of } D_\alpha)).$$

where a subclass D of a partial order P is a P-*section* if every extension of a condition in D is in D.

[2]There is also an equivalent one-sorted formulation in which the only objects are classes and sets are defined as being classes which are elements of other classes. For reasons of clarity we will use the two-sorted version throughout the paper.

quantification over sets as well as classes. This results in major differences between the two theories which can be seen for example in their relation to ZFC: GB is a conservative extension of ZFC, meaning that every sentence about sets that can be proved in GB can already be proved in ZFC and so GB cannot prove "new" theorems about ZFC. MK on the other hand can do just that, in particular MK implies CON(ZFC)[3] and so MK is not conservative over ZFC. The consistency strength of MK is strictly stronger than that of ZFC but lies below that of ZFC + there is an inaccessible cardinal as $\langle V_\kappa, V_{\kappa+1} \rangle$ for κ inaccessible, is a model for MK in ZFC.

As said above we choose MK (and not GB) as underlying theory to define class forcing. The reason lies mainly in the fact that within MK we can show the Definability Lemma for class forcing without having to restrict the forcing notion whereas in GB this would not be possible. We use the following axiomatization of MK:

A) Set Axioms:

1. Extensionality for sets: $\forall x \forall y (\forall z (z \in x \leftrightarrow z \in y) \rightarrow x = y)$.
2. Pairing: For any sets x and y there is a set $\{x, y\}$.
3. Infinity: There is an infinite set.
4. Union: For every set x the set $\bigcup x$ exists.
5. Power set: For every set x the power set $P(x)$ of x exists.

B) Class Axioms:

1. Foundation: Every nonempty class has an $\in$-minimal element.
2. Extensionality for classes: $\forall z (z \in X \leftrightarrow z \in Y) \rightarrow X = Y$.
3. Replacement: If a class F is a function and x is a set, then $\{F(z) : z \in x\}$ is a set.
4. Class-Comprehension:

$$\forall X_1 \ldots \forall X_n \exists Y \ Y = \{x : \varphi(x, X_1, \ldots, X_n)\}$$

where φ is a formula containing class parameters in which quantification over both sets and classes are allowed.
5. Global Choice: There exists a global class well-ordering of the universe of sets.

There are different ways of axiomatizing MK, one of them is obtained by using the Limitation of Size Axiom instead of Global Choice and Replacement. Limitation of Size is an axiom that was introduced by von Neumann and says that for every $C \in \mathcal{M}$, C is a proper class if and only if there is a one-to-one function from the universe of sets to C, i.e. all the proper classes have the same size. The two axiomatizations are equivalent: Global Choice and Replacement follow from Limitation of size and

---

[3]This is because in MK we can form a Satisfaction Predicate for V and then by reflection we get an elementary submodel V_α of V. But any such V_α models ZFC.

vice versa.[4] A nontrivial argument shows that Limitation of Size does not follow from Replacement plus Local Choice.

In the definition of forcing we will use the following induction and recursion principles:

Proposition 1 (Induction) *Let (Ord, R) be well-founded and $\varphi(\alpha)$ a property of an ordinal α. Then it holds that*

$$\forall \alpha \in Ord\,((\forall \beta \in Ord\,(\beta\,R\,\alpha \rightarrow \varphi(\beta))) \rightarrow \varphi(\alpha)) \rightarrow \forall \alpha \in Ord\,\varphi(\alpha)$$

Proof Otherwise, as R is well-founded, there exists an R-minimal element α of Ord such that $\neg\varphi(\alpha)$. That is a contradiction.   □

Proposition 2 (Recursion) *For every well-founded binary relation R on Ord and every formula $\varphi(X, Y)$ satisfying $\forall X\,\exists!\,Y\,\varphi(X, Y)$, there is a unique binary relation S on $Ord \times V$ such that for every $\alpha \in Ord$ it holds that $\varphi(S_{<\alpha}, S_\alpha)$, where $S_\alpha = \{x \mid (\alpha, x) \in S\}$ and $S_{<\alpha} = \{(\beta, x) \in S \mid \beta R \alpha\}$.*

Proof By induction on α it holds that for each γ there exists a unique binary relation S^γ on $Ord_{<\gamma} \times V$, where $Ord_{<\gamma} = \{\beta \in Ord \mid \beta R \gamma\}$, such that $\varphi(S^\gamma_{<\alpha}, S^\gamma_\alpha)$ holds for all $\alpha R \gamma$. Then it follows from Class-Comprehension that we can take $S = \bigcup_{\gamma \in Ord} S_\gamma$.   □

3   Generics, Names and the Extension

To lay out forcing in MK we have to redefine the basic notions like names, interpretation of names etc. to arrive at the definition of the forcing extension. As we work in a two-sorted theory we will define these notions for sets and classes respectively. Let us start with the definition of the forcing notions and its generics. We use the notation $(X_1, \ldots, X_n) \in C$ to mean $X_i \in C$ for all i.

Definition 3 Let $P \in C$ and $\leq_P \in C$ be a partial ordering with greatest element 1^P. We call $(P, \leq_P) \in C$ an (M, C)-forcing and often abbreviate it by writing P. With the above convention $(P, \leq_P) \in C$ means that P and $\leq_P$ are in C.

$G \subseteq P$ is P-generic over (M, C) if

1. G is compatible: If $p, q \in G$ then for some r, $r \leq p$ and $r \leq q$.
2. G is upwards closed: $p \geq q \in G \rightarrow p \in G$.
3. $G \cap D \neq \emptyset$ whenever $D \subseteq P$ is dense, $D \in C$.

---

[4]This is because Global Choice is equivalent with the statement that every proper class is bijective with the ordinals.

Note that from now on we will assume M to be countable (and transitive) and $\mathcal{C}$ to be countable to ensure that for each $p \in P$ there exists G such that $p \in G$ and G is P-generic.

We build the hierarchy of names for sets and classes in the following way (we will use capital greek letters for class-names and lower case greek letters for set-names):

Definition 4

$\mathcal{N}_0^s = \emptyset$.

$\mathcal{N}_{\alpha+1}^s = \{\sigma : \sigma \text{ is a subset of } \mathcal{N}_\alpha^s \times P \text{ in } M\}$.

$\mathcal{N}_\lambda^s = \bigcup\{\mathcal{N}_\alpha^s : \alpha < \lambda\}$, if λ is a limit ordinal.

$\mathcal{N}^s = \bigcup\{\mathcal{N}_\alpha^s : \alpha \in ORD(M)\}$ is the class of all set-names of P.

$\mathcal{N} = \{\Sigma : \Sigma \text{ is a subclass of } \mathcal{N}^s \times P \text{ in } \mathcal{C}\}$.

Note that the $\mathcal{N}_\alpha^s$ (for $\alpha > 0$) are in fact proper classes (and indeed $\mathcal{N}$ is a hyperclass) and therefore Definition 4 is an inductive definition of a sequence of proper classes of length the ordinals. The fact that with this definition we stay inside $\mathcal{C}$ follows from Proposition 2.

Lemma 5

a) If $\alpha \leq \beta$ then $\mathcal{N}_\alpha^s \subseteq \mathcal{N}_\beta^s$.
b) $\mathcal{N}^s \subseteq \mathcal{N}$.

Proof

a) By induction on β. For $\beta = 0$ there is nothing to prove.

Successor step $\beta \to \beta + 1$. Assume $\mathcal{N}_\alpha^s \subseteq \mathcal{N}_\beta^s$ for all $\alpha \leq \beta$. Let $\tau \in \mathcal{N}_\alpha^s$ for some $\alpha < \beta + 1$. Then we know by assumption that $\tau \in \mathcal{N}_\beta^s$. So by Definition 4 there is some $\gamma < \beta$ such that $\tau = \{\langle \pi_i, p_i \rangle \mid i \in I\}$ where for each $i \in I$, $\pi_i \in \mathcal{N}_\gamma^s$ and $p_i \in P$. By assumption $\pi_i \in \mathcal{N}_\beta^s$ for all $i \in I$ and so $\tau \in \mathcal{N}_{\beta+1}^s$.

Limit step λ. Assume $\mathcal{N}_\alpha^s \subseteq \mathcal{N}_\beta^s$ for all $\alpha \leq \beta < \lambda$. But by Definition 4, $\sigma \in \mathcal{N}_\lambda^s$ iff $\sigma \in \mathcal{N}_\beta^s$ for some $\beta < \lambda$ and so it follows that $\mathcal{N}_\alpha^s \subseteq \mathcal{N}_\lambda^s$ for all $\alpha \leq \lambda$.

b) By Definition 4, $\Sigma \in \mathcal{N}$ iff Σ is a subclass of $\mathcal{N}^s \times P$ iff for every $\langle \tau, p \rangle \in \Sigma$, $\tau \in \mathcal{N}^s$ and $p \in P$ iff for every $\langle \tau, p \rangle \in \Sigma$ there is an ordinal α such that $\tau \in \mathcal{N}_\alpha^s$ and $p \in P$. Let $\sigma \in \mathcal{N}^s$, i.e. there is an ordinal β such that $\sigma \in \mathcal{N}_\beta^s$. Then it holds that for every $\langle \tau, p \rangle \in \sigma$ there is an ordinal $\alpha < \beta$ such that $\tau \in \mathcal{N}_\alpha^s$ and $p \in P$. So $\sigma \in \mathcal{N}$.

$\square$

We define the interpretations of set- and class-names recursively.

Definition 6

$\sigma^G = \{\tau^G : \exists p \in G(\langle \tau, p \rangle \in \sigma)\}$ for $\sigma \in \mathcal{N}^s$.

$\Sigma^G = \{\sigma^G : \exists p \in G(\langle \sigma, p \rangle \in \Sigma)\}$ for $\Sigma \in \mathcal{N}$.

According to the definitions above we define the extension of an MK model (M, C) to be the extension of the set part and the extension of the class part:

Definition 7 $(M, C)[G] = (M[G], C[G]) = (\{\sigma^G : \sigma \in N^s\}, \{\Sigma^G : \Sigma \in N\}).$

Definition 8 If P is a partial order with greatest element 1^P, we define the canonical P-names of $x \in M$ and $C \in C$:

$\check{x} = \{\langle \check{y}, 1^P \rangle \mid y \in x\}.$
$\check{C} = \{\langle \check{x}, 1^P \rangle \mid x \in C\}.$

From these definitions the basic facts of forcing follow easily:

Lemma 9 Let $M = \langle M, C \rangle$ be a model of MK, where M is a transitive model of ZFC and C the family of classes of M. Then it holds that:

a) $\forall x \in M(\check{x} \in N^s \wedge \check{x}^G = x)$ and $\forall C \in C(\check{C} \in N \wedge \check{C}^G = C).$
b) $(M, C) \subseteq (M, C)[G]$ in the sense that $M \subseteq M[G]$ and $C \subseteq C[G].$
c) $G \in (M, C)[G]$, i.e. $G \in C[G]$
d) $M[G]$ is transitive and $Ord(M[G]) = Ord(M).$
e) If (N, C') is a model of MK, $M \subseteq N$, $C \subseteq C'$, $G \in C'$ then $(M, C)[G] \subseteq (N, C').$

Proof

a) Using Definitions 6 and 8 we can easily show this by induction.
b) follows immediately from 1.
c) Let $\Gamma = \{\langle \check{p}, p \rangle : p \in P\}$. Then this is a name for G as $\Gamma^G = \{\check{p}^G \mid p \in G\} = \{p \mid p \in G\} = G.$
d) It follows from Definition 6 and Definition 7 that $M[G]$ is transitive. For every $\sigma \in N^s$ the rank of σ^G is at most rank σ, so $Ord(M[G]) \subseteq Ord(M).$
e) For each name $\Sigma \in N$, $\Sigma \in (M, C)$ and therefore $\Sigma \in (N, C')$. As $G \in C'$ the interpretation of Σ in $(M, C)[G]$ is the same as in $(N, C').$

$\square$

4 Definability and Truth Lemmas

We will define the forcing relation and show that it is definable in the ground model and how it relates to truth in the extension. The main focus will be the Definability Lemma, since it now is possible to prove that it holds for all forcing notions in contrast to A-definable class forcings in a ZFC setting (see [3]). Note that when we talk about a formula $\varphi(x_1, \ldots, x_m, X_1, \ldots, X_n)$ we mean φ to be a second-order formula that allows second-order quantification and we always assume the model (M, C) to be countable.

Definition 10 Suppose p belongs to P, $\varphi(x_1, \ldots, x_m, X_1, \ldots, X_n)$ is a formula, $\sigma_1, \ldots, \sigma_m$ are set-names and $\Sigma_1, \ldots, \Sigma_n$ are class-names. We write $p \Vdash$

$\varphi(\sigma_1, \ldots, \sigma_m, \Sigma_1, \ldots, \Sigma_n)$ iff whenever $G \subseteq P$ is P-generic over $(M, \mathcal{C})$ and $p \in P$, we have $(M, \mathcal{C})[G] \models \varphi(\sigma_1^G, \ldots, \sigma_m^G, \Sigma_1^G, \ldots, \Sigma_n^G)$.

Lemma 11 (Definability Lemma) *For any φ, the relation "$p \Vdash \varphi(\sigma_1, \ldots, \sigma_m, \Sigma_1, \ldots, \Sigma_n)$" of $p, \vec{\sigma}, \vec{\Sigma}$ is definable in $(M, \mathcal{C})$.*

Lemma 12 (Truth Lemma) *If G is P-generic over $(M, \mathcal{C})$ then*

$$(M, \mathcal{C})[G] \models \varphi(\sigma_1^G, \ldots, \sigma_m^G, \Sigma_1^G, \ldots, \Sigma_n^G) \Leftrightarrow \exists p \in G \, (p \Vdash \varphi(\sigma_1, \ldots, \sigma_m, \Sigma_1, \ldots, \Sigma_n)).$$

Following the approach of set forcing we introduce a new relation $\Vdash^*$ and prove the Definability and Truth Lemma for this $\Vdash^*$. Then we will show that $\Vdash^*$ equals the intended forcing relation $\Vdash$.

The definition of $\Vdash^*$ consists of ten cases: six cases for atomic formulas, where the first two are for set-names, the second two for the "hybrid" of set- and class-names and the last two for class-names, one for $\wedge$ and $\neg$ respectively and two quantifier cases, one for first-order and one for second-order quantification. By splitting the cases in this way we can see very easily that it is enough to prove the Definability Lemma for set-names only (case one and two in the Definition) and then infer the general Definability Lemma by induction.

Definition 13 $D \subseteq P$ is *dense below p* if $\forall q \leq p \, \exists r \, (r \leq q, r \in D)$.

Definition 14 Let σ, γ, π be elements of $\mathcal{N}^s$ and Σ, Γ elements of $\mathcal{N}$.

1. $p \Vdash^* \sigma \in \gamma$ iff $\{q : \exists \langle \pi, r \rangle \in \gamma$ such that $q \leq r, q \Vdash^* \pi = \sigma\}$ is dense below p.
2. $p \Vdash^* \sigma = \gamma$ iff for all $\langle \pi, r \rangle \in \sigma \cup \gamma$, $p \Vdash^* (\pi \in \sigma \leftrightarrow \pi \in \gamma)$.
3. $p \Vdash^* \sigma \in \Sigma$ iff $\{q : \exists \langle \pi, r \rangle \in \Sigma$ such that $q \leq r, q \Vdash^* \pi = \sigma\}$ is dense below p.
4. $p \Vdash^* \sigma = \Sigma$ iff for all $\langle \pi, r \rangle \in \sigma \cup \Sigma$, $p \Vdash^* (\pi \in \sigma \leftrightarrow \pi \in \Sigma)$.
5. $p \Vdash^* \Sigma \in \Gamma$ iff $\{q : \exists \langle \pi, r \rangle \in \Gamma$ such that $q \leq r, q \Vdash^* \pi = \Sigma\}$ is dense below p.
6. $p \Vdash^* \Sigma = \Gamma$ iff for all $\langle \pi, r \rangle \in \Sigma \cup \Gamma$, $p \Vdash^* (\pi \in \Sigma \leftrightarrow \pi \in \Gamma)$.
7. $p \Vdash^* \varphi \wedge \psi$ iff $p \Vdash^* \varphi$ and $p \Vdash^* \psi$.
8. $p \Vdash^* \neg \varphi$ iff $\forall q \neg \leq p \, (\neg q \Vdash^* \varphi)$.
9. $p \Vdash^* \forall x \varphi$ iff for all σ, $p \Vdash^* \varphi(\sigma)$.
10. $p \Vdash^* \forall X \varphi$ iff for all Σ, $p \Vdash^* \varphi(\Sigma)$.

We have to show that $\Vdash^*$ is definable within the ground model. For this it is enough to concentrate on the first two of the above cases, because we can reduce the definability of the $\Vdash^*$-relation for arbitrary second-order formulas to its definability for atomic formulas $\sigma \in \tau, \sigma = \tau$, where σ and τ are set-names. The rest of the cases then follow by induction. So let us restate Lemma 11 for the case of $\Vdash^*$ and set-names:

Lemma 15 (Definability Lemma for the Atomic Cases of Set-Names) *The relation "$p \Vdash^* \varphi(\sigma, \tau)$" is definable in $(M, \mathcal{C})$ for $\varphi = $ "$\sigma \in \tau$" and $\varphi = $ "$\sigma = \tau$".*

Proof We will show by induction[5] on $\beta \in ORD$ that there are unique classes $X_\beta, Y_\beta \subseteq \beta \times M$ which define the $\Vdash^*$-relation for the first two cases of Definition 14 in the following way: for all $\alpha < \beta$, $R_\alpha = (X_\beta)_\alpha$, $S_\alpha = (Y_\beta)_\alpha$ where $(X_\beta)_\alpha = \{x \mid \langle \alpha, x \rangle \in X_\beta\}$ and

$$R_\alpha = \{(p, \sigma, \in, \tau) \mid p \in P, \sigma \text{ and } \tau \text{ are set } P\text{-names}, \quad (\star)$$
$$\text{rank}(\sigma) \text{ and rank}(\tau) < \alpha, \text{ for all } q \leq p$$
$$\text{there is } q' \leq q \text{ and } \langle \pi, r \rangle \in \tau \text{ such that}$$
$$q' \leq r \text{ and } (q', \pi, =, \sigma) \in S_\alpha\}$$

and

$$S_\alpha = \{(p, \sigma, =, \tau) \mid p \in P, \sigma \text{ and } \tau \text{ are set } P\text{-names}, \quad (\star\star)$$
$$\text{rank}(\sigma) \text{ and rank}(\tau) < \alpha,$$
$$\text{for all } \langle \pi, r \rangle \in \sigma \cup \tau \text{ such that}$$
$$(p, \pi, \in, \sigma) \in R_\alpha \text{ iff } (p, \pi, \in, \tau) \in R_\alpha\}$$

To show that X_β and Y_β are definable we will define the classes R_α and S_α at each step by recursion on the tupel (p, σ, e, τ) according to the following well-founded partial order on $P \times \mathcal{N}^s \times \{`` \in ", `` = "\} \times \mathcal{N}^s$.

Definition 16 Suppose $(p, \sigma, e, \tau), (q, \sigma', e', \tau') \in P \times \mathcal{N}^s \times \{`` \in ", `` = "\} \times \mathcal{N}^s$. Say that $(q, \sigma', e', \tau') < (p, \sigma, e, \tau)$ if

- $max(rank(\sigma'), rank(\tau')) < max(rank(\sigma), rank(\tau))$, or
- $max(rank(\sigma'), rank(\tau')) = max(rank(\sigma), rank(\tau))$, and $rank(\sigma) \geq rank(\tau)$ but $rank(\sigma') < rank(\tau')$, or
- $max(rank(\sigma'), rank(\tau')) = max(rank(\sigma), rank(\tau))$, and $rank(\sigma) \geq rank(\tau) \leftrightarrow rank(\sigma') \geq rank(\tau')$, and e is "=" and e' is "$\in$".

Note that clause 1 and 2 of Definition 14 always reduce the $<$-rank of the members of $P \times \mathcal{N}^s \times \{`` \in ", `` = "\} \times \mathcal{N}^s$.

"Successor step $\beta \rightarrow \beta + 1$." We know that there are unique classes X_β, Y_β such that for all $\alpha < \beta$, $R_\alpha = (X_\beta)_\alpha$, $S_\alpha = (Y_\beta)_\alpha$ and $(\star)$ and $(\star\star)$ hold. We want to show that there are unique classes $X_{\beta+1}, Y_{\beta+1}$ such that for all $\alpha < \beta + 1$, $R_\alpha = (X_{\beta+1})_\alpha$, $S_\alpha = (Y_{\beta+1})_\alpha$ and $(\star)$ and $(\star\star)$ hold. So let for all $\alpha < \beta$ $(X_{\beta+1})_\alpha = (X_\beta)_\alpha = R_\alpha$ and $(Y_{\beta+1})_\alpha = (Y_\beta)_\alpha = S_\alpha$ and define $(X_{\beta+1})_\beta = R_\beta$ and $(Y_{\beta+1})_\beta = S_\beta$ uniquely as follows:

---

[5]To show how this induction works in the context of a class-theory we will not simply use Propositions 1 and 2, but rather give the complete construction.

A) $(p,\sigma,\text{“}\in\text{”},\tau) \in R_\beta$ if and only if for all $q \leq p$ there is $q' \leq q$ and $\langle \pi, r \rangle \in \tau$ such that $q' \leq r$ and $(q',\pi,\text{“}=\text{”},\sigma) \in S_\beta$.

B) $(p,\sigma,\text{“}=\text{”},\tau) \in S_\beta$ if and only if for all $\langle \pi, r \rangle \in \sigma \cup \tau$: $(p,\pi,\text{“}\in\text{”},\sigma) \in R_\beta$ iff $(p,\pi,\text{“}\in\text{”},\tau) \in R_\beta$.

These definitions clearly satisfy ($\star$) and ($\star\star$) and to see that they are indeed inductive definitions over the well-order defined in Definition 16, we consider the following three cases for each of the definitions A) and B):

1. $\text{rank}(\sigma) < \text{rank}(\tau)$
2. $\text{rank}(\tau) < \text{rank}(\sigma)$
3. $\text{rank}(\sigma) = \text{rank}(\tau)$

Ad A.1: $(q',\pi,\text{“}=\text{”},\sigma) < (p,\sigma,\text{“}\in\text{”},\tau)$ because $\text{rank}(\sigma), \text{rank}(\pi) < \text{rank}(\tau)$ (first clause of Denfition 16).

Ad A.2: $(q',\pi,\text{“}=\text{”},\sigma) < (p,\sigma,\text{“}\in\text{”},\tau)$ because $\max(\text{rank}(\pi), \text{rank}(\sigma)) = \max(\text{rank}(\sigma), \text{rank}(\tau))$ and $\text{rank}(\sigma) \geq \text{rank}(\tau)$ and $\text{rank}(\pi) < \text{rank}(\sigma)$ (second clause of Definition 16).

Ad A.3: $(q',\pi,\text{“}=\text{”},\sigma) < (p,\sigma,\text{“}\in\text{”},\tau)$ because $\max(\text{rank}(\pi), \text{rank}(\sigma)) = \max(\text{rank}(\sigma), \text{rank}(\tau))$ and $\text{rank}(\sigma) \geq \text{rank}(\tau)$ and $\text{rank}(\pi) < \text{rank}(\sigma) = \text{rank}(\tau)$ (second clause of Definition 16).

Ad B.1: $(p,\pi,\text{“}\in\text{”},\sigma) < (p,\sigma,\text{“}=\text{”},\tau)$ because $\text{rank}(\sigma), \text{rank}(\pi) < \text{rank}(\tau)$ and $(p,\pi,\text{“}\in\text{”},\tau) < (p,\sigma,\text{“}=\text{”},\tau)$ because $\max(\text{rank}(\pi), \text{rank}(\tau)) = \max(\text{rank}(\sigma), \text{rank}(\tau))$ and $\text{rank}(\sigma) < \text{rank}(\tau)$ and $\text{rank}(\pi) < \text{rank}(\tau)$ (third clause of Definition 16).

Ad B.2: $(p,\pi,\text{“}\in\text{”},\sigma) < (p,\sigma,\text{“}=\text{”},\tau)$ because of the second clause of Definition 16 and $(p,\pi,\text{“}\in\text{”},\tau) < (p,\sigma,\text{“}=\text{”},\tau)$ because $\text{rank}(\pi), \text{rank}(\tau) < \text{rank}(\sigma)$.

Ad B.3: $(p,\pi,\text{“}\in\text{”},\sigma) < (p,\sigma,\text{“}=\text{”},\tau)$ and $(p,\pi,\text{“}\in\text{”},\tau) < (p,\sigma,\text{“}=\text{”},\tau)$ because $\max(\text{rank}(\pi), \text{rank}(\tau)) = \max(\text{rank}(\sigma), \text{rank}(\tau))$ and $\text{rank}(\sigma) \geq \text{rank}(\tau)$ and $\text{rank}(\pi) < \text{rank}(\sigma), \text{rank}(\tau)$ (both second clause of Definition 16).

"Limit step λ." We know that for every $\beta < \lambda$ there are unique classes X_β, Y_β such that for all $\alpha < \beta$, $R_\alpha = (X_\beta)_\alpha$, $S_\alpha = (Y_\beta)_\alpha$ and ($\star$) and ($\star\star$) hold. We have to show that there are unique classes $X_\lambda, Y_\lambda \subseteq \lambda \times M$, λ limit, such that for all $\beta < \lambda$, $R_\beta = (X_\lambda)_\beta$, $S_\beta = (Y_\lambda)_\beta$ and ($\star$) and ($\star\star$) hold respectively. We define the required classes as follows:

$$\langle \alpha, x \rangle \in X_\lambda \leftrightarrow \exists \langle \langle R_\gamma, S_\gamma \rangle \mid \gamma \leq \alpha \rangle \exists X, Y((\forall \gamma \leq \alpha((X)_\gamma = R_\gamma \text{ and }$$
$$(Y)_\gamma = S_\gamma \text{ and they satisfy } (\star) \text{ and } (\star\star) \text{ resp.}) \wedge$$
$$(x \in (X)_\gamma \text{ for some } \gamma \leq \alpha))$$

$$\langle \alpha, x \rangle \in Y_\lambda \leftrightarrow \exists \langle \langle R_\gamma, S_\gamma \rangle \mid \gamma \leq \alpha \rangle \exists X, Y((\forall \gamma \leq \alpha((X)_\gamma = R_\gamma \text{ and }$$
$$(Y)_\gamma = S_\gamma \text{ and they satisfy } (\star) \text{ and } (\star\star) \text{ resp.}) \wedge$$
$$(x \in (Y)_\gamma \text{ for some } \gamma \leq \alpha))$$

From the proof of the successor step we see that the sequence $\langle\langle R_\gamma, S_\gamma\rangle \mid \gamma \leq \alpha\rangle$ is unique for every $\alpha < \lambda$ and therefore X_λ, Y_λ are also unique. This definition is possible only in Morse-Kelly with its version of Class-Comprehension and not in Gödel-Bernays, because we are quantifying over class variables (in fact we only need Δ_1^1 Class-Comprehension). □

The general Definability Lemma now follows immediately from this Lemma and Definition 14. We now turn to the Truth Lemma.

In the following a capital greek letter denotes a name from $\mathcal{N}$ (and therefore can be a set- or a class-name), whereas a lower case greek letter is a name from $\mathcal{N}^s$ (and therefore can only be a set-name).

Lemma 17

a) If $p \Vdash^* \varphi$ and $q \leq p$ then $q \Vdash^* \varphi$
b) If $\{p \mid q \Vdash^* \varphi\}$ is dense below p then $p \Vdash^* \varphi$.
c) If $\neg p \Vdash^* \varphi$ then $\exists q \leq p(q \Vdash^* \neg\varphi)$.

Proof

a) By induction on φ: Let φ be $\Sigma \in \Gamma$, then by Definition 4 $D = \{q' : \exists\langle\pi, r\rangle \in \Gamma$ such that $q' \leq r, q' \Vdash^* \pi = \Sigma\}$ is dense below p. Then for all $q \leq p$, D is also dense below q and therefore $q \Vdash^* \varphi$. The other cases follow easily.
b) By induction on φ. Let φ be $\Sigma \in \Gamma$ and $\{q \mid q \Vdash^* \Sigma \in \Gamma\}$ is dense below p. From Definition 14 it follows that $\{q \mid \{s : \exists\langle\pi, r\rangle \in \Gamma$ such that $s \leq r, s \Vdash^* \pi = \Sigma\}$ is dense below $q\}$ is dense below p and from a well-known fact it follows that $D = \{s : \exists\langle\pi, r\rangle \in \Gamma$ such that $s \leq r, s \Vdash^* \pi = \Sigma\}$ is dense below p. Again by Definition 14 we get as desired $p \Vdash^* \Sigma \in \Gamma$.

   The other cases follow easily; for the case of negation we will use the fact that if $\{p \mid q \Vdash^* \neg\varphi\}$ is dense below p then $\forall q \leq p(\neg q \Vdash^* \varphi)$, using a).
c) follows directly from b).

□

Now, the proofs for the Truth Lemma and $\Vdash^* = \Vdash$ follow similarly to the proofs in set forcing (note that a name $\Sigma \in \mathcal{N}$ can also be a set-name and therefore we don't need to mention the cases for set-names explicitly):

Lemma 18 (Truth Lemma)   *If G is P-generic then*

$$(M, \mathcal{C})[G] \models \varphi(\Sigma_1^G, \ldots, \Sigma_m^G) \Leftrightarrow \exists p \in G\,(p \Vdash^* \varphi(\Sigma_1, \ldots, \Sigma_m)).$$

Proof By induction on φ.

$\Sigma \in \Gamma$. "$\rightarrow$" Assume $\Sigma^G \in \Gamma^G$ then choose a $\langle\pi, r\rangle \in \Gamma$ such that $\Sigma^G = \pi^G$ and $r \in G$. By induction there is a $p \in G$ with $p \leq r$ and $p \Vdash^* \pi = \Sigma$. Then for all $q \leq p$, $q \Vdash^* \pi = \Sigma$ and by Definition 4 $p \Vdash^* \Sigma \in \Gamma$.

"$\leftarrow$": Assume $\exists p \in G(p \Vdash^* \Sigma \in \Gamma)$. Then $\{q : \exists\langle\pi, r\rangle \in \tau$ such that $q \leq r, q \Vdash^* \sigma = \pi\} = D$ is dense below p and so by genericity $G \cap D \neq \emptyset$. So there is a $q \in G, q \leq p$ such that $\exists\langle\pi, r\rangle \in \Gamma$

with $q \leq r, q \Vdash^* \pi = \Sigma$. By induction $\pi^G = \Sigma^G$ and as $r \geq q, r \in G$ and therefore $\pi^G \in \Gamma^G$. So $\Sigma^G \in \Gamma^G$.

$\Sigma = \Gamma$. "$\rightarrow$" Assume $\sigma^G = \Gamma^G$. Then for all $\langle \pi, r \rangle \in \Sigma \cup \Gamma$ with $r \in G$ it holds that $\pi^G \in \Sigma^G \leftrightarrow \pi^G \in \Gamma^G$. Let $D = \{p \mid \text{either } p \Vdash^* \Sigma = \Gamma \text{ or for some } \langle \pi, r \rangle \in \Sigma \cup \Gamma, p \Vdash^* \neg(\pi \in \Sigma \leftrightarrow \pi \in \Gamma)\}$. Then D is dense: By contradiction, let $q \in P$ and assume that there is no $p \leq q$ such that $p \in D$. But if there is no $p \leq q$ such that for some $\langle \pi, r \rangle \in \Sigma \cup \Gamma, p \Vdash^* \neg(\pi \in \Sigma \leftrightarrow \pi \in \Gamma)\}$ then by Lemma 17 $q \Vdash^* (\pi \in \Sigma \leftrightarrow \pi \in \Gamma)$ for all $\langle \pi, r \rangle \in \Sigma \cup \Gamma$ and therefore $q \Vdash^* \Sigma = \Gamma$. So there is a $p \leq q$ such that $p \in D$. Since the filter G is generic, there is a $p \in G \cap D$. If $p \Vdash^* \neg(\pi \in \Sigma \leftrightarrow \pi \in \Gamma)\}$ for some $\langle \pi, r \rangle \in \Sigma \cup \Gamma$ then by induction $\neg(\pi^G \in \Sigma^G \leftrightarrow \pi^G \in \Gamma^G)$ for some $\langle \pi, r \rangle \in \Sigma \cup \Gamma$. But this is a contradiction to $\Sigma^G = \Gamma^G$ and so $P \Vdash^* \Sigma = \Gamma$.

"$\leftarrow$" Assume that there is $p \in G$ ($p \Vdash^* \Sigma = \Gamma$). By Definition 4 it follows that for all $\langle \pi, r \rangle \in \Sigma \cup \Gamma$ $P \Vdash^* (\pi \in \Sigma \leftrightarrow \pi \in \Gamma)$. Then by induction $\pi^G \in \Sigma^G \leftrightarrow \pi^G \in \Gamma^G$ for all $\langle \pi, r \rangle \in \Sigma \cup \Gamma$. So $\Sigma^G = \Gamma^G$.

$\varphi \wedge \psi$ "$\rightarrow$" Assume that $(M, \mathcal{C})[G] \models \varphi \wedge \psi$ iff $(M, \mathcal{C})[G] \models \varphi$ and $(M, \mathcal{C})[G] \models \psi$. Then by induction $\exists p \in G$ $P \Vdash^* \varphi$ and $\exists q \in G, q \Vdash^* \psi$ and we know that $\exists r \in G (r \leq p$ and $r \leq q)$ such that $r \Vdash^* \varphi$ and $r \Vdash^* \psi$ and so by Definition 4 $r \Vdash^* \varphi \wedge \psi$.

"$\leftarrow$" Assume $\exists p \in G, p \Vdash^* \varphi \wedge \psi$, then $p \Vdash^* \varphi$ and $p \Vdash^* \psi$. So $(M, \mathcal{C})[G] \models \varphi$ and $(M, \mathcal{C})[G] \models \psi$ and therefore $(M, \mathcal{C})[G] \models \varphi \wedge \psi$.

$\neg\varphi$ "$\rightarrow$" Assume that $(M, \mathcal{C})[G] \models \neg\varphi$. $D = \{p \mid p \Vdash^* \varphi$ or $p \Vdash^* \neg\varphi\}$ is dense (using Lemma 17 and Definition 4). Therefore there is a $p \in G \cap D$ and by induction $p \Vdash^* \neg\varphi$.

"$\leftarrow$" Assume that there is $p \in G$ such that $p \Vdash^* \neg\varphi$. If $(M, \mathcal{C}) \models \varphi$ then by induction hypothesis there is a $q \in G$ such that $q \Vdash^* \varphi$. But then also $r \Vdash^* \varphi$ for some $r \leq p, q$ and this is a contradiction because of Definition 4. So $(M, \mathcal{C}) \models \neg\varphi$.

$\forall X\varphi$ "$\rightarrow$" Assume that $(M, \mathcal{C})[G] \models \forall X\varphi$. Following the lines of the "$\rightarrow$"-part of the proof for $\Sigma = \Gamma$, there is a dense $D = \{p \mid \text{either } p \Vdash^* \forall X\varphi$ or for some $\sigma, p \Vdash^* \neg\varphi(\sigma)\}$. By induction we show that the second case is not possible and so it follows that $p \Vdash^* \forall X\varphi$.

"$\leftarrow$" By induction.

$\square$

Lemma 19 $\Vdash^* = \Vdash$

Proof $p \Vdash^* \varphi(\sigma_1, \ldots, \sigma_n) \rightarrow p \Vdash \varphi(\sigma_1, \ldots, \sigma_n)$ follows directly from the Truth Lemma. For the converse we use Lemma 17 c) and note that we assumed the existence of generics. Then from $\neg p \Vdash^* \varphi(\sigma_1, \ldots, \sigma_n)$ it follows that for some $q \leq p, q \Vdash^* \neg\varphi(\sigma_1, \ldots, \sigma_n)$ and so $\neg p \Vdash \varphi(\sigma_1, \ldots, \sigma_n)$. $\square$

5 The Extension Fulfills the Axioms

We have shown that in MK we can prove the Definability Lemma without restricting the forcing notion as we have to do when working with A-definable class forcing in ZFC (see [3]). Unfortunately we do not have the same advantage when proving the preservation of the axioms. For example, when proving the Replacement Axiom we have to show that the range of a set under a class function is still a set and this does not hold in general for class forcings. In [3] two properties of forcing notions are introduced, namely pretameness and tameness. Pretameness is needed to prove the Definability Lemma and show that all axioms except Power Set are preserved. For the Power Set Axiom this restriction needs to be strengthened to tameness. Let us give the definitions in the MK context:

Definition 20 (Pretameness) $D \subseteq P$ is predense $\leq p \in P$ if every $q \leq p$ is compatible with an element of D.

P is pretame if and only if whenever $\langle D_i \mid i \in a \rangle$ is a sequence of dense classes in $\mathcal{M}$, $a \in M$ and $p \in P$ then there exists a $q \leq p$ and $\langle d_i \mid i \in a \rangle \in M$ such that $d_i \subseteq D_i$ and d_i is predense $\leq q$ for each i.

Definition 21 $q \in P$ meets $D \subseteq P$ if q extends an element in D.

A predense $\leq p$ partition is a pair (D_0, D_1) such that $D_0 \cup D_1$ is predense $\leq p$ and $p_0 \in D_0, p_1 \in D_1 \to p_0, p_1$ are incompatible. Suppose $\langle (D_0^i, D_1^i) \mid i \in a \rangle$, $\langle (E_0^i, E_1^i) \mid i \in a \rangle$ are sequences of predense $\leq p$ partitions. We say that they are equivalent $\leq p$ if for each $i \in a$, $\{q \mid q$ meets $D_0^i \leftrightarrow q$ meets $E_0^i\}$ is dense $\leq p$. When $p = 1^P$ we omit $\leq p$.

To each sequence of predense $\leq p$ partitions $\vec{D} = \langle (D_0^i, D_1^i) \mid i \in a \rangle \in M$ and G is P-generic over $\langle M, \mathcal{C} \rangle$, $p \in G$ we can associate the function

$$f_{\vec{D}}^G : a \to 2$$

defined by $f(i) = 0 \leftrightarrow G \cap D_0^i \neq \emptyset$. Then two such sequences are equivalent $\leq p$ exactly if their associated functions are equal, for each choice of G.

Definition 22 (Tameness) P is tame iff P is pretame and for each $a \in M$ and $p \in P$ there is $q \leq p$ and $\alpha \in ORD(M)$ such that whenever $\vec{D} = \langle (D_0^i, D_1^i) \mid i \in a \rangle \in M$ is a sequence of predense $\leq q$ partitions, $\{r \mid \vec{D}$ is equivalent $\leq r$ to some $\vec{E} = \langle (E_0^i, E_1^i) \mid i \in a \rangle$ in $V_\alpha^M\}$ is dense below q.

Theorem 23 *Let $(M, \mathcal{C})$ be a model of MK. Then, if G is P-generic over $(M, \mathcal{C})$ and P is tame then $(M, \mathcal{C})[G]$ is a model of MK.*

Proof Extensionality and Foundation follow because $M[G]$ is transitive (see Lemma 9 d); axioms 2 and 3 from Definitions 4 and 6. For Pairing, let σ_1^G, σ_2^G be such that $\sigma_1, \sigma_2 \in \mathcal{N}^s$. Then the interpretation of the name $\sigma = \{\langle \sigma_1, 1^P \rangle, \langle \sigma_2, 1^P \rangle\}$ in the extension gives the desired $\sigma^G = \{\sigma_1^G, \sigma_2^G\}$. Infinity follows because ω exists in (M, C) and the notion of ω is absolute to any model, $\omega \in (M, C)[G]$. Union follows as in the set forcing case.

Replacement This follows as in [3] from the property of pretameness and we give the proof to make clear where the property of pretameness is needed: Suppose that $F : \sigma^G \to M[G]$. Then for each σ_0 of rank $< \text{rank}\,\sigma$ the class $D(\sigma_0) = \{p \mid \text{for some } \tau, q \Vdash \sigma_0 \in \sigma \to F(\sigma_0) = \tau\}$ is dense below p, for some $p \in G$ which forces that F is a total function on σ. We now use pretameness to "shrink" this class to a set: so for each $q \leq p$ there is an $r \leq q$ and $\alpha \in Ord(M)$ such that $D_\alpha(\sigma_0) = \{s \mid s \in V_\alpha^M \text{ and for some } \tau \text{ of rank } < \alpha, s \Vdash \sigma_0 \in \sigma \to F(\sigma_0) = \tau\}$ is predense $\leq r$ for each σ_0 of rank $< \text{rank}\,\sigma$. Then it follows by genericity that there is a $q \in G$ and $\alpha \in Ord(M)$ such that $q \leq p$ and $D_\alpha(\sigma_0)$ is predense $\leq q$ for each σ_0 of rank $< \text{rank}\,\sigma$. So let $\pi = \{\langle \tau, r \rangle \mid \text{rank}\,\tau < \alpha, r \in V_\alpha^M, r \Vdash \tau \in \text{ran}(F)\}$ and then it follows that $\text{ran}(F) = \pi^G \in M[G]$.

Power Set This follows from tameness as shown in [3].

Class-Comprehension Let $\Gamma = \{\langle \sigma, p \rangle \in \mathcal{N}^s \times P \mid p \Vdash \varphi(\sigma, \Sigma_1, \ldots, \Sigma_n)\}$. Because of the Definability Lemma, we know that $\Gamma \in \mathcal{N}$. By Definitions 4 and 6, $\Gamma^G = \{\sigma^G \mid \exists p \in G(\langle \sigma, p \rangle \in \Gamma)\}$ and we need to check that this equals the desired $Y = \{x \mid (\varphi(x, \Sigma_1^G, \ldots, \Sigma_n^G))^{(M, C)[G]}\}$. So let $\sigma^G \in \Gamma^G$. Then by the definition of Γ^G we know that $p \Vdash \varphi(\sigma, \Sigma_1, \ldots, \Sigma_n)$ and because of the Truth Lemma it follows that $(M, C)[G] \models \varphi(\sigma^G, \Sigma_1^G, \ldots, \Sigma_n^G)$. For the converse, let $x \in Y$. By the Truth Lemma, $\exists p \in G(p \Vdash \varphi(\pi, \Sigma_1, \ldots, \Sigma_n)$, where π is a name for x. By definition of $\Gamma, \langle \pi, p \rangle \in \Gamma$.

Global Choice Let $<_M$ denote the well-order of M and let σ_x, σ_y be the least names for some $x, y \in M[G]$. As the names are elements of M, we may assume that $\sigma_x <_M \sigma_y$. So we define the relation $<_G$ in $M[G]$ using M and $<_M$ as parameters, so that $x <_G y$ iff $\sigma_x <_M \sigma_y$ for the corresponding least names of x and y. Let $R = \{(x, y) \mid x, y \in M[G] \text{ and } x <_G y\}$. Then by Class-Comprehension the class R exists. □

Friedman [3] gives us a simple sufficient condition for tameness that translates directly into the context of MK:

Definition 24 For regular, uncountable $\kappa > \omega$, P is κ-distributive if whenever $p \in P$ and $\langle D_i \mid i < \beta \rangle$ are dense classes, $\beta < \kappa$ then there is a $q \leq p$ meeting each D_i (p meets D if $p \leq q \in D$ for some q).

P is tame below κ if the tameness conditions hold for P with the added restriction that $Card(a) < \kappa$.

Lemma 25 *If P is κ-distributive then P is tame below κ.*

Proof Analogous to set forcing.[6] □

---

[6]See [3, p. 37].

6   Laver's Theorem

In the following we will give an example which shows that a fundamental theorem that holds for set forcing can be violated by tame class forcings.

Laver's Theorem (see [5]) shows that for a set-generic extension $V \subseteq V[G]$, $V \models ZFC$ with the forcing notion $P \in V$ and G P-generic over V, V is definable in $V[G]$ from parameter $V_{\delta+1}$ (of V) and $\delta = |P|^+$ in $V[G]$. This result makes use of the fact that every such forcing extension has the approximation and cover properties as defined in [4] and relies on certain results for such extensions.

In general, the same does not hold for class forcing. In fact there are class forcings such that the ground model is not even second-order definable from set-parameters:

Theorem 26 *There is an MK-model* $(M, \mathcal{C})$ *and a first-order definable, tame class forcing* $\mathbb{P}$ *with* G $\mathbb{P}$*-generic over* $(M, \mathcal{C})$ *such that the ground model* M *is not definable with set-parameters in the generic extension* $(M, \mathcal{C})[G]$.

Proof We are starting from L. For every successor cardinal α, let P_α be the forcing that adds one Cohen set to α: P_α is the set of all functions p such that

$$dom(p) \subset \alpha, \quad |dom(p)| < \alpha, \quad ran(p) \subset \{0, 1\}.$$

Let P be the Easton product of the P_α for every successor α: A condition $p \in P$ is a function $p \in L$ of the form $p = \langle p_\alpha : \alpha \text{ successor cardinal}\rangle \in \Pi_{\alpha \text{ succ.}} P_\alpha$ (p is stronger then q if and only if $p \supset q$) and p has Easton support: for every inaccessible cardinal κ, $|\{\alpha < \kappa \mid p(\alpha) \neq \emptyset\}| < \kappa$. Then P is the forcing which adds one Cohen set to every successor cardinal.

Let $\mathbb{P} = P \times P = \Pi_{\alpha \text{ succ.}} P_\alpha \times \Pi_{\alpha \text{ succ.}} P_\alpha$ be the forcing that adds simultaneously two Cohen sets to every successor cardinal.[7] Note that $\Pi_{\alpha \text{ succ.}} P_\alpha \times \Pi_{\alpha \text{ succ.}} P_\alpha$ is isomorphic to $\Pi_{\alpha \text{ succ.}} P_\alpha \times P_\alpha$. Let G be $\mathbb{P}$-generic. Then $G = \Pi_{\alpha \text{ succ.}} G_0(\alpha) \times G_1(\alpha)$ and we let $G_0 = \Pi_{\alpha \text{ succ.}} G_0(\alpha)$ and $G_1 = \Pi_{\alpha \text{ succ.}} G_1(\alpha)$ with G_0, G_1 P-generic over L. We consider the extension $L[G_0] \subseteq L[G_0][G_1]$ and we will show, that $L[G_0]$ is not definable in $L[G_0][G_1]$ from parameters in $L[G_0]$.

The reason that we cannot apply Laver's and Hamkins' results of [5] to this extension is that it does not fulfill the δ approximation property[8]: As the forcing adds a new set to every successor, the δ approximation property cannot hold at successor cardinals δ: the added Cohen set is an element of the extension and a subset of the ground model and all of its $< \delta$ approximations are elements of the ground model but the whole set is not.

---

[7]It follows by a standard argument that $\mathbb{P}$ is pretame (and indeed tame) over $(M, \mathcal{C})$, see [3].

[8]A pair of transitive classes $M \subseteq N$ satisfies the δ *approximation property* (with $\delta \in Card^N$) if whenever $A \subseteq M$ is a set in N and $A \cap a \in M$ for any $a \in M$ of size less than δ in M, then $A \in M$. For models of set theory equipped with classes, the pair $M \subseteq N$ satisfies the δ *approximation property for classes* if whenever $A \subseteq M$ is a class of N and $A \cap a \in M$ for any a of size less than δ in M, then A is a class of M.

Note that the forcing is weakly homogeneous, i.e. for every $p, q \in \mathbb{P}$ there is an automorphism π on $\mathbb{P}$ such that $\pi(p)$ is compatible with q. This is because every P_α is weakly homogeneous (let $\pi(p) \in P_\alpha$ such that $dom(\pi(p)) = dom(p)$ and $\pi(p)(\lambda) = q(\lambda)$ if $\lambda \in dom(p) \cap dom(q)$ and $\pi(p)(\lambda) = p(\lambda)$ otherwise, then π is order preserving and a bijection) and therefore also P is weakly homogeneous (define π componentwise using the projection of p to p_α). Similar for $P \times P$.

To show that $L[G_0]$ is not definable in $L[G_0][G_1]$ with parameters, assume to the contrary that there is a set-parameter a_0 such that $L[G_0]$ is definable by the second-order formula $\varphi(x, a_0)$ in $L[G_0][G_1]$ from a_0. Let α be such that $a_0 \in L[G_0 \restriction \alpha, G_1 \restriction \alpha]$. Now consider $a = G_0(\alpha^+)$, the Cohen set which is added to α^+ in the first component of $\mathbb{P}$. a is P_{α^+}-generic over $L[G_0 \restriction \alpha, G_1 \restriction \alpha]$ and as a is an element of $L[G_0]$ the formula φ holds for a. So we also know that there is a condition $q \in G$ such that $q \Vdash \varphi(\dot{a}, a_0)$.

Now we construct another generic $G^* = G_0^* \times G_1^*$ which produces the same extension but also an element for which φ holds and which is not an element of $L[G_0]$. This new generic adds the same sets as G, but we switch G_0 and G_1 at α^+ so that the set added by $G_1(\alpha^+)$ is now added in the new first component G_0^*. However we have to make sure that the new generic respects q so that φ is again forced in the extension. We achieve this by fixing the generic G on the length of $q(\alpha^+)$ (we can assume that the length is the same on G_0 and G_1).

It follows that $q \in G_0^* \times G_1^*$ and because of weakly homogeneity $G_0^* \times G_1^*$ is generic and $L[G_0][G_1] = L[G_0^*][G_1^*]$. Because of the construction of G^*, the formula $\varphi(x, a_0)$ holds for the set $b = G_0^*(\alpha^+)$ but b is not an element of $L[G_0]$. That is a contradiction! □

We have seen that there are different ways of approaching class forcing, namely on the one hand as definable from a class parameter A in a ZFC model (M, A) and on the other hand in the context of an MK model $(M, \mathcal{C})$. That presents us with three notions of genericity: set-genericity, A-definable class genericity and class-genericity. One of the questions that arises now is in which way we can define the next step in this "hierarchy" of genericity. To answer this question, Sy Friedman and the author of this paper are currently working on so-called hyperclass forcings in a variant of MK, i.e. forcings in which the conditions are classes (see [1]). We will show in which context such forcings are definable and which application they have to class-theory.

Acknowledgements I want to thank the Austrian Academy of Sciences for their generous support through their Doctoral Fellowship Program. I would also like to thank my doctoral advisor Sy David Friedman for his insightful comments on previous drafts of this paper and his kind support. This paper was prepared as part of the project: "The Hyperuniverse: Laboratory of the Infinite" of the JTF grant ID35216.

References

1. C. Antos, S.D. Friedman, Hyperclass forcing in Morse-Kelley class theory. J. Symb. Logic **82**(2), 549–575 (2017)
2. R. Chuaqui, Internal and forcing models for an impredicative theory of classes. Dissertationes Mathematicae 176 (1980)
3. S.D. Friedman, *Fine Structure and Class Forcing*. de Gruyter Series in Logic and Its Applications, vol. 3 (Walter de Gruyter, New York, 2000)
4. J.D. Hamkins, Extensions with the approximation and cover properties have no new large cardinals. Fundam. Math. **108**(3), 257–277 (2003)
5. R. Laver, Certain very large cardinals are not created in small forcing extensions. Ann. Pure Appl. Logic **149**, 1–6 (2007)

Hyperclass Forcing in Morse-Kelley Class Theory

Carolin Antos and Sy-David Friedman

Abstract In this article we introduce and study hyperclass-forcing (where the conditions of the forcing notion are themselves classes) in the context of an extension of Morse-Kelley class theory, called MK**. We define this forcing by using a symmetry between MK** models and models of ZFC^{-} plus there exists a strongly inaccessible cardinal (called SetMK**). We develop a coding between β-models $\mathcal{M}$ of MK** and transitive models M^+ of SetMK** which will allow us to go from $\mathcal{M}$ to M^+ and vice versa. So instead of forcing with a hyperclass in MK** we can force over the corresponding SetMK** model with a class of conditions. For class-forcing to work in the context of ZFC^{-} we show that the SetMK** model M^+ can be forced to look like $L_{\kappa^*}[X]$, where κ^* is the height of M^+, κ strongly inaccessible in M^+ and $X \subseteq \kappa$. Over such a model we can apply definable class forcing and we arrive at an extension of M^+ from which we can go back to the corresponding β-model of MK**, which will in turn be an extension of the original $\mathcal{M}$. Our main result combines hyperclass forcing with coding methods of Beller et al. (Coding the universe. Lecture note series. Cambridge University Press, Cambridge, 1982) and Friedman (Fine structure and class forcing. de Gruyter series in logic and its applications, vol 3, Walter de Gruyter, New York, 2000) to show that every β-model of MK** can be extended to a minimal such model of MK** with the same ordinals. A simpler version of the proof also provides a new and analogous minimality result for models of second-order arithmetic.

Originally published in C. Antos, S.D. Friedman, Hyperclass Forcing in Morse-Kelley Class Theory. J. Symb. Logic **82**(2), 549–575 (2017).

C. Antos (✉)
Zukunftskolleg/Department of Philosophy, University of Konstanz, Konstanz, Germany
e-mail: carolin.antos-kuby@uni-konstanz.de

S.-D. Friedman
Kurt Gödel Research Center for Mathematical Logic, University of Vienna, Vienna, Austria

© Springer International Publishing AG 2018
C. Antos et al. (eds.), *The Hyperuniverse Project and Maximality*,
https://doi.org/10.1007/978-3-319-62935-3_2

1 Introduction

When considering forcing notions with respect to their size, there are three different types: the original version of forcing, where the forcing notion is a set, called set forcing; forcing in ZFC, where the forcing notion is a class, called definable class forcing and class forcing in Morse-Kelley class theory (MK). In this article we consider a fourth type which we call definable hyperclass forcing and give applications for this forcing in the context of Morse-Kelley class theory, where hyperclass forcing denotes a forcing with class conditions. We will define hyperclass forcing indirectly by using a correspondence between certain models of MK and models of a version of ZFC$^-$ (minus PowerSet) and show that we can define definable hyperclass forcing by going to the related ZFC$^-$ model and using definable class forcing there.

Two problems arise when considering definable class forcing in ZFC: the forcing relation might not be definable in the ground model and the extension might not preserve the axioms. As an example consider $Col(\omega, ORD)$ with conditions $p : n \to Ord$ for $n \in \omega$ which adds a cofinal sequence of length ω in the ordinals. Here Replacement fails.[1] These problems were addressed in a general way by the second author in [4] where class forcings are presented which are definable (with parameters) over a model $\langle M, A \rangle$ where M is a transitive model of ZFC, $A \subseteq M$ and Replacement holds in M for formulas mentioning A as a unary predicate. Two properties of the forcing notion are introduced, pretameness and tameness and it is shown that for a pretame forcing notion the Definability Lemma holds and Replacement is preserved and that tameness (which is a strengthening of pretameness) is equivalent to the preservation of the Power Set axiom. In this article we will adjust this approach to definable class forcing in ZFC$^-$. Pretameness is defined as follows:

Definition 1 A forcing notion P is pretame iff whenever $\langle D_i | i \in a \rangle$, $a \in M$, is an $\langle M, A \rangle$-definable sequence of dense classes and $p \in P$ then there is $q \leq p$ and $\langle d_i | i \in a \rangle \in M$ such that $d_i \subseteq D_i$ and d_i is predense $\leq q$ for each i.

For definable hyperclass forcing we will work in the context of Morse-Kelley class theory, by which we mean a theory with a two-sorted language, i.e. the object are sets and classes and we have corresponding quantifiers for each type of object. We denote the classes by upper case letters and sets by lower case letters, the same will hold for class-names and set-names and so on. Hence atomic formulas for the $\in$-relation are of the form "$x \in X$" where x is a set-variable and X is a set- or class-variable. The models $\mathcal{M}$ of MK are of the form $\langle M, \in, \mathcal{C} \rangle$, where M is a transitive model of ZFC, $\mathcal{C}$ the family of classes of $\mathcal{M}$ (i.e. every element of $\mathcal{C}$ is a subset of M) and $\in$ is the standard $\in$ relation (from now on we will omit mentioning this relation). We use the following axiomatization of MK:

---

[1] A detailed analyses on how even the Definability Lemma for class forcings can fail can be found in [7].

A) Set Axioms:

1. Extensionality for sets: $\forall x \forall y (\forall z (z \in x \leftrightarrow z \in y) \rightarrow x = y)$.
2. Pairing: For any sets x and y there is a set $\{x, y\}$.
3. Infinity: There is an infinite set.
4. Union: For every set x the set $\bigcup x$ exists.
5. Power set: For every set x the power set $P(x)$ of x exists.

B) Class Axioms:

1. Foundation: Every nonempty class has an $\in$-minimal element.
2. Extensionality for classes: $\forall z (z \in X \leftrightarrow z \in Y) \rightarrow X = Y$.
3. Replacement: If a class F is a function and x is a set, then $\{F(z) : z \in x\}$ is a set.
4. Class-Comprehension:

$$\forall X_1 \ldots \forall X_n \exists Y \ Y = \{x : \varphi(x, X_1, \ldots, X_n)\}$$

   where φ is a formula containing class parameters in which quantification over both sets and classes are allowed.
5. Global Choice: There exists a global class well-ordering of the universe of sets.

Class forcing in MK was defined by the first author in [2]. Here the Definability Lemma holds for unrestricted forcing notions, but for the preservation of the axioms we still need pretameness and tameness.

The structure of this article will be as follows: First, we define the correspondence between certain models of a version of MK and ZFC$^-$ and show that this correspondence is indeed a coding between a variant of MK and certain models of the ZFC$^-$ which allows us to go back and forth between them. Then we define definable hyperclass forcing and show how the problems of definable class forcing in the setting of ZFC$^-$ can be handled. We conclude the chapter by giving an example of definable hyperclass forcing by showing that every β-model of a variant of MK can be extended to a minimal β-model of the same variant of MK with the same ordinals.

2 Coding Between MK* and SetMK*

In the context of ZFC we can talk about definable class forcings as done in [4], where we deal directly with the class forcing notion as it is definable from a class predicate. Here we want to develop a way of defining definable hyperclass forcings in MK, i.e. forcings with class conditions, but we will choose an indirect approach, which will allow us to reduce the technical problems as much as possible to the context of definable class forcing. So instead of talking directly about hyperclasses, we will use a correspondence between models of a variant of MK (called MK*) and models of a variant of ZFC$^-$ (called SetMK*). We get an idea of how such a model of SetMK* looks by considering the following model of MK: $\langle V_\kappa, V_{\kappa+1} \rangle$ where κ is strongly inaccessible. Similar to this model we will show how to define a model of SetMK* with a strongly inaccessible cardinal κ which is the largest cardinal such that the sets of the MK* model are elements of V_κ and the classes are elements of $V_{\kappa*}$, where κ^* is the height of the SetMK* model. We will then force over such a model with a definable class forcing which will give us an extension of the SetMK* model. From this extension we can then go back to a model of MK* and this is the definable hyperclass-generic extensions of the original MK* model.

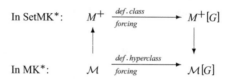

In the following we will describe how we can go from MK* to SetMK* and vice versa and show that the basic properties of class forcing over a model of SetMK* hold. Then we give an application of definable hyperclass forcing regarding minimal models of MK**.

But before we develop the relation between these models further we will impose a restriction on the models we are considering.

Definition 2 A model $\mathcal{M}$ of Morse-Kelley class theory is a β-model of MK if a class is well-founded in $\mathcal{M}$ if and only if it is true that the class is well-founded.

We introduce this restriction for two reasons: First, we will define a coding which allows us to go from a β-model of MK* to a transitive model of SetMK* and this coding only works in the intended way if we know that every well-founded class in the model is really well-founded (see Sect. 2). Secondly we will prove a theorem about minimal models and such a notion only makes sense if we work with minimal β-models. So from now on, we will always talk about β-models of (variants of) MK.

The associated model of set theory will be a model of ZFC$^-$ (i.e. minus the Power Set Axiom) where we understand such a model to include the Collection (or Bounding) Principle.[2] To ensure this we have to add the Class-Bounding Principle, a "class version" of the Bounding Principle, and we call the resulting axiomatic system MK*:

---

[2]Note that in ZFC minus Power Set the Bounding Principle does not follow from Replacement. This is used in [8], where he showed that in ZF$^-$ the different formulations of the Axiom of Choice are not equivalent. As for MK, work done in [5] shows that for example ultrapower constructions don't work without first adding a version of Class-Bounding. For more information see [6].

Definition 3 The axioms of MK* consist of the axioms of MK plus the Class-Bounding Axiom

$$\forall x \exists A \, \varphi(x, A) \rightarrow \exists B \, \forall x \exists y \, \varphi(x, (B)_y)$$

where $(B)_y = \{z \mid (y, z) \in B\}$.

Note that as we have Global Choice, this is equivalent to AC_∞:

$$\forall x \exists A \, \varphi(x, A) \rightarrow \exists B \, \forall x \varphi(x, (B)_x).$$

Equivalently, SetMK* will include the set version of Bounding (here called Set-Bounding):

$$\forall x \in a \, \exists y \, \varphi(x, y) \rightarrow \exists b \, \forall x \in a \, \exists y \in b \, \varphi(x, y)$$

As we will show in the proof of Theorem 8 and the proof of Theorem 13, Set-Bounding in SetMK* follows from Class-Bounding in MK* and vice versa.

We are now going to show how to translate the theory of MK* to a first-order set theory SetMK*. The axioms of SetMK* are:

1. ZFC$^-$ (including Set-Bounding).
2. There is a strongly inaccessible cardinal κ.
3. Every set can be mapped injectively into κ.

We can construct a transitive model M^+ of SetMK* out of any β-model (M, C) of MK* by taking all sets which are coded by a pair (M_0, R), where M_0 belongs to C and R is a binary relation within C. We will show that M^+ is the unique model of SetMK* with largest cardinal κ such that $M = V_\kappa^{M^+}$ and the elements of C are the subsets of M in M^+.

To describe the coding between SetMK* and MK* we will define what a coding pair (M_0, R) is and what it means for a coding pair (M_0, R) to code a set x in a model of SetMK*. In the coding below we work with relations which are classes, i.e. objects of rank ORD which provide isomorphic copies of the membership relation on the transitive closure of x for a set x which may have rank greater than ORD (see Definitions 5 and 7).

Definition 4 A pair (M_0, R) is a coding pair in the β-model $\mathcal{M} = (M, C)$ if M_0 is an element of C with a distinguished element a, $R \in C$ and R is a binary relation on M_0 with the following properties:

a) $\forall z \in M_0 \exists! n$ such that z has R-distance n from a, i.e. there is an R-chain $(zRz_{n-1}R \ldots Rz_1 Ra)$,
b) if $x, y, z \in M_0$ with $y \neq z$, yRx, zRx then $(M_0, R) \upharpoonright y$ is not isomorphic to $(M_0, R) \upharpoonright z$, where $(M_0, R) \upharpoonright y$ denotes the R-transitive closure below y (i.e. y together with all elements which are connected to y via an R-chain), respectively for z,

c) if $y, z \in M_0$ are on level n (i.e. have the same R-distance n from a) and $y \neq z$ then $vRy \rightarrow \neg(vRz)$,

d) R is well-founded.

Note that in the definition of the codes in $(M, \mathcal{C})$ we need the assumption that $(M, \mathcal{C})$ is a β-model as for a class to code a set in M^+ it has to be well-founded not only in the MK model but "in the real world".

The meaning of the definition becomes clearer when we view the coding pair as a tree T whose nodes are exactly the distinct elements of M_0, the top node is a and R is the extension relation of the tree. A tree T' with top node a' is a subtree of T if a' is a node of T and T' contains all T-nodes (not only immediately) below a'. If T' is a subtree of T such that a' lies directly below a then T' is called a direct subtree of T. Then property b) states that for every node x distinct direct subtrees are not isomorphic and property c) implies that the trees below two distinct points on the same level are disjoint (and not only on the next level).

The idea behind the coding pairs is, that every coding pair will define a unique set x in the SetMK* model. Note that at the same time every x in M^+ can correspond to different coding pairs in $\mathcal{M}$.

In the following we will give some intuition on what such a correspondence between coding pairs in $\mathcal{M}$ and sets in M^+ should look like: Every $x \in M^+$ is coded by a tree T_x where x is associated to the top node a_x of T_x, the elements $y \in x$ are associated to the nodes on the first level below a_x so that every node on this level gives rise to a subtree T_y which codes y so that the elements of y are associated to the nodes on the second level below a_x and so on:

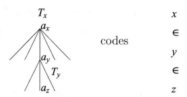

Note that there are only countably many levels but a level can have class many elements. If two elements a_y, a_z have the same R_x predecessor (i.e. are connected to the same node on the previous level) their subtrees T_y, T_z will never be isomorphic and therefore don't code the same element of M^+ (by property b) of Definition 4). But it can happen that there are isomorphic subtrees on different levels or on the same level but not connected to the same node on the level above. This can be made clear in the following two examples: First let $y \in x$, $v \in y$ and $w \in y$ and $v \in w$. Then there are two isomorphic trees T_v and T'_v both coding v but on different levels:

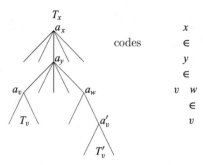

Secondly let $v \in w$, $v \in y$ and $w, y \in x$. Again there are two isomorphic trees T_v and T'_v coding v but this time on the same level:

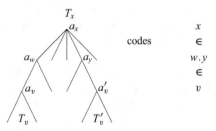

To show that Definition 4 indeed defines a coding, we have to show that there is a correspondence between x and its coding pair. As we want to include non-transitive sets we will work with $(TC(\{x\}), \in)$ (note that we used the transitive closure of $\{x\}$ rather than the transitive closure of x as the transitive closure of two different sets could be the same). As we have seen, the coding tree will have a lot of isomorphic subtrees, for example many different pairs $(a_i, \{\})$ coding the empty set. So the tree T_x itself will not be isomorphic to $(TC(\{x\}), \in)$ and we will have to collapse (M_x, R_x) to a structure $(M_x, R_x)/ \approx$ in which we have identified all these isomorphic subtrees. We define this quotient of the coding pair in the following way:

Definition 5 For a coding pair (M_0, R), let $[a] = \{b \in M_0 \mid (M_0, R) \upharpoonright b$ isomorphic to $(M_0, R) \upharpoonright a\}$ be the equivalence class of all the top nodes of subtrees of the coding tree T which are isomorphic to the subtree T_a (here $(M_0, R) \upharpoonright b$ denotes the "sub-coding pair" which is the subtree T_b as detailed in Definition 4). By Global Choice let $\tilde{a}$ be a fixed representative of this class. Then let $\tilde{M_0} = \{\tilde{a} \mid a \in M_0\}$ and define the relation $\tilde{R}$ as follows: $\tilde{a}\tilde{R}\tilde{b}$ iff $\exists a_0, b_0$ such that $a_0 \in [a]$ and $b_0 \in [b]$ and $a_0 R b_0$.

Note that if $a_0 \approx a_1$ and $b_0 R a_0$ then there is b_1 with $b_1 R a_1$ such that $b_0 \approx b_1$ as the isomorphism between T_{a_0} and T_{a_1} will restrict to the trees T_{b_0} and T_{b_1}.

The following example shows how this quotient structure looks for a possible coding tree of the set 3:

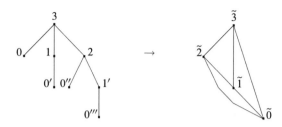

As one can see, the resulting structure $(\tilde{M}_3, \tilde{R}_3)$ is then isomorphic to $(TC(\{3\}), \in)$. In the following we will show that this construction works in general:

Lemma 6 *Let (M_0, R) be a coding pair. Then the quotient structure $(\tilde{M}_0, \tilde{R})$ as defined in Definition 5 is extensional and well-founded.*

Proof By Class-Comprehension $\tilde{R} \in \mathcal{C}$ and $\tilde{R}$ is well-founded as we can always find an R-minimal element a, build the equivalence class $[a]$ and find its representative $\tilde{a}$. Then $\tilde{a}$ is $\tilde{R}$ minimal as otherwise there exists $\tilde{a}'$ such that $\tilde{a}'\tilde{R}\tilde{a}$ and therefore there is $a_0' \in [a']$ such that $a_0' R a$.

To show that $\tilde{R}$ is extensional, let $\tilde{y}, \tilde{z} \in \tilde{M}_0$ with $\tilde{y} \neq \tilde{z}$ and assume that they have the same extension $\{\tilde{x} \mid \tilde{x}\tilde{R}\tilde{y}\} = \{\tilde{x} \mid \tilde{x}\tilde{R}\tilde{z}\}$. Going back to (M_0, R) this means that the elements of the related equivalence classes $[y], [z]$ have the same isomorphism types of children, i.e. for every $x_0, y_0, z_0 \in M_0$ with $x_0 R y_0$, $y_0 \in [y]$ and $z_0 \in [z]$ we can find x_1 with $x_1 R z_0$ such that $x_0, x_1 \in [x]$. By using property b) of Definition 4 it follows that the $[y] = [z]$, because we do not have multiplicities in (M_0, R), i.e. isomorphic subtrees that are connected to the same R-predecessor. It follows that $\tilde{y} = \tilde{z}$. □

Note that the quotient structure always has a fixed top node which is the representative of the equivalence class of the distinguished node of (M_0, R), which has the distinguished node as its only element.

It follows from Mostowski's Theorem that there is a unique transitive structure with the $\in$-relation that is isomorphic to $(\tilde{M}_0, \tilde{R})$. This structure then has the form $(TC(\{x\}), \in)$ for a unique set x.

Definition 7 A coding pair (M_x, R_x) is called a coding pair for x, if x is the unique set such that $(\tilde{M}_x, \tilde{R}_x)$ is isomorphic to $(TC(\{x\}), \in)$.

In the following we will use this coding to associate a transitive model of SetMK* to each β-model of MK* and vice versa.

Theorem 8 *Let $\mathcal{M} = (M, \mathcal{C})$ be a β-model of MK* and*

$$M^+ = \{x \mid \text{there is a coding pair } (M_x, R_x) \text{ for } x\}$$

Then M^+ is the unique, transitive set that obeys the following properties:

a) $M^+ \models$ SetMK,*
b) $\mathcal{C} = P(M) \cap M^+$,
c) $M = V_\kappa^{M^+}$, κ is the largest cardinal in M^+ and strongly inaccessible in M^+.

The coding between $\mathcal{M}$ and M^+ is the key to prove the theorem. So before proving this theorem we will prove two useful fact about the coding.

As we have seen there can be more than one coding pair for an $x \in M^+$. Of course these coding pairs are isomorphic because they are all built according to Definition 4 but we also would like to know that they are isomorphic in $\mathcal{M}$. For elements of M^+ that can be coded by sets in $\mathcal{M}$ this is trivial but for elements that are coded by proper classes we have to show the following:

Lemma 9 (Coding Lemma 1) *Let $\mathcal{M} = (M, \mathcal{C})$ be a transitive β-model of MK^*. Let $N_1, N_2 \in \mathcal{C}$ and R_1, R_2 be well-founded binary relations in $\mathcal{C}$ such that (N_1, R_1) and (N_2, R_2) are coding pairs as described in Definition 4. Then if there is an isomorphism between (N_1, R_1) and (N_2, R_2) there is such an isomorphism in $\mathcal{C}$.*

Proof Let T_1, T_2 be the coding trees associated to the coding pairs (N_1, R_1), (N_2, R_2). Assume to the contrary that there is an isomorphism between T_1 and T_2 but not one in $\mathcal{C}$. It follows that the tree below the top node of T_1 is isomorphic to the tree below the top node of T_2, but there is no such isomorphism in $\mathcal{C}$. Then, as T_1 and T_2 are well-founded we can choose a T_1-minimal node a_1 of T_1 such that for some node a_2 of T_2 the tree U_1 (the tree T_1 below and including a_1) is isomorphic to U_2 (the tree T_2 below and including a_2) but there is no isomorphism in $\mathcal{C}$. Because of the minimality of a_1 we know that for every node $a_{1,i}$ of U_1 just below a_1 and every node $a_{2,j}$ of U_2 just below a_2, if $U_{1,i}$ is isomorphic to $U_{2,j}$ then there is an isomorphism in $\mathcal{C}$. Moreover the property "$U_{1,i}, U_{2,j}$ are ismorophic" is expressible in $(M, \mathcal{C})$.

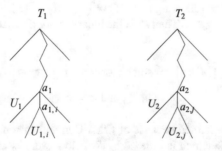

Now we can apply the Class Bounding Principle of MK^* to get a class B so that for each $a_{1,i}, a_{2,j}$ for which $U_{1,i}, U_{2,j}$ are isomorphic, $(B)_c$ is such an isomorphism for some set c. Using the global well-order of M we can choose a unique $c(a_{1,i}, a_{2,j})$ for each relevant pair $\langle a_{1,i}, a_{2,j} \rangle$ and combine the isomorphisms $(B)_{c(a_{1,i}, a_{2,j})}$ to get an isomorphism between U_1 and U_2 in $\mathcal{C}$, which is a contradiction. ☐

So all coding trees of the same element of M^+ are isomorphic in $\mathcal{C}$. For the converse it is obvious that two isomorphic coding trees code the same element in M^+ as they give rise to the same $(\tilde{M}_x, \tilde{R}_x)$.

The next lemma shows that we are able to see something of the coding in M^+:

Lemma 10 (Coding Lemma 2) *For all $x \in M^+$ there is a one-to-one function $f \in M^+$ such that $f : x \to M_x$, where (M_x, R_x) is a coding pair for x.*

Proof Let T_x be a coding tree for x and for each $y \in x$ let T_y is the subtree of T_x with top node a_y lying just below the top node of T_x such that T_y codes y. Note that the choice of a_y is unique after having fixed the tree T_x.

To show that $f = \{\langle y, a_y \rangle \mid y \in x\}$ belongs to M^+, we have to find a coding tree for f. Firstly we construct a coding tree $T_{\langle y, a_y \rangle}$ for every $\langle y, a_y \rangle$ with $y \in x$. As a_y is a set in M, it is a set in M^+ and therefore coded by some T_{a_y}. So we can build $T_{\langle y, a_y \rangle}$ by connecting the trees T_y and T_{a_y}. To make sure that the relation $R_{\langle y, a_y \rangle}$ on the new tree is well-defined we can relabel the nodes of the tree T_{a_y} and so we get the following picture:

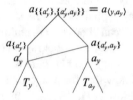

In this way we code every pair $\langle y, a_y \rangle$ with $y \in x$ and we can now join all the codes to code f.

Let (M_f, R_f) be the following pair: $M_f = \bigcup_{z \in x} M_{\langle z, a_z \rangle} \cup \{a_f\}$ where $a_f \in M$ and $a_f \notin M_{\langle z, a_z \rangle}$ for every $z \in x$. Then R_f is the binary relation which is defined using R_x as parameter:

$$R_f = \{\langle v, w \rangle \mid \text{ for some } y \in x \text{ either } \langle v, w \rangle \in R_{\langle y, a_y \rangle} \text{ or }$$

$$v = a_{\langle y, a_y \rangle} \text{ and } w = a_f\}$$

M_f and R_f are well-defined because of Class Comprehension in MK^* and so f is coded by the tree T_f which is ordered by R_z below every a_z and by putting $a_{\langle y, a_y \rangle}$ below a_f otherwise.

□

Now we give the proof of Theorem 8.

Proof

a) We show that if $\mathcal{M}$ is a β-model of MK^* then $M^+ \models SetMK^*$. The first step is proving that M^+ satisfies ZFC^- with Set-Bounding.

Observe that M^+ is transitive: Let $x \in M^+$. Then for every $y \in x$ there is a coding tree for y (namely the corresponding subtree of T_x). Therefore $y \in M^+$ and so $x \subseteq M^+$. From transitivity it follows that Extensionality and Foundation hold in M^+; Infinity follows as $\omega \in M^+$.

Pairing: Let x, y be coded by T_x, T_y respectively. Then $\{x, y\}$ is coded by the tree:

Union: Let x be coded by T_x:

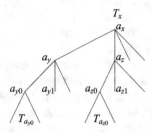

The obvious way to code $\bigcup x$ would be to join the $a_{y0}, a_{y1}, \ldots, a_{z0}, a_{z1}, \ldots$ together by one top node $a_{\bigcup x}$. But in general this is not a coding tree by reasons of isomorphism: Our coding trees have the property that subtrees which are connected to the same node on the next level above are all pairwise non-isomorphic. In this case that means that the trees $T_{a_y}, T_{a_z}, \ldots$ are pairwise non-isomorphic, as are the trees $T_{a_{y0}}, T_{a_{y1}}, \ldots$ and the trees $T_{a_{z0}}, T_{a_{z1}}, \ldots$ and so on. But, as we explained before, it can happen that some of the $T_{a_{yi}}$ are isomorphic to, for example, some of the $T_{a_{zj}}$. So if we connect these trees by a top node the resulting tree would have isomorphic subtrees connected by the same node on the next level and therefore would not be a coding tree. This problem can easily be resolved by taking equivalence classes of the subtrees of T_x from the second level below a_x (where two trees are equivalent if the are isomorphic). Then we take a representative from each equivalence class and connect them to the top node $a_{\bigcup x}$ (as before, this is possible by Class Comprehension in MK^* and Coding Lemma 1).

To prove Comprehension and Bounding we need to take a closer look at how formulas in M^+ translate to formulas in $\mathcal{M}$:

Lemma 11 *For each first-order formula φ there is a formula ψ of second-order class theory such that for all $x_1, \ldots, x_n \in M^+$, $M^+ \models \varphi(x_1, \ldots, x_n)$ if and only if $\mathcal{M} \models \psi(c_1, \ldots, c_n)$ for any choice of codes $c_1, \ldots, c_n$ for $x_1, \ldots, x_n$.*

Proof The proof is by induction over the complexity of the formula φ. For the first atomic case assume that $M^+ \models y \in x$. Let c_x and c_y be codes for x and y respectively and let T_x, T_y be the associated coding trees. As we know that $y \in x$ it follows that there is a direct subtree $T_{y'}$ of T_x such that $T_{y'}$ is a coding tree for y ("direct subtree" means a subtree whose top node lies just below the top node of the original tree). As $T_{y'}$ and T_y are both codes for y they are isomorphic and by Coding Lemma 1 we know that they are isomorphic in $\mathcal{M}$. So $\mathcal{M} \models$ "c_y is isomorphic to a direct subtree of c_x," and this therefore is the desired ψ.

For the second atomic case assume that $M^+ \models y = x$. Let c_x and c_y be codes for x and y respectively. As $y = x$, c_y is also a code for x and again by Coding Lemma 1 we know that the codes are isomorphic in $\mathcal{M}$ thus giving us the desired ψ.

The cases of $\neg \varphi$, $\varphi_1 \wedge \varphi_2$ follow easily by using the induction hypothesis. For the quantifier case consider $M^+ \models \forall x \varphi$. By induction hypothesis let ψ be the second-order formula associated to φ. Then $M^+ \models \forall x \varphi$ translates to $\mathcal{M} \models \forall c$, if c is a code then $\psi(c)$. $\qquad \square$

Comprehension Let $a, x_1, \ldots, x_n \in M^+$ and let $\varphi(x, x_1, \ldots, x_n, a)$ be any first-order formula. We will show that $b = \{x \in a : M^+ \models \varphi(x, x_1, \ldots, x_n, a)\}$ in an element of M^+ by using Class Comprehension in $\mathcal{M}$ to find the corresponding $B \in \mathcal{C}$ and build from it a coding tree for b.

Let $T_{x_1}, \ldots, T_{x_n}, T_a$ be codes for the corresponding elements of M^+ and let ψ be the formula corresponding to φ provided by Lemma 11. Assume that b is non-empty, i.e. that there is x_0 in a such that φ holds. Therefore there is a c_0 such that $\psi(c_0, T_{x_1}, \ldots, T_{x_n}, T_a)$ holds. Let c be a variable that varies over the level directly below the top level of T_a so that each $T_{a(c)}$ denotes a direct subtree of T_a. Then by Class Comprehension there is a class B such that if $\psi(T_{a(c)}, T_{x_1}, \ldots, T_{x_n}, T_a)$ holds then $(B)_c$ is the direct subtree $T_{a(c)}$ of T_a and if not then $(B)_c$ is T_{c_0}.

So let T_b be the coding tree with top node a_b and whose direct subtrees are all of the $(B)_c$:

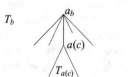

$$T_b \qquad \overset{a_b}{\underset{T_{a(c)}}{\wedge}} a(c) \qquad \qquad \text{ordered by } R_{a(c)}$$

Then T_b codes $b \in M^+$ with $b = \{x \in a : \varphi(x, x_1, \ldots, x_n, a)\}$.

Bounding We have to show that for $a \in M^+$ and φ a first-order formula

$$M^+ \models \forall x \in a \, \exists y \, \varphi(x, y) \rightarrow \exists b \, \forall x \in a \, \exists y \in b \, \varphi(x, y).$$

So assume that $\forall x \in a \, \exists b \, \varphi(x, y)$. Let T_y, T_a be coding trees for y and a respectively and let ψ be the second-order formula corresponding to φ provided by Lemma 11. By Class-Bounding in MK^* we know that

$$\exists B \, \forall T_x \text{ direct subtree of } T_a \exists y' \, \psi(T_x, (B)_{y'}),$$

where $(B)_{y'} = \{z \mid (y', z) \in B\}$. By Class Comprehension we can join together all the section $(B)_{y'}$ which are coding trees $T_{(B)_{y'}}$ to obtain a tree T_b with top node a_b such that the $T_{(B)_{y'}}$ are the direct subtrees of T_b. It follows that in $\mathcal{M}$ there is a tree

T_b such that for every tree T_x subtree of T_a there is a $T_{(B)_y}$ direct subtree of T_b such that $\psi(T_x, T_{(B)_y})$ and the tree T_b gives us the desired b in M^+.

Replacement Follows from Comprehension and Bounding.

Choice We have to show that every element of M^+ can be well-ordered (we aim for the strongest version of the axiom of Choice in a set-theory without Power Set (see [8]). So let $x \in M^+$ and let T_x be a coding tree for x with top node a_x. We know that the direct subtrees T_y of T_x code the elements y of x and their top nodes a_y are elements of M. As we have a well-order of M we can well-order the class $B = \{a_y \mid a_y$ is the top node of a direct subtree T_y of $T_x\}$. We call this well-order W. Now we can build a tree for every pair $\langle a_y, a_z \rangle \in W$ by using the trees T_y, T_z analogous as we did in the proof of Coding Lemma 2:

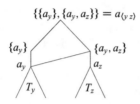

So for every $\langle a_y, a_z \rangle \in W$ we get a coding tree for the pair $\langle y, z \rangle$ with $y, z \in x$. As we have shown in the proof of Coding Lemma 2 we can now join together the trees by a single top node a_w using Class Comprehension. We now get a tree T_w which is a coding tree for an element w of M^+ and w is a well-order of x.

Remark 12 The next two results below (b and c) will show, that there even is a global choice function for the sets in $V_\kappa^{M^+}$ for κ an inaccessible cardinal, as there is a class which well-orders M and we will show that every class in $\mathcal{C}$ is an element of M^+.

b) We have to show that $\mathcal{C} = P(M) \cap M^+$. So assume that $X \in \mathcal{C}$ and $y \in X$. Then $y \in M$ and so can be coded by the following tree: y is the top node of the tree T_y. On the first level below the top node there are nodes for every element of y which are named by pairwise different elements z_i of $M \setminus \{y\}$. On the first level below such an z_i there are nodes for every element in z_i named by pairwise different elements v_j of $M \setminus \{y, z_i\}$ and so on. So T_y is a coding tree for y and therefore $y \in M^+$. This can be done for all $y \in X$ and by Class Comprehension the trees T_y can be connected to a tree T_X with top node a_X. Then the pair (M_X, R_X) gives a code for X with $M_X = \bigcup_{y \in X} M_y \cup \{a_X\}$ and

$$R_X = \{\langle v, w \rangle \mid \text{for some } y \in X \text{ either } \langle v, w \rangle \in R_y \text{ or } v = a_y \text{ and } w = a_X\}$$

Therefore $X \in M^+$.

For the converse, let $x \in M^+$ and $x \subseteq M$. Then there exists a coding pair (M_x, R_x) of x such that $(\tilde{M}_x, \tilde{R}_x) \cong (TC\{x\}, \in)$ (see Lemma 6). As $(\tilde{M}_x, \tilde{R}_x)$ is in C, has rank $Ord(M)$ and we can build $TC(\{x\})$ by transfinite induction from $(\tilde{M}_x, \tilde{R}_x)$, we can decode x in C and so $x \in C$.

c) Now we will show that there is a strongly inaccessible cardinal κ in M^+ which is the largest cardinal in M^+ and the elements of M (the sets in $\mathcal{M}$) are exactly the elements of $V_\kappa^{M^+}$.

Let κ be $Ord(M)$. Then as $\kappa \subseteq M$ and $\kappa \in M^+$ it follows from b) that κ is a class in C. Let $f : \beta \to \kappa$ with β is a ordinal less than κ be a function in C. From the Class Bounding Principle it follows that f is bounded in κ. So κ is regular in $\mathcal{M}$ and therefore regular in M^+. Moreover, again by b), any subset of an ordinal β of M which belongs to M^+ is a class in C and indeed a set in M, so the power set of β in M^+ equals the power set of β in M and so κ is strongly inaccessible. It follows that if $x \in M$ then $x \in V_\kappa^{M^+}$. For the converse let $x \in V_\kappa^{M^+}$ and let (M_x, R_x) be a coding pair and T_x the associate coding tree for x. By Coding Lemma 2 any coding tree of a set is a set, so T_x is an element of M. Clause 3 of the axioms of SetMk* follows directly from Coding Lemma 2 and so κ is the largest cardinal in M^+.

That M^+ is unique follows from its construction: Let M^{++} be another such model of SetMK* (i.e. it is transitive, $C = P(M) \cap M^{++}$ and $M = V_\kappa^{M^{++}}$ with κ largest cardinal in M^{++} and strongly inaccessible cardinal in M^{++}). Then M^+ and M^{++} have the same largest cardinal κ, they have the same subsets of κ and as every set in both models can be coded by a subset of κ they are the same.

This concludes the proof of Theorem 8. $\qquad\square$

The converse of Theorem 8 follows by the corresponding axioms in the SetMK* model:

Theorem 13 *Let N be a transitive model of SetMK* that has a strongly inaccessible cardinal κ that is the largest cardinal, let $C = P(M) \cap N$ and M is defined to be V_κ^N. Then $\mathcal{M} = (M, C)$ is a β-model of MK* and the model M^+ derived from $\mathcal{M}$ by Theorem 8 equals N.*

Proof We have to show that (M, C) fulfills the axioms of MK*: Extensionality, Pairing, Infinity, Union, Power Set, and Foundation follow directly by the corresponding axioms of SetMK*. By the definition of M and C it follows that every set is a class and elements of classes are sets.

For the remaining axioms, note that there is an easy converse for Lemma 11: For each formula φ of second-order class theory there is a first-order formula ψ such that for all $x_1, \ldots, x_n \in M$, $\mathcal{M} \models \varphi(x_1, \ldots, x_n)$ if and only if $N \models \psi(x_1, \ldots, x_n)$. This holds because by assumption all elements of $\mathcal{M}$ are elements of C or M and therefore elements of N and so φ and ψ are the same where the statement that x is a set in $\mathcal{M}$ translates to $x \in V_\kappa^N$ and the statement that X is a class in $\mathcal{M}$ translates to $X \in P(M) \cap N$. So for Class Comprehension we have to show that the following holds:

$$\forall X_1 \ldots \forall X_n \exists Y \; Y = \{x : \varphi(x, X_1, \ldots, X_n)\}$$

where φ is a formula containing class parameters in which quantification over both sets and classes is allowed. By the definition of M and C this statement is exactly the Comprehension Axiom of N where ψ is the first-order formula corresponding to φ: $y = \{x \in V_\kappa^N : N \models \psi(x, x_1, \ldots, x_n, V_\kappa^N)\}$.

For Class Bounding we have to show:

$$\forall x\, \exists A\, \varphi(x, A) \rightarrow \exists B\, \forall x\, \exists y\, \varphi(x, (B)_y)$$

where $(B)_y = \{z \mid (y, z) \in B\}$. So assume that $\forall x\, \exists A\, \varphi(x, A)$ holds in $\mathcal{M}$. Then translating this to N we know by Set-Bounding that

$$\forall x \in V_\kappa^{M^+} \exists A \in P(M) \cap M^+\, \psi(x, A) \rightarrow \exists b\, \forall x \in V_\kappa^{M^+} \exists y \in b\, \psi(x, y)$$

where ψ is the first-order formula corresponding to φ. By Set-Comprehension we can form a set b_0 from b such that $b_0 = \{y \mid y \in b \wedge y \subseteq V_\kappa^N\}$. Then there is a function $f \in N$ from V_κ^N onto b_0 (as b_0 has size less or equal κ) and so f is also an element of $\mathcal{M}$. The we can define the class $(B)_z = \{w \mid w \in f(z)\}$ and therefore also $B = \{(z, w) \mid z \in V_\kappa^N \wedge w \in f(z)\}$. So Class-Bounding holds.

For Global Choice we have to show that there is a well-ordering of M. We know that every element of N can be well-ordered and so V_κ^N can be well-ordered. The well-order is therefore an element of C.

(M, C) has to be a β-model: Any well-founded relation in (M, C) corresponds to a well-founded relation in N and because N is a transitive model of ZF^-, well-foundedness is absolute (we can define a rank function into the "real" ordinals which witnesses the well-foundedness in V).

Finally when we build the M^+ of $\mathcal{M}$ according to Theorem 8, M^+ and N are both transitive, have the same largest cardinal κ and the same subsets of κ and are therefore equal. □

Remark 14 We can also use this switching between models of MK* and SetMK* for class-forcing: Instead of doing class-forcing over MK* we go to SetMK* and do a set-forcing there. Note that by doing this indirect version of class-forcing we don't lose the tameness requirement for the forcing: Assume the class-forcing is not tame (as for example a forcing which collapses the universe to ω). Then we go to $M^+ \models SetMK^*$ and force with the associated set-forcing. But such a forcing destroys the inaccessibility of κ and therefore the preservation of PowerSet in the MK* extension $\mathcal{M}[G]$.

Corollary 15

$$M^+ = \bigcup_{C \in \mathcal{C}} L_{\kappa^*}(C).$$

where κ^* is the height of M^+ and

$$L_0(C) = TC(\{C\})$$
$$L_{\beta+1}(C) = Def(L_\beta(C))$$
$$L_\lambda(C) = \bigcup_{\beta<\lambda} L_\beta(C), \ \lambda \ limit.$$

Proof Let $x \in M^+$. Then there is a coding pair (M_x, R_x) for x such that $(\tilde{M}_x, \tilde{R}_x)$ is isomorphic to $(TC(\{x\}), \in)$. As $\tilde{M}_x$ and $\tilde{R}_x$ are elements of C we can code the pair $(\tilde{M}_x, \tilde{R}_x)$ by a class $C_x \in C$. As C_x is an element of M^+, $L_{\kappa^*}(C_x)$ is an inner model in M^+. But now we can decode x in $L_{\kappa^*}(C_x)$ as we can build $(TC(\{x\}))$ by transfinite induction from $(\tilde{M}_x, \tilde{R}_x)$. So $x \in L_{\kappa^*}(C_x)$.

For the converse, let $x \in \bigcup_{C \in C} L_{\kappa^*}(C)$, i.e. there is an $C_x \in C$ such that $x \in L_{\kappa^*}(C_x)$. As $L_{\kappa^*}(C_x)$ is an inner model of $\mathcal{M}$, x is an element of C and by Theorem 8 b) it is an element of M^+. $\qquad\qquad\qquad\qquad\qquad\qquad\qquad\qquad\qquad\qquad\qquad\qquad\qquad\qquad\qquad\Box$

3 Hyperclass Forcing and Forcing in SetMK**

In the last section we have seen how to move back and forth between a model of MK* and its associated SetMK* model. Now we will use this relation between a model of class theory and a model of set theory to define hyperclass forcing. A hyperclass is a collection whose elements are classes. The key idea is that instead of trying to formalize forcing for a definable hyperclass forcing notion, we can go to the associated model of SetMK* where the forcing notion is now a class and so we force with a definable class forcing there and then go back to a new MK* model. First let us define the relevant notions:

Definition 16 Let $\mathcal{M} = (M, C)$ be a model of MK* and for $\mathbb{P} \subseteq C$ let $(\mathbb{P}, \leq) = \mathbb{P}$ be an $\mathcal{M}$-definable partial ordering with a greatest element $1^{\mathbb{P}}$. $P, Q \in \mathbb{P}$ are compatible if for some R, $R \leq P$ and $R \leq Q$. A definable hyperclass $D \subseteq \mathbb{P}$ is dense if $\forall P \exists Q (Q \leq P$ and $Q \in D)$. Then a $\mathbb{G} \subseteq C$ is called a $\mathbb{P}$-generic hyperclass over $\mathcal{M}$ iff $\mathbb{G}$ is a pairwise compatible, upward-closed subcollection of $\mathbb{P}$ which meets every dense subcollection of $\mathbb{P}$ which is definable over $\mathcal{M}$.

We will assume that for each $P \in \mathbb{P}$ there exists $\mathbb{G}$ such that $P \in \mathbb{G}$ and $\mathbb{G}$ is $\mathbb{P}$-generic over $\mathcal{M}$ (this is always possible if the model $\mathcal{M}$ is countable).

To define the structure $(M, C)[\mathbb{G}]$ where $\mathbb{G}$ is a $\mathbb{P}$-generic hyperclass over (M, C) we will use Theorem 8 and Proposition 13. By Theorem 8 we go to the model $M^+ \models SetMK^*$. As $\mathbb{P}$ is a subcollection of C in $\mathcal{M}$ it becomes a subclass of $P(M) \cap M^+$ and is an M^+-definable class, $\mathbb{G}$ remains a pairwise compatible, upward-closed subclass of $\mathbb{P}$ which meets every dense subclass of $\mathbb{P}$ which is definable over M^+ and therefore is definable class-generic over M^+. Then we define names, their interpretation and the extension of M^+ as usual: A $\mathbb{P}$-name in M^+ is a set in M^+ consisting of pairs (τ, p) where τ is a $\mathbb{P}$-name in M^+ and p belongs to $\mathbb{P}$ (as we are in the set model we now denote the elements of $\mathbb{P}$ with lower-case letters). Then $\mathcal{N} = \cup \{\mathcal{N}_\alpha \mid \alpha \in Ord(M^+)\}$ is the collection of all names where $\mathcal{N}_0 = \emptyset$, $\mathcal{N}_{\alpha+1} = \{\sigma \mid \sigma \text{ is a subset of } \mathcal{N} \times P \text{ in } M^+\}$ and $\mathcal{N}_\lambda = \cup \{\mathcal{N}_\alpha \mid \alpha < \lambda\}$ for a limit ordinal λ. For a $\mathbb{P}$-name σ its interpretation is $\sigma^{\mathbb{G}} = \{\tau^{\mathbb{G}} \mid p \in \mathbb{G} \text{ for some } (\tau, p) \in \sigma\}$. Then $M^+[\mathbb{G}]$ is the set of all such $\tau^{\mathbb{G}}$. Finally we can define the extension of $\mathcal{M}$:

Definition 17 Let $\mathcal{M} = (M, \mathcal{C})$ be a β-model of MK^*, $\mathbb{P}$ be a definable hyperclass forcing and $\mathbb{G} \subseteq \mathbb{P}$ be a $\mathbb{P}$-generic hyperclass over $\mathcal{M}$. Let M^+ be the model of $SetMK^*$ associated to $\mathcal{M}$ by Theorem 8 and assume that $M^+[\mathbb{G}] \models SetMK^*$ with largest cardinal κ with $M^+[\mathbb{G}]$ transitive. Then $\mathcal{M}[\mathbb{G}] = (M, \mathcal{C})[\mathbb{G}]$ is the β-model of MK^* derived from $M^+[\mathbb{G}]$ by Theorem 13, whose sets are the elements of $V_\kappa^{M^+[\mathbb{G}]}$ and whose classes are the subsets of $V_\kappa^{M^+[\mathbb{G}]}$ in $M^+[\mathbb{G}]$, where κ is the largest cardinal of $M^+[\mathbb{G}]$ and is strongly inaccessible. Such a model is called a definable hyperclass-generic outer model of $\mathcal{M}$.

This definition assumes that the definable class-forcing $\mathbb{P}$ again produces a model of $SetMK^*$ with the same largest cardinal κ where κ is strongly inaccessible (we say in short that $\mathbb{P}$ does not change κ). Unfortunately the assumption that $SetMK^*$ is preserved is not as straightforward as it might seem. Definable class-forcing was developed by Friedman [4]. There the concept of pretameness and tameness of a forcing notion is introduced and it is shown that such a forcing has a definable forcing relation and preserves the axioms. In the case of $SetMK^*$ we now have the added problem that we are not forcing over a model of full ZFC but rather over ZFC$^-$, i.e. without the Power Set Axiom. This can cause problems when we use concepts like the hierarchy of the V_α, for example to prove that pretame class-forcings preserve the Replacement (or in our case the Set-Bounding) Axiom. So we cannot simply transfer the results of [4] but have to prove the Definability Lemma and the preservation of the axioms again without making use of the Power Set Axiom.

To define definable class-forcing in $SetMK^*$ first note that the following still holds: Let M^+ be a transitive model of $SetMK^*$, P be a M^+-definable forcing notion and G P-generic over M^+. Then $M^+[G]$ is transitive and $Ord(M^+[G]) = Ord(M^+)$. It follows from the definition of the interpretation of names and the definition of $M^+[G]$ that if $y \in \sigma^G$ then $y = \tau^G$ for some $\tau \in TC(\sigma)$ and therefore $M^+[G]$ is transitive. Furthermore for every $x \in Ord(M^+)$ there exists a name σ for x (i.e. $x = \sigma^G$ as defined above) with name-rank of $\sigma =$ the least $\alpha \in Ord(M)$ such that $\sigma \in \mathcal{N}_{\alpha+1}$ and by induction the von Neumann rank of σ^G is at most the name rank of σ. So we know that if "new" sets are added by the forcing they have size at most the "old" sets from M^+ and so $Ord(M^+[G]) \subseteq Ord(M^+)$.

We will first treat the case where we already assume that the forcing relation is definable and P is a pretame class-forcing and then show how we can ensure that in general pretame class-forcings preserve the axioms and the Definability Lemma holds.

Proposition 18 *Let M^+ be a model of SetMK* and let P be a pretame definable class-forcing over M^+ that does not change κ and whose forcing relation is definable. Let $G \subseteq P$ be definable class-generic over M^+. Then $M^+[G]$ is a model of SetMK*.*

Proof Extensionality, Pairing, Comprehension, Infinity, Foundation and Choice still hold by the proof for definable class-forcing over full ZFC. We have to show that Set-Bounding holds in $M^+[G]$, i.e.

$$M^+[G] \models \forall x \in a\, \exists y\, \varphi(x, y) \to \exists b\, \forall x \in a\, \exists y \in b\, \varphi(x, y)$$

Let σ be a name for a. We can extend any p for which $p \Vdash \forall x \in \sigma\, \exists y\, \varphi(x, y)$ to force that there is an isomorphism between σ and an ordinal α (by using AC) and so we can assume without loss of generality that σ is $\check{\alpha}$ where $\alpha \in Ord$ and therefore $p \Vdash \forall x < \alpha\, \exists y\, \varphi(x, y)$. Then for such a fixed p and for each $x < \alpha$ we can define by the Definability of the forcing relation $D_x = \{q \leq p \mid \exists \tau\, q \Vdash \varphi(x, \tau)\}$ where D_x is dense below p. By pretameness there is a $q \leq p$ and $\langle d_x \mid x < \alpha \rangle \in M^+$ such that for all $x < \alpha$, d_x is pretense $\leq q$ and by genericity there is such a q in G. Then we know that for all pairs $\langle x, r \rangle$ where $x < \alpha$ and $r \in d_x$ there is τ such that $r \Vdash \varphi(x, \tau)$. By the Set-Bounding principle in M^+ we get a set $T \in M^+$ such that $\forall (x, r)$ with $r \in d_x\, \exists \tau \in T$ such that $r \Vdash \varphi(x, \tau)$. Finally let π be a name for $\{\tau^G \mid \tau \in T\}$, i.e. $\pi = \{\langle \tau, 1^{\mathbb{P}} \rangle \mid \tau \in T\}$. Then, because the generic below q hits every d_x, $\varphi(x, \tau)$ will hold for some $\tau \in T$. It follows that $q \Vdash \forall x < \alpha\, \exists y \in \pi\, \varphi(x, y)$. Then Union follows with the use of Set-Bounding. □

With this proposition we have shown that in a model of MK* we can force with a definable hyperclass-forcing $\mathbb{P}$ and preserve MK*, provided $\mathbb{P}$ translates to a pretame class-forcing in SetMK* which preserves the inaccessibility of κ and whose forcing relation is definable. But in practice we don't usually know if the forcing relation is definable, even if we know that P is pretame due to the absence of a suitable hierarchy (like the V-hierarchy which suffices when forcing over ZF-models). So we will introduce a preparatory forcing which does not add any new sets but converts the SetMK* model M^+ into a model of the form $L_\alpha[A]$ for some generic class predicate $A \subseteq ORD$ preserving SetMK* (relative to A). This will allow us to use the relativized L hierarchy and therefore adapt the proof of the Definability Lemma for a pretame class-forcing and the fact that it preserves the axioms.

With this proposition we have shown that in a model of MK* we can force with a definable hyperclass-forcing $\mathbb{P}$ and preserve MK*, provided $\mathbb{P}$ translates to a pretame class-forcing in SetMK* which preserves the inaccessibility of κ and whose forcing relation is definable. But in the setting of SetMK* we don't usually know if the forcing relation is definable, even if we know that $\mathbb{P}$ is pretame, due to the absence of a suitable hierarchy. In the standard case of proving the Definability Lemma for pretame class forcing in models (M, A) of ZF we define a function by induction using the fact that we have the V-hierarchy to refer to the "least possible stage" where something occurs (see [4, p. 34ff]). Similarly, in the proof that the forcing preserves the axioms we construct predense sets associated to certain stages in the V-hierarchy. To be able to adapt these proofs to the case of ZF^- we will introduce a preparatory forcing which does not add any new sets but converts the SetMK* model M^+ into a model of the form $L_\alpha[A]$ for some generic class predicate $A \subseteq ORD$ preserving SetMK* (relative to A). This will allow us to use the relativized L-hierarchy instead of the V-hierarchy to prove the Definability Lemma and verify that the axioms are preserved.

Such a preparatory forcing presents us with two difficulties: first we have to show that its forcing relation is definable and the forcing is pretame, so that we can infer from Proposition 18 that it preserves the axioms. Secondly we have to show that the predicate A, that was added by the forcing, can be coded into a subset of κ so as to avoid problems when going back to the MK* model.

To prove the pretameness of such a forcing we have to add a new axiom to SetMK*, namely a variant of Dependent Choice. To ensure that this axiom holds in M^+, we will add its class version to MK* and show that it is transformed to the appropriate set version using the coding introduced in the last section.

Definition 19 Let MK** consist of the axioms of MK* plus Dependent Choice for Classes (we denote this with DC_∞):

$$\forall \vec{X} \exists Y \varphi(\vec{X}, Y) \rightarrow \forall X \exists \vec{Z} \, (Z_0 = X \wedge \forall i \in ORD \, \varphi(\vec{Z} \upharpoonright i, Z_i))$$

where $\vec{X}$ is an α-length sequence of classes for some $\alpha \in ORD$, $\vec{Z}$ is an ORD-length sequence of classes and $Z \upharpoonright i$ is the sequence of the "previously chosen" $Z_j, j < i$.

In the resulting SetMK** model M^+, DC_∞ becomes a form of κ-Dependent Choice:

$$\forall \vec{x} \exists y \varphi(\vec{x}, y) \rightarrow \forall x \exists \vec{z} \, (z_0 = x \wedge \forall i < \kappa \, \varphi(\vec{z} \upharpoonright i, z_i))$$

where $\vec{x}$ is a $< \kappa$-length sequence of sets, $\vec{z}$ is a κ-length sequences of sets and $z \upharpoonright i$ is the sequence of the "previously chosen" $z_j, j < i$.

The coding between MK** and SetMK** works exactly as in the MK* case, we only have to prove that it transforms DC_∞ into DC_κ and vice versa.

Proposition 20

1. Let $\mathcal{M} = (M, \mathcal{C})$ be a β-model of MK**. Then we can define a model

$$M^+ = \{x \mid \text{there is a coding pair } (M_x, R_x) \text{ that codes } x\}$$

Then M^+ is the unique, transitive set that obeys the following properties:

a) $M^+ \models SetMK^{**}$,
b) $\mathcal{C} = P(M) \cap M^+$,
c) $M = V_\kappa^{M^+}$, κ is the largest cardinal in M^+ and strongly inaccessible in M^+.

2. Let M^+ be a model of SetMK** that has a strongly inaccessible cardinal κ, let $\mathcal{C} = P(M) \cap M^+$ and $M = V_\kappa^{M^+}$. Then $\mathcal{M} = (M, \mathcal{C})$ is a model of MK**.

Proof For 1.: Using the proof of Theorem 8 it only remains to show that M^+ is a model of κ-Dependent Choice, where κ is strongly inaccessible in M^+: $M^+ \models \forall \vec{x} \exists y \varphi(\vec{x}, y) \rightarrow \forall \vec{x} \exists \vec{z}(z_0 = x \land \forall i < \kappa \, \varphi(\vec{z} \upharpoonright i, z_i))$ where $\vec{x}, \vec{z}$ are κ-length sequences. So assume that $M^+ \models \forall \vec{x} \exists y \varphi(\vec{x}, y)$. From what we have show above, we know that $\vec{x}$ is an ordinal length sequence of elements in $\mathcal{M}$ and also y is an element of $\mathcal{M}$ (as these can be classes we will write them with upper case letters in $\mathcal{M}$). Let ψ be the second-order formula associated to φ, i.e. ψ is the formula that says exactly the same as φ only that its variables can be classes. Then by DC_∞ we have that $\forall \vec{X} \exists Y \psi(\vec{X}, Y) \rightarrow \forall \vec{X} \exists \vec{Z}(Z_0 = X \land \forall i \in ORD \, \psi(\vec{Z} \upharpoonright i, Z_i))$ where $\vec{X}, \vec{Z}$ are sequences of classes with ordinal length and $Z \upharpoonright i$ is the sequence of the previously "chosen" $Z_j, j < i$. As before all the classes mentioned here are elements of M^+ where $\vec{Z}$ is a κ-length sequence and so we have proven the κ-Dependent Choice.

For 2.: Again we only have to proof the case of DC_∞ and this is an direct analog to the proof of the Comprehension Axiom in the proof of Proposition 13.   □

Lemma 21 *Let M^+ be a model of SetMK** with largest cardinal κ and P be an M^+-definable class forcing notion. Then if P is $\leq \kappa$-closed it is $\leq \kappa$-distributive.*

Proof Let $p \in P$ and $\langle D_i \,|\, i < \beta \rangle$ is an M^+ definable sequence of dense classes, $\beta \leq \kappa$, and we want to show that there is a $q \leq p$ meeting each D_i (q meets D_i if $q \leq q_i \in D_i$ for some q_i). As we have shown that P is $\leq \kappa$-closed we want to construct a descending sequence $p_0 \geq p_1 \geq \ldots \geq p_i \geq \ldots$ ($i < \beta$) with $p_i \in D_i$ for all $i < \beta$. Here we need the SetMK** version of the Dependent Choice Axiom we added to MK*: Recall that κ-Dependent Choice says that $\forall \vec{x} \exists y \varphi(\vec{x}, y) \rightarrow \forall \vec{x} \exists \vec{z}(z_0 = x \land \forall i < \kappa \, \varphi(\vec{z} \upharpoonright i, z_i))$ where $\vec{x}$ is a $< \kappa$-length sequence of sets, $\vec{z}$ is a κ-length sequences of sets and $z \upharpoonright i$ is the sequence of the previously "chosen" $z_j, j < i$. If we take $\varphi(\vec{x}, y)$ to mean that "$\vec{x}$ is a descending sequence of conditions, $x_i \in D_i$ for $i < \text{length} \vec{x}$, y is a lower bound for $\vec{x}$ and $y \in D_{length_{\vec{x}}}$" then we know that we can find a descending sequence $p_0 \geq p_1 \geq \ldots \geq p_i \geq \ldots (i < \beta)$ with $p_i \in D_i$ for all $i < \beta$ such that there is an $q \in P$ with $q \leq p$ and $q \leq p_i$ for all $i < \beta$ and so q meets all D_i.   □

Theorem 22 *Let M^+ be a model of SetMK** with largest cardinal κ and let κ^* denote the height of M^+. Then there is an M^+-definable forcing P such that the Definability Lemma holds and P is pretame, which adds a class predicate $A \subseteq \kappa^*$ such that $M^+ = L_{\kappa^*}[A]$ and $(M^+, A) \models$ SetMK** relativized to A.*

Proof Let $P = \{p : \beta \rightarrow 2 \,|\, \beta < \kappa^*, p \in M^+\}$ and let G be P-generic over M^+. Let $\bigcup G = g : \kappa^* \rightarrow 2$ and $A = \{\gamma < \kappa^* \,|\, g(\gamma) = 1\}$. Note that G is an amenable predicate, i.e. $G \cap a$ belongs to M^+ for every $a \in M^+$ and P is $\leq \kappa$-closed, as for every $\lambda \leq \kappa$ and every descending sequence $p_0 \geq p_1 \geq \ldots \geq p_i \geq \ldots (i < \lambda)$ there is $q = \bigcup_{i < \lambda} p_i \in P$ such that $\forall i < \lambda \, q \leq p_i$.

To show that the forcing relation is definable in the ground model, we will concentrate on the atomic cases "$p \Vdash \sigma \in \tau$" and "$p \Vdash \sigma = \tau$". Then the other cases follow by induction. For $p \Vdash \sigma \in \tau$ first consider the case where the length of

p is larger then the ranks of σ and τ (i.e. there is an γ such that $rank\,\sigma, rank\,\tau < \gamma$ and $Dom(p) > \gamma$). Then the question if $\sigma^G \in \tau^G$ is already decided by p, meaning that $\sigma^G \in \tau^G$ exactly when $\sigma^p \in \tau^p$ with $\tau^p = \{\pi^p \mid \langle \pi, q \rangle \in \tau, p \leq q\}$ as p "has no holes" and therefore a condition that extends p will never change the decisions made below the length of p. This now defines the forcing relation because P doesn't add any new sets and therefore σ^p and τ^p are already elements of the ground model. If p is not large enough to decide if σ^G is an element of τ^G, then we have to check that every q that extends p decides that this is the case so we get the definition "$p \Vdash \sigma \in \tau \leftrightarrow \forall q \leq p\,(|q| > rank\,\sigma, rank\,\tau \rightarrow \sigma^q \in \tau^q)$". The definitions for the "$=$" case can be given the same way and so the forcing is definable. The Truth Lemma then follows from Definability by the usual arguments.

Next we want to show that P is pretame: As P is $\leq \kappa$-closed, we know by Lemma 21 that P is $\leq \kappa$-distributive. Then P is also pretame for sequences of dense classes of length $\leq \kappa$ and therefore P is pretame.

We have shown that P doesn't add any new sets to the extension but a subclass $A \subset \kappa^*$. So the forcing just reorganizes M^+ and adds A as a predicate. Then every set of ordinals from M^+ is copied into an interval of the generic and so every set of ordinals and therefore also every set is coded by A. Also as A adds no new sets it holds that $L_{\kappa^*}[A] \subseteq M^+$. It follows that $M^+[G] = L_{\kappa^*}[A]$ and therefore already $M^+ = L_{\kappa^*}[A]$.

It remains to show that $(M^+, A) \models (SetMK^{**})^A$, i.e. $SetMK^{**}$ holds for formulas which can mention A as a predicate. As P preserves the strongly inaccessibility of κ it follows by Proposition 18 that $M^+[G] \models SetMK^*$ and that means that $(M^+, A) \models SetMK^*$. But as the Comprehension and Bounding can mention the generic this implies that $(M^+, A) \models (SetMK^*)^A$. For the DC_κ note that by adding A we now have a global well-order of the extension. That means that if we have a $< -\kappa$ sequence $\vec{x}$ in $M^+[G]$ such that $\forall \vec{x}\,\exists y\,\varphi(\vec{x}, y)$ and we want to find a κ-length sequence $\vec{z}$ such that $\forall x\exists \vec{z}\,(z_0 = x \wedge \forall i < \kappa\,\varphi(\vec{z} \restriction i, z_i))$ we can just take z_i to be least so that $\varphi(\vec{z} \restriction i, z_i)$ for each i. $\qquad\square$

As our ultimate goal is to go back to an MK^{**} model, we want to show that the predicate A can be coded into a subset of κ:

Theorem 23 *Let (M^+, A) be a model of $SetMK^{**}$ relativized to a predicate A, with largest cardinal κ and let κ^* denote the height of (M^+, A), where A is the generic predicate added by the forcing P in Theorem 22 and $M^+ = L_{\kappa^*}[A]$. Then we can force that there is a $X \subseteq \kappa$ such that $L_{\kappa^*}[A] \subseteq L_{\kappa^*}[X]$, $SetMK^{**}$ is preserved and κ remains strongly inaccessible.*

Proof To get A definable in $M^+[X]$, for some $X \subseteq \kappa$, we want to use an almost disjoint forcing which codes the predicate A into such an X. The forcing will be along the following lines: we will need to define a family S of almost disjoint sets (i.e. for $x, y \subseteq \kappa$, x and y are almost disjoint if $x \cap y$ is bounded in κ) A_β which we will use to code the predicate $A \subseteq \kappa^*$ into an X. We will define A_β to be the least subset of κ (i.e. least in the canonical well-order of $L_{\kappa^*}[A \cap \beta]$) in $L_{\kappa^*}[A \cap \beta]$ which is distinct from the $A_{\bar{\beta}}$ for $\bar{\beta} < \beta$. The idea is that we can decode A in $L_{\kappa^*}[X]$ if we know the A_β's. But as A is a proper class we don't know that we can always find such distinct A_β's. So we will have to assume that the cardinality of β is at most κ not only in $L_{\kappa^*}[A]$ but also in $L_{\kappa^*}[A \cap \beta]$ because now to find an A_β distinct from each $A_{\bar\beta}$, $\bar\beta < \beta$, we can list these $A_{\bar\beta}$'s as $\langle A_i \mid i < \kappa \rangle$ and obtain A_β by diagonalization. To fulfill that assumption however we have to "reshape" A into a predicate A' that has the property that if $\beta < \kappa^*$ then the cardinality of β is $\leq \kappa$ in $L_{\kappa^*}[A' \cap \beta]$. Then we can code A as the even part of A' to get $(M^+, A') \models (SetMK^{**})^{A'}$ and finally code A' by a subset of κ.

So the proof consists of two steps: First we have to show that we can reshape A and then we have to force with an almost disjoint forcing to show that the reshaped predicate A' can be coded into a subset of κ, preserving SetMK** in each step.

Step 1: We add a reshaped predicate A' over $(L_{\kappa^*}[A], A)$ by the following forcing:

$$P = \{p : \beta \to 2 \mid \kappa \leq \beta < \kappa^*, \forall \gamma \leq \beta \, (L_{\kappa^*}[A \cap \gamma, p \restriction \gamma] \models |\gamma| \leq \kappa)\}$$

The main obstacle is to show that P is definably-distributive, i.e. we have to show that for a $p \in P$ and (M^+, A)-definable sequences of dense classes of set-length $\langle D_i \mid i < \alpha \rangle$ for all $\alpha \leq \kappa$, there is a $q \leq p$ meeting each D_i with $q \in P$.

Claim 24 P is definably-distributive.

Proof Note that it suffices to show definable-distributivity for κ; so we consider an (M^+, A)-definable sequence of dense classes $\langle D_i \mid i < \kappa \rangle$. We want to define a descending sequence of conditions $p \geq p_0 \geq p_1 \geq \ldots$ where $p_i \geq q$, $q \in P$ and $p_{i+1} \in D_i$ for each $i < \kappa$. To show that the p_i are indeed conditions we have to show that $L_{\kappa^*}[A \cap \gamma, p_i \restriction \gamma] \models |\gamma| \leq \kappa$ for every $\gamma \leq |p_i|$. In the following we will use the fact that a condition is always extendible to any length $< \kappa^*$: $\forall p \, \forall \beta < \kappa^* \, \exists q \leq p, |q| \geq \beta, q \in P$. This holds because there is an $x \subseteq \kappa$ such that β is coded by x and $p * x \in P$ and has length $|p| + \kappa$. If this is still below β we can lengthen p further by a sequence of 0's: $q = p * x * \vec{0}$. This will again be an element of P as we know from the information in the code x of β that the ordinals will collapse.

First, we assume that the sequence of dense classes is Σ_1-definable, i.e. $\{(q, i) \mid q \in D_i\}$ is Σ_1-definable with parameter.

As we have seen that every condition is extendible, we can extend p to catch a parameter $x \in L_{|p|}[A]$ such that the sequence of the D_i is Σ_1-definable with parameter x. Let p_0 be this extension of p. Then, as we have Global Choice, we can consider the $<_{(M^+, A)}$-least pair (q_0, w_0) such that $q_0 \leq p_0$ and w_0 witnesses "$q_0 \in D_0$". Then we choose p_1 such that p_1 is a condition which extends q_0 such that $w_0 \in L_{|p_1|}[A \cap |p_1|]$. Now we define p_2 in the same way: Choose (q_1, w_1) such that $q_1 \leq p_1$ and w_1 witnesses "$q_1 \in D_1$". Then let $p_2 \leq q_1$ such that $w_1 \in L_{|p_2|}[A \cap |p_2|]$. Define the rest of the successor cases (p_{n+1}, w_{n+1}) similarly.

For the first of the limit cases, let $p_\omega = \bigcup_{n < \omega} p_n$ and we claim that $p_\omega \in P$. So we have to show that $\forall \gamma \leq |p_\omega|$, γ collapses to κ using only $A \cap \gamma$ and $p_\omega \restriction \gamma$. We know that if $\gamma < |p_\omega|$ then $\gamma < |p_n|$ for some n. So we only have to consider the case where $\gamma = |p_\omega|$. It follows from the construction of the p_n's that the sequence $\langle p_n \mid n < \omega \rangle$ is definable over $L_{|p_\omega|}[A \cap |p_\omega|, p_\omega]$ and is a cofinal sequence in p_ω, i.e.

it converges to p_ω. Then also the sequence of the lengths of the p_n's, $\langle |p_n| \mid n < \omega \rangle$ is definable over $L_{|p_\omega|}[A \cap |p_\omega|, p_\omega]$ and converges to $|p_\omega|$. As we know that $|p_n|$ collapses to κ for every $n < \omega$, we know that in $L_{|p_\omega|}[A \cap |p_\omega|, p_\omega]$ $|p_\omega|$ definably collapses to κ. So $L_{|p_\omega|+1}[A \cap |p_\omega|, p_\omega] \models |p_\omega|$ is collapsed to κ. The other limit cases can be handled in the same way.

Now we go to the Σ_2-definable case. Note that we cannot simply copy the construction of the p_n-sequence because the witness q_{n+1} we need for the definition of the next p_{n+1} will now be a solution to a Π_1-statement and will therefore not be absolute in the other models. But we know that for $V = L_{\kappa^*}[A]$ it holds that $\forall \alpha < \kappa^* \, \exists \beta \leq \kappa^*, \alpha < \beta$ such that $L_\beta[A]$ is Σ_n-elementary in $L_{\kappa^*}[A]$. This holds because for a pair α, n we can take the Σ_n-Skolem Hull N of α in $L_{\kappa^*}[A]$. Then in M we have a solution for every Σ_n-property with parameters $< \alpha$, M is transitive and bounded by Class-Bounding. Then there is a $\beta \leq \kappa^*$ such that M is equal to $L_\beta[A]$.

So we can always find models that are Σ_1-elementary submodels of (M^+, A) in which we can carry out the definition of the sequence of conditions: As before we choose for every $n < \omega$ a pair (q_n, w_n) such that $q_n \leq p_n$ such that w_n witnesses "$q_n \in D_n$" and then let $p_{n+1} \leq q_n$ such that $w_n \in L_{|p_{n+1}|}[A \cap |p_{n+1}|, p_{n+1}]$ and $L_{|p_{n+1}|}[A \cap |p_{n+1}|, p_{n+1}]$ is an Σ_1-elementary submodel of $L_{\kappa^*}[A]$. This also holds in the limit case by using the same construction we did for the Σ_1 case where again the model $L_{|p_\omega|+1}[A \cap |p_\omega|, p_\omega]$ is an Σ_1-elementary submodel of $L_{\kappa^*}[A]$. The same can be done for all the Σ_m-definable cases. $\square$

Now that we know that P is $\leq \kappa$-distributive, we know that P is $\leq \kappa$-pretame and therefore $(M^+, A, A') \models (SetMK^{**})^{A, A'}$ (similar to proof of Theorem 22 by using Proposition 18 and the fact that there is a global well-order of the extension). Then we can code A to be the even part of A' and we get a model $(M^+, A') \models (SetMK^{**})^{A'}$. It remains to show that A' can be coded into a subset of κ.

Step 2: Code A' into $X \subseteq \kappa$. As we know that A' is reshaped we can define a collection of sets $S = \langle A_\beta \mid \beta < \kappa^* \rangle$ in the following way: let A_β be the least $B \subseteq \kappa$ in $L_{\kappa^*}[A' \cap \beta]$ such that $B \notin \{A_{\bar\beta} \mid \bar\beta < \beta\}$. S can be turned into a collection $S' = \langle A'_\beta \mid \beta < \kappa^* \rangle$ of almost disjoint sets A'_β by mapping every set to the set of codes of its proper initial segments: $B \subseteq \kappa$ is mapped to $B' = \{\text{Code}\,(B \cap \alpha) \mid \alpha < \kappa\} \subseteq \kappa$. Then for two distinct subsets B and C of κ, $|B' \cap C'| < \kappa$ and therefore they are almost disjoint. We want to show that we can code A' by a subset X of κ by showing that $X \cap A'_\beta$ is bounded if and only if $\beta \in A'$. This can be done by a forcing Q with the conditions (g, S) where $S \subseteq A'$, $|S| < \kappa$ and g is an element of $^{<\kappa}2$. Extension is defined by: $(g, S) \geq (h, T)$ iff h extends g, $S \subseteq T$ and if $\beta \in S$ and $h(\gamma) = 1$ for a $\gamma \in A'_\beta$ then $g(\gamma) = 1$. Note that two conditions with the same first component $\langle g, S \rangle$ and $\langle g, T \rangle$ are compatible because we can always find a common extension $\langle g, S \cup T \rangle$. Thus a function which maps every element of a definable antichain into its first component is injective (as otherwise the conditions would be compatible). So we have injectively mapped a definable class to a set as there are only κ many first components. By Bounding such a function exists as a set and so Q is set-c.c., i.e. every definable antichain is only set-sized. Then Q is pretame, as every definable dense class can be seen as an antichain. Now let G be a Q-generic, $G_0 = \bigcup\{g \mid (g, S) \in G\}$ and $X = \{\gamma \mid G_0(\gamma) = 1\}$. we argue that we can find the almost disjoint sets in $L_{\kappa^*}[X]$ because A' is reshaped and therefore it holds for any β that $|\beta| \leq \kappa$ in $L_{\kappa^*}[A' \cap \beta]$. So after X has decoded $A' \cap \beta$ it can find A'_β and then continue the decoding in the following way: $\beta \in A'$ if there is an $(g, S) \in G$ with $\beta \in S$ and by the definition of extension if $G_0(\gamma) = 1$ for a $\gamma \in A'_\beta$ then $g(\gamma) = 1$. So $X \cap A'_\beta = \{\gamma \mid g(\gamma) = 1\} \cap A'_\beta$ and that is bounded and therefore we have a code of A' by X via

$$X \cap A'_\beta \text{ is bounded if and only if } \beta \in A'.$$

As this forcing is κ-closed (i.e. closed for $< \kappa$ sequences), κ stays regular and therefore strongly inaccessible and by Proposition 18 SetMK* is preserved and by Proposition 18 SetMK* is preserved.   □

We have seen how definable hyperclass-forcing can be carried out over a model $\mathcal{M}$ of MK**: First we go to the related SetMK** model M^+ (Theorem 8). Then in order to be able to force over this model, we change M^+ to a model $L_{\kappa^*}[A]$ for a generic predicate A (Theorem 22). Finally we showed how to code A into a subset $X \subseteq \kappa$ to avoid having an undefinable predicate once we go back to the extension of the original MK** model (Theorem 23). At this point we can force with any desirable pretame definable class-forcing over $L_{\kappa^*}[X]$, go back to MK** and get the desired definable hyperclass-forcing over MK**.

So we have given a template which allows us to do definable hyperclass-forcing over MK**. In the following we will show how to use this template to produce minimal β-models of MK**.

4   Minimal β-Models of MK**

As an application of definable hyperclass forcing we will show that every β-model of MK** can be extended to a minimal β-model of MK** via the use of SetMK** models. Here a minimal model $M(S)$ of SetMK** is the least transitive model of SetMK** containing a real S and equivalently a minimal β-model $\mathcal{M}(S)$ of MK** is the least β-model of MK** containing a real S.[3] For that we will use and modify the template developed in the last section: We start with an arbitrary β-model $\mathcal{M} = (M, \mathcal{C})$ of MK** and from that we get the corresponding model M^+ of SetMK** (by Theorem 8) with $M = V_\kappa^{M^+}$ and $\mathcal{C} = P(M) \cap M^+$ where κ is strongly inaccessible in M^+. Let κ^* denote the height of M^+ and apply Theorem 22 to arrive at $M^+ = L_{\kappa^*}[A]$ where $A \subseteq \kappa^*$ and (M^+, A) satisfies SetMK$(**)$ relative to A. We now

---

[3]We can see here that it is vital to restrict ourselves to β-models in order to talk about minimal models of MK by comparing this to the situation in ZFC: There it also only makes sense to talk about minimal models containing a real for well-founded models (and not for ill-founded models). So by making the transformation from MK to SetMK we have to restrict ourselves to β-models.

show that we can extend M^+ to a minimal model of SetMK** and then go back to an MK** model, which will be a minimal β-model of MK**.

Theorem 25 *Every β-model of MK** can be extended to a minimal β-model of MK** with the same ordinals.*

Proof First we use the template described above to arrive at the model $L_{\kappa^*}[A]$ and then we will code the predicate A into a subset of κ by using Theorem 23 with a small modification in the "reshaping" forcing. Instead of forcing that each $\gamma < \kappa^*$ collapses in $L_{\kappa^*}[A \cap \gamma, p \upharpoonright \gamma]$, we will force it to already collapse instantly in the next level, i.e. in $L_{\gamma+1}[A \cap \gamma, p \upharpoonright \gamma]$. So the forcing will be:

$$P = \{p : \beta \to 2 \mid \kappa \le \beta < \kappa^*, \forall \gamma \le \beta \, (L_{\gamma+1}[A \cap \gamma, p \upharpoonright \gamma] \models |\gamma| \le \kappa)\}$$

The proof that P is definably-distributive then works in exactly the same way. As in Theorem 23 we can code A to be the even part of the predicate A' added by the reshaping forcing which in turn can be coded into an $X \subseteq \kappa$ by an almost disjoint forcing. This gives us that there are no SetMK** models containing X of height between κ and κ^*: In the reshaping forcing we destroyed the Replacement axiom level by level relative to A and in the almost disjoint coding we can now choose the codes instantly level-by-level (i.e. every code for γ appears in $L_{\gamma+1}[X]$). So A' can be recovered level-by-level from X and therefore Replacement is also destroyed level-by-level relative to X. We arrive at a SetMK** model $L_{\kappa^*}[X]$, with $X \subseteq \kappa$, which is the least transitive ZFC$^-$ model containing X (again κ remains regular and indeed strongly inaccessible, because the almost disjoint coding is κ-closed).

We will extend this to a minimal model of SetMK** in two steps: First we extend $L_{\kappa^*}[X]$ to a model $L_{\kappa^*}[Y]$ such that no cardinal $\bar{\kappa} < \kappa^*$ can serve as a "source" for a SetMK** model (i.e. is the largest cardinal of a SetMK^{**} model containing $Y \cap \bar{\kappa}$) and second we show that we can add a real S such that in $L_{\kappa^*}[S]$ there are no SetMK** models containing S below κ^*. Then it only remains to show that from $L_{\kappa^*}[S]$ we can go back to a minimal β-model of MK**.

Step 1: With the modification of Theorem 23, we have shown that there are no SetMK** models containing X between κ and κ^*. But it could still be that there exist cardinals below κ which are sources for SetMK** models. We will destroy these cardinals by shooting a club through a "fat-stationary" set which has no such cardinals and then force all limit cardinals to belong to this club.

So let $S = \{\bar{\kappa} < \kappa \mid \bar{\kappa}$ is a limit cardinal and for all $\bar{\beta} > \bar{\kappa}$, if $L_{\bar{\beta}}[X \cap \bar{\kappa}] \models ZFC^-$ then $L_{\bar{\beta}}[X \cap \bar{\kappa}] \not\models \bar{\kappa}$ is strongly inaccessible$\}$.

Definition 26 S is *fat-stationary* if for every club C in $L_{\kappa^*}[X]$, $S \cap C$ contains closed subsets of any order type less than κ.

We prove the following:

Lemma 27 *S is fat-stationary and there is a κ-distributive (i.e. $< \kappa$ distributive) forcing of size κ that adds a club $C \subseteq S$.*

Proof First we will show that S is stationary with respect to clubs in $L_{\kappa^*}[X]$. So suppose C is a club in $L_\alpha[X]$ for an $\alpha < \kappa^*$. We build an increasing sequence $\langle M_n \mid n < \omega \rangle$ of sufficiently elementary submodels of $L_\alpha[X]$ in the following way: Let M_0 be the Σ_1-Skolem Hull of $\omega \cup \{X, C\}$ in $L_\alpha[X]$. Then $C \in M_0$ and $\kappa_0 = sup(M_0 \cap \kappa)$ is a cardinal. Next, let M_1 be the Σ_1-Skolem Hull of $\kappa_0 + 1 \cup \{X, C\}$ in $L_\alpha[X]$ and $\kappa_1 = sup(M_1 \cap \kappa)$. Repeat this construction for all $n < \omega$. Then this sequence of elementary submodels is definable over $M_\omega = \bigcup_{n<\omega} M_n$ and $\kappa_\omega = sup_{n<\omega}\kappa_n < \kappa$ is a cardinal in C as C is closed, unbounded in κ. Also κ_ω is an element of S because if $L_{\bar\alpha}[X \cap \kappa_\omega]$ is the transitive collapse of M_ω then there are no ZFC$^-$ models containing $X \cap \kappa_\omega$ of height $< \bar\alpha$ (by elementarity), of height $= \bar\alpha$ because $\langle \kappa_n \mid n < \omega \rangle$ is definable over it (and so κ_ω becomes definably singular) and any ZFC$^-$ model containing $X \cap \kappa_\omega$ of height $> \bar\alpha$ sees that κ_ω has cofinality ω (as the κ_n-sequence is an element of it).

To show that S is fat-stationary we can use the same proof as for stationarity except one uses a longer δ-sequence of elementary submodels, for δ a limit cardinal less than κ.

Now for the second part of the Lemma we can force with a set-forcing to add a club. Here we will closely follow the proof of the ZFC version of this claim, as proven in [1] (see there for more details). Let $Q = \{p \mid p$ is a closed, bounded subset of $S\}$ be a forcing notion ordered by end-extensions: $q \leq p$ iff $p = q \cap (sup(p) + 1)$. For G Q-generic over $L_{\kappa^*}[X]$ let $C = \bigcup G$. Then C is closed and unbounded and a subset of S. To show that Q is κ-distributive we have to show that for every $\tau < \kappa$ and sequence $\mathcal{D} = \langle D_i \mid i \in \tau \rangle$ of open, dense subsets of Q, $\bigcap_{i<\tau} D_i$ is dense in Q. Now we can define a sequence of elementary substructures $\langle M_\alpha \mid \alpha < \kappa \rangle$ of $L_{\kappa^*}[X]$ such that $c_\alpha = M_\alpha \cap \kappa$ is an ordinal and $\langle c_\alpha \mid \alpha < \kappa \rangle$ is an increasing and continuous sequence cofinal in κ. Let E be the collection of the c_α, $\alpha < \kappa$. Because S is fat-stationary, $S \cap E$ contains a closed subset A of order-type $\tau + 1$. Then in the model M_α, with $\alpha = sup(A)$, we can define an increasing sequence $\langle p_i \mid i < \tau \rangle$, such that $p_i \in Q$ and $p_{i+1} \in D_i \cap M_\alpha$. We can define $p_\tau = \bigcup_{i<\tau} p_i \cup \{\alpha\}$ and this will be in $\bigcap_{i<\tau} D_i$. Note that this (set-) forcing is an element of $L_{\kappa^*}[X]$ and therefore preserves ZFC$^-$. Furthermore, as this forcing doesn't add sets of size $< \kappa$, κ stays strongly inaccessible and SetMK* is preserved because of Proposition 18.   □

Let X' be the join of X with the club we added. Then $X' \subseteq \kappa$ and the resulting model is $L_{\kappa^*}[X']$.

Lemma 28 *We can force all limit cardinals to belong to C with a forcing of size κ such that κ remains strongly inaccessible.*

Proof Enumerate C as follows: $C = \langle \bar\kappa_i \mid i < \kappa \rangle$. We may assume that each $\bar\kappa_i$ is a strong limit cardinal (as κ is strongly inaccessible we can thin out C). Then we can build an Easton product of collapses, where we collapse every $\bar\kappa_{i+1}$ to the successor of $\bar\kappa_i$ and therefore ensure that all limit cardinals below κ are limits of cardinals in C and therefore are themselves in C.

So for $i < \kappa$ consider $Col_i(\bar\kappa_i^+, \bar\kappa_{i+1})$, where the conditions are functions p with $dom(p) \subset \bar\kappa_i^+$, $|dom(p)| < \bar\kappa_i^+$ and $range(p) \subset \bar\kappa_{i+1}$. Cardinals below $\bar\kappa_i^+$ and above $\bar{\kappa}_{i+1}^{\bar{\kappa}_i}$ are preserved (the size of the forcing is $\bar{\kappa}_{i+1}^{\bar{\kappa}_i}$) and in the extension we have a function which maps $\bar{\kappa}_i^+$ onto $\bar{\kappa}_{i+1}$.

Now we can build the Easton product (product with Easton support) of these collapses for every $i < \kappa$: A condition p in this forcing is a function such that $p = \langle p_i \,|\, i < \kappa \rangle \in \Pi_{1 < \kappa} \mathrm{Col}_i(\bar{\kappa}_i^+, \bar{\kappa}_{i+1})$ and the forcing is ordered by end-extension. p has Easton support, i.e. for every inaccessible cardinal λ, $|\,\{\alpha < \lambda \,|\, p(\alpha) \neq \emptyset\}\,| < \lambda$. As usual with Easton Products the forcing notion P can be split into two parts $P(\leq \lambda) = \Pi_{1 \leq \lambda} \mathrm{Col}_i(\bar{\kappa}_i^+, \bar{\kappa}_{i+1})$ and $P(> \lambda) = \Pi_{\lambda < 1 < \kappa} \mathrm{Col}_i(\bar{\kappa}_i^+, \bar{\kappa}_{i+1})$ for every regular cardinal λ. For this reason and as each $\bar{\kappa}_i$ is a strong limit, each collapse from $\bar{\kappa}_{i+1}$ to $\bar{\kappa}_i^+$ will not be affected by the other collapses and κ remains regular and strong limit. Furthermore, as this forcing is in $L_{\kappa^*}[X']$ (it is of size κ) it preserves SetMK**.

Because of the unboundedness of C, every limit cardinal is also a limit of cardinals in C and therefore, as C is closed, it is an element of C. □

We conclude Step 1 by choosing X'' to be the join of X' and the above Easton product. Then we arrive at a model $L_{\kappa^*}[X'']$ with $X'' \subseteq \kappa$ such that for every cardinal $\bar{\kappa} < \kappa^*$ there is no model of ZFC$^-$ containing $X'' \cap \bar{\kappa}$ in which $\bar{\kappa}$ is inaccessible and therefore $\bar{\kappa}$ is not a source for a SetMK** model.

Step 2: We want to extend the results from the last step to hold for all ordinals, i.e. for all ordinals $\alpha < \kappa^*$ there is no SetMK** model of height $< \kappa^*$ containing a real S in which α is strongly inaccessible. This makes use of Jensen coding and a result about admissibility spectra which is connected to it. We will use these results as black boxes and will only state the main definitions and theorems here:

Theorem 29 (Jensen Coding) *Suppose that $\langle M, A \rangle$ is a transitive model of ZFC, i.e. M is a transitive model of ZFC, $A \subseteq M$ and Replacement holds in M for formulas mentioning A as a unary predicate. Then there is an $\langle M, A \rangle$-definable class forcing P such that if $G \subseteq P$ is P-generic over $\langle M, A \rangle$, then:*

a) *$\langle M[G], A, G \rangle \models ZFC$.*
b) *For some $R \subseteq \omega$, $M[G] \models V = L[R]$ and $\langle M[G], A, G \rangle \models A, G$ are definable from the parameter R.*

The very elaborate proof of this result uses Jensen's fine structure theory and, very roughly, the forcing involved consists of three components: an almost disjoint coding at successor cardinals, a variation thereof at limit cardinals and a reshaping forcing.[4]

Definition 30 Let T be the theory of ZF without Power Set and with Replacement restricted to Σ_1 formulas. Then $\Lambda(R)$ for a real R denotes the admissibility spectrum of R and is defined as the class of all ordinals α such that $L_\alpha[R] \models T$, i.e. the class of all R-admissible ordinals.

---

[4]An detailed account of this can be found in [3], a simplified version of the proof can be found in [4].

Theorem 31 (S.-D. Friedman)[5] *Suppose φ is Σ_1 and $L \models \varphi(\kappa)$ whenever κ is an L-cardinal. Then there exists a real $R <_L 0^{\sharp}$ such that $\Lambda(R) \subseteq \{\alpha \mid L \models \varphi(\alpha)\}$ and R is cardinal preserving over L.*

We will use these theorems to prove the following lemma:

Lemma 32 *We can extend the model $L_{\kappa^*}[X'']$ to be of the form $L_{\kappa^*}[S]$ for a real S such that $L_{\kappa^*}[S] \models SetMK^{**}$ and whenever $\bar{\alpha} < \kappa^*$ is an ordinal there is no model of $SetMK^{**}$ of height $< \kappa^*$ containing S in which $\bar{\alpha}$ is strongly inaccessible.*

Proof First we add a real R to the resulting model of Step 1 and get a model $L_{\kappa^*}[R] \models SetMK^{**}$. This can be done by using Jensen coding over the model $L_{\kappa^*}[X'']$. Although we start from a model of ZFC$^-$ rather than ZFC our model is of the form $L_{\kappa^*}[X'']$ and therefore we can use the standard pretameness argument for Jensen coding to show that ZFC$^-$ is preserved.[6] Also, κ will still be inaccessible in the extension because Jensen coding preserves inaccessibles.[7] Note that the result from Step 1 still holds: In $L_{\kappa^*}[R]$ we have that if $\bar{\kappa} < \kappa^*$ is a cardinal then there is no transitive model of SetMK** containing R in which $\bar{\kappa}$ is inaccessible as otherwise there would have been such a model containing $X'' \cap \bar{\kappa}$ as the latter is coded by R in $L_{\bar{\kappa}}[R]$.

Now we use Theorem 31 relativized to the real R to produce a new real S such that this holds for ordinals $\bar{\kappa}$. Theorem 31 works in the context of ZFC$^-$ for the same reasons as for Jensen coding. Note that $L_{\kappa^*}[R] \models \varphi(\bar{\kappa})$ for every $L_{\kappa^*}[R]$-cardinal $\bar{\kappa}$ where $\varphi(\alpha)$ is the following Σ_1 property with parameter R: "Either $L_{\alpha}[R] \models$ there is a largest cardinal or there is $\beta > \alpha$ such that $L_{\beta}[R] \models \alpha$ is singular and for all γ with $\alpha < \gamma < \beta$, $L_{\gamma}[R] \not\models ZFC^-$". This property says that either α is a successor or we can "see" the singularity of α before we see a ZFC$^-$ model for which it could be a source. Then by Theorem 31 there exists a real S generic over $L_{\kappa^*}[R]$ such that $L_{\kappa^*}[S] \models SetMK^{**}$ and $\Lambda(S) \subseteq \{\alpha \mid L[R] \models \varphi(\alpha)\}$. As α which is inaccessible in a model of ZFC$^-$ containing S is S-admissible, we get the desired property for all ordinals.   □

We now have a minimal model $L_{\kappa^*}[S]$ of SetMK**, i.e. the least transitive model of SetMK** containing S. It only remains to show that by going back to MK** we arrive at a minimal β-model of MK^{**}. To see that consider the model $(L_{\kappa}[S], \mathcal{C})$ where $\mathcal{C}$ consists of the subsets of $L_{\kappa}[S]$ in $L_{\kappa^*}[S]$. This is a β-model of MK** by Proposition 13 and it is the least such model containing S because otherwise there exists a β-model $(N, \mathcal{C}') \subset (L_{\kappa}[S], \mathcal{C})$, $(N, \mathcal{C}') \models MK^{**}$ containing S that would give rise to a model N^+ of SetMK**. If we then go to the $L[S]$ of N^+ we arrive at a model $L_{\alpha}[S]$ for some $\alpha < \kappa^*$ which is a model of SetMK**. This is a contradiction to the minimality of $L_{\kappa^*}[S]$.   □

---

[5]See [4], Theorem 7.5, p. 142.

[6]See [4, Chapter 4].

[7]This follows from an property called diagonal distributivity (see [4], p. 37).

5 Further Work and Open Questions

These results opens up a wider area of further research and related open questions.

In the definition of definable hyperclass forcing we used the restriction to β-models of MK^* to make the coding of a transitive $SetMK^*$ model work. It would be interesting to investigate what happens if we drop this restriction:

Question How can definable hyperclass forcing be defined for an arbitrary model of MK^{**}?

Dropping the β-model assumption for the coding would mean to work only internally in the MK^{**} model and restricting ourselves to just coding pairs. We are confident that this can be done, but there are many details to be worked out.

In Theorem 22 we introduced a preparatory forcing to convert the $SetMK^*$ model M^+ into a model of the form $L_\alpha[A]$ for some generic class predicate $A \subseteq ORD$. This entails a string of further modifications to our original setup: We add the Dependent Choice axiom, arriving at the theory $SetMK^{**}$, and extend the model M^+ to a model where the predicate A is coded into a subset of κ (see Theorem 23). We can ask if and how these modifications could be circumvented:

Question Can we avoid the use of the preparatory forcing by restricting the original MK^* model and/or the hyperclass forcing $\mathbb{P}$?

At the moment, let us just remark that if the classes already carry a "good" wellorder, i.e. a wellorder "$\prec$" such that for each X there is a class $\prec_X$ such that the predecessors of X in $\prec$ are the $(\prec_X)_i$, $i < Ord$, and $X \mapsto \prec_X$ is 2nd order definable, then there is no need for the preparation (or for MK^{**}, MK^* is enough) for then M^+ will already have the necessary definable hierarchy for class forcing. This will happen if the given MK^* model is already minimal, or more generally, if M^+ satisfies $V = L[X]$ for some subset X of κ.

In this paper we consider three variants of the axioms of Morse-Kelley; the standard form MK, the extension via Class-Bounding, here called MK^* and the additional extension with Dependent Choice, called MK^{**}. The obvious question presents itself, which is how they are related:

Question Assuming just the consistency of MK, are there models of MK that don't satisfy MK^* and models of MK^* that don't satisfy MK^{**}?

Another fruitful topic is the analogy between Morse-Kelley and second-order arithmetic.

Question What results and questions can be transferred from the context of Morse-Kelley class theory to second-order arithmetic and vice versa?

As an example for this transfer, let us consider the question of minimal β-models of MK^{**}. It can be translated to minimal β-models of second-order arithmetic (plus Dependent Choice) in the following way:

Theorem 33 *Every β-model of second-order arithmetic with Dependent Choice can be extended to a minimal β-model of second-order arithmetic with Dependent Choice with the same ordinals.*

Proof outline Starting with a β-model of second-order arithmetic we can go to a related model of ZFC$^-$ where the inaccessible cardinal κ is now simply $\aleph_0$. Then the question about models below the largest cardinal becomes trivial and we can concentrate on the case of eliminating models of ZFC$^-$ between $\aleph_0$ and the height of the model α^*. First we change the ZFC$^-$ model to a model $L_{\alpha^*}[A]$ in a way analogous to Theorem 22. Then we can adapt the proof of Theorem 23 in a similar way as we did in the beginning of the proof of Theorem 25: we code the predicate A into $X \subseteq \aleph_0$ with an almost disjoint forcing, where we first reshape A to a predicate A'. In this reshaping forcing we destroy the Replacement axiom level-by-level relative to A and therefore it is also destroyed level-by-level relative to X. We arrive at a model $L_{\alpha^*}[X]$, with $X \subseteq \aleph_0$, which is the least transitive ZFC$^-$ model containing X and from this we can go back to a minimal β-model of second-order arithmetic. $\qquad\square$

Of course, definable hyperclass forcing is not the last step in considering a hierarchy of forcing notions via their size. One could ask further:

Question What would a general hyperclass forcing look like and in which context can it be developed (a hypercass theory)? What would a hyperhyperclass forcings look like, i.e. a forcing where conditions are hyperclasses?

Here we developed a further step in this hierarchy after set forcing, definable class forcing and class-forcing in MK. We hope that it will serve as a basis for further fruitful research.

Acknowledgements The first author wants to thank the Austrian Academy of Sciences for their generous support through their Doctoral Fellowship Program. Both authors are grateful to the John Templeton Foundation for its generous support through Grant ID 35216, which supported the preparation of this article.

References

1. U. Abraham, S. Shelah, Forcing closed unbounded sets. J. Symb. Log. **48**(3), 643–657 (1983)
2. C. Antos, Class forcing in class theory, in *The Hyperuniverse Project and Maximality*, ed. by C. Antos, S.-D. Friedman, R. Honzik, C. Ternullo (Springer, Heidelberg, 2017). https://doi.org/10.1007/978-3-319-62935-3_1
3. A. Beller, R. Jensen, P. Welch, *Coding the Universe*. Lecture note Series (Cambridge University Press, Cambridge, 1982)
4. S.-D. Friedman, *Fine Structure and Class Forcing*. de Gruyter Series in Logic and Its Applications, vol. 3 (Walter de Gruyter, New York, 2000)
5. V. Gitman, J.D. Hamkins, Kelley-Morse set theory and choice principles for classes (in preparation)
6. V. Gitman, J. Hamkins, T.A. Johnstone, What is the theory ZFC without powerset? Math. Log. Q. **62**(4–5), 391–406 (2016)
7. P. Holy, R. Krapf, P. Lücke, A. Njegomir, P. Schlicht, Class forcing, the forcing theorem and Boolean completions. J. Symb. Logic **81**(4), 1500–1530 (2016)
8. A. Zarach, Unions of ZF$^-$-models which are themselves ZF$^-$-models, in *Logic Colloquium '80 (Prague, 1980)*. Studies in Logic and the Foundations of Mathematics, vol. 108 (North-Holland, Amsterdam, 1982), pp. 315–342

Multiverse Conceptions in Set Theory

Carolin Antos, Sy-David Friedman, Radek Honzik, and Claudio Ternullo

Abstract We review different conceptions of the set-theoretic multiverse and evaluate their features and strengths. In Sect. 1, we set the stage by briefly discussing the opposition between the 'universe view' and the 'multiverse view'. Furthermore, we propose to classify multiverse conceptions in terms of their adherence to some form of mathematical realism. In Sect. 2, we use this classification to review four major conceptions. Finally, in Sect. 3, we focus on the distinction between actualism and potentialism with regard to the universe of sets, then we discuss the Zermelian view, featuring a 'vertical' multiverse, and give special attention to this multiverse conception in light of the hyperuniverse programme introduced in Arrigoni-Friedman (Bull Symb Logic 19(1):77–96, 2013). We argue that the distinctive feature of the multiverse conception chosen for the hyperuniverse programme is its utility for finding new candidates for axioms of set theory.

1 The Set-Theoretic Multiverse

1.1 Introduction

Recently, a debate concerning the set-theoretic multiverse has emerged within the philosophy of set theory, and it is plausible to expect it to remain at centre stage for a long time to come. The 'multiverse' concept was originally triggered by the independence phenomenon in set theory, whereby set-theoretic statements such as

Originally published in C. Antos, S.-D. Friedman, R. Honzik, C. Ternullo, Multiverse conceptions in set theory. Synthese **192**(8), 2463–2488 (2015).

C. Antos • S.-D. Friedman • C. Ternullo (✉)
KGRC, Vienna, Austria
e-mail: claudio.ternullo@univie.ac.at

R. Honzik
KGRC, Vienna, Austria

Charles University, Prague, Czech Republic

© Springer International Publishing AG 2018
C. Antos et al. (eds.), *The Hyperuniverse Project and Maximality*,
https://doi.org/10.1007/978-3-319-62935-3_3

CH (and many others) can be shown to be independent from the ZFC axioms by using different *models* (universes). The collection of *some* or *all* of these models constitutes the set-theoretic multiverse.

While the existence of a set-theoretic multiverse is a well-known mathematical fact, it is far from clear how one should properly conceive of it, and in what sense the 'multiverse phenomenon' bears on our experience and conceptions of sets. Furthermore, as is frequent in the philosophy of mathematics, the issue of the nature of the multiverse intersects other no less prominent issues, concerning the nature of mathematical objectivity, ontology and truth. For instance, in what sense may the set-theoretic multiverse imply a revision of our 'standard' conception of truth? What are universes within the multiverse like and how should they be selected? Is the multiverse a merely transient phenomenon or can it legitimately claim to represent the ultimate set-theoretic ontology? Is there still any chance for the 'universe view' to prevail notwithstanding the existence of the multiverse? These are only a few examples of the philosophical issues one may want to examine and, in what follows, some of these questions will be addressed. Our goal is twofold: to provide an account of the positions at hand in a systematic way, and to present our own theory of the multiverse.

In order to fulfill the first goal, in Sect. 2, we will give an overview of some of the available conceptions, whereas in Sect. 3 we will introduce one further conception, which befits the goals of the hyperuniverse programme. Our focus will be more on philosophical features than on mathematical details, although, sometimes, a more accurate mathematical account will inevitably have to be given.

1.2   The 'Universe View' and the 'Multiverse View'

Let us preliminarily explain, in general terms, what a 'multiverse view' amounts to. In particular, we clarify how it can and should be contrasted to a 'universe view'.

The universe of sets, V, is the cumulative hierarchy of all sets, starting with the empty set and iterating, along the ordinals, the power-set operation at successor stages and the union operation at limit stages. The ZFC axioms will be our reference axioms and we know that these axioms are unable to specify many relevant properties of the universe. For instance, the axioms do not tell us what the size of the continuum is or whether there exist measurable cardinals (provided they are consistent with ZFC). In fact, different versions of V, obtained through model-theoretic constructions, are compatible with the axioms. Set-theorists work with lots of these constructions: set-generic or class-generic extensions (obtained through forcing), inner models, models built using, for instance, ultrafilters, ultraproducts, elementary embeddings and so on. Some of these are sets, others are classes, some satisfy CH and some do not, some satisfy $\diamond$ and some do not, and so on. A huge variety of combinatorial possibilities comes with the study of set-theoretic models and the bulk of contemporary set theory consists in studying, classifying and producing models not only of ZFC, but also of some of its extensions.

Two immediate questions may arise: how should we interpret such a situation in relation to the uncritical assumption that V denotes a 'fixed' entity, a determinate object? What is the relationship between the universe and the multiverse? The two possible responses to such questions yield what we may see as the two basic possible philosophical alternatives at hand: the universe view and the multiverse view.

The former conception is characterised as follows: there is a definite, unique, 'ultimate' set-theoretic structure which captures all true properties of sets. Its supporters are aware that the first-order axioms of set theory are satisfied by different structures, but from this fact they only infer that the currently known axioms are not sufficient to describe *the* universe in full. They may think that set-theoretic indeterminacy will be significantly reduced by adding new axioms which will provide us with a more determinate picture of the universe, but they may also believe that we will never reach a complete understanding of the universe, after all.

On the other hand, the multiverse view can be characterised in the following way: there is a wide realm of models which satisfy the axioms, all of which contain relevant, sometimes alternative, pieces of information about sets. Each of these, or, at least, some of these represent all equally legitimate universes of sets and, accordingly, there is no unique universe, nor should there be one. The multiverse view supporter believes that the absence of a unique reference of the set-theoretic axioms will not and cannot be repaired: set theory is about different realms of sets, each endowed with properties which differentiate it from another.

In light of current set-theoretic practice, both conceptions are legitimate and tenable, and both are problematic. It is easy to see why. In very rough terms, the universe view supporter owes us an account of how, notwithstanding the existence of one single conceptual framework, we can think in a perfectly coherent way of different alternative frameworks. If she thinks that such alternative frameworks are not definitive, then she has to explicate why they are epistemically reliable (that is, why they give us 'true' knowledge about sets). The multiverse view supporter, on the other hand, owes us an account of how, notwithstanding the existence of multiple frameworks, one can always imagine each of them as being 'couched' within V. Granted, such frameworks may well be *mutually* incompatible, but, surely, they must *all* be compatible with V.

1.3 A Proposed Classification

There are a lot more nuanced versions of each of the two positions. In order to address more closely what we believe are the most relevant ones, we want to propose a systematic way to group them. We will add one further criterion of differentiation, that of their commitment to some form of *realism*. In plain terms, commitment to realism measures how strongly each conception holds that the universe or the multiverse exist *objectively*.

It is fairly customary in the contemporary philosophy of mathematics to differentiate realists in ontology from realists in truth-value, and we will pre-eminently focus here on *realism in ontology*.[1] Thus, the universe view or multiverse view may split into two further positions, according to whether one is a realist or a non-realist universe view or multiverse view supporter. Such a differentiation yields, in the end, four positions, each of which, we believe, has had some tradition in the philosophy of mathematics or has been influential in, or relevant to the current debate in the foundations of set theory.

A realist universe view may alternatively be described as that of a Gödelian platonist. Although Gödel's views may have changed during his lifetime, it seems rather plausible to construe his many references to the reality of sets in the context of a realist (platonist) universe view.[2]

But one could be a universe view supporter without believing in the external existence of the universe. One may, for instance, view the universe view as only 'practically' confirmed on the basis of some specific mathematical results. For instance, Maddy's 'thin realist' would presumably hold that the universe view is preferable as long as it better fits set theory's first and foremost purpose of producing a 'unified' arena wherein all mathematics can be carried out.[3]

The realist multiverse view supporter fosters a peculiar strain of realism, based on the assumption that there are different, alternative, 'platonistically' existing concepts of sets instantiated by different, alternative universes or, alternatively, that there are different universes, which correspond to alternative concepts of set. As known, such a conception has been set forth and articulated in full as a new version of platonism known as *full-blooded platonism* (FBP).[4] Within the context of the debate we are interested in, this conception has been recently advocated by Hamkins, and we will devote substantial efforts to examining its features.

Finally, the non-realist multiverse view supporter is someone who does not believe in the existence of universes and, in particular, does not believe in the existence of a single universe. To someone with these inclinations, the multiverse is a 'practical' phenomenon, so to speak, with which one should deal as with

---

[1]For the distinction and its conceptual relevance within the philosophy of mathematics, see Shapiro's introduction to Shapiro [34] or Shapiro [33].

[2]See, for instance, the following oft-quoted passage in his Cantor paper: "It is to be noted, however, that on the basis of the point of view here adopted [that is, the 'platonistic conception', *our note*], a proof of the undecidability of Cantor's conjecture from the accepted axioms of set theory (in contradistinction, e.g., to the proof of the transcendency of π) would by no means solve the problem. For if the meaning of the primitive terms of set theory as explained on page 262 and in footnote 14 are accepted as sound, it follows that the set-theoretical concepts and theorems describe some well-determined reality in which Cantor's conjecture must be either true or false" ([13], in [14], p. 260). For an account of the development of Gödel's conceptions, see, for instance, Wang's books [38, 39] and also van Atten-Kennedy [37].

[3]For the full characterisation of Maddy's 'thin' realist, see Maddy [27, 28]. A possible middle ground between a realist and a non-realist universe view has been described by Putnam in his [31]. A 'moderate realist', as featured there, would be someone who does not buy into full-blown platonism but who, at the same time, still believes to be able to find evidence in favour of some sort of 'ultimate' universe.

[4]See, especially, Balaguer [2, 3].

any other fact of mathematical practice. This position seems to represent the basic 'uncommitted' viewpoint, but may be further elaborated using, for instance, the formalist viewpoint. A typical representative of this attitude is Shelah, whose position we will describe in Sect. 2.

As we said above, one may be a realist in truth-value and a non-realist in ontology and vice versa, one may be both and one may be neither. The introduction of one further criterion of differentiation, namely the commitment to realism in truth-value, in principle, would give us further available positions, but the four conceptions briefly described above are sufficient to cover the whole spectrum of the existing conceptions, at least for now.[5]

2 Multiverse Conceptions

We will now proceed to examine multiverse conceptions in detail. We will, in turn, present four positions, before introducing ours. As anticipated, Hamkins and Shelah will be our featured representatives of, respectively, a realist and a non-realist multiverse view. We will also be scrutinising two further positions, whose authors, it seems to us, are, in fact, universe view supporters, and these are Woodin's and Steel's. Incidentally, this fact testifies to the essential non-rigidity of set-theorists' stances in the practical arena: the universe view and the multiverse view are variously advocated or rejected philosophically, but the universe and the multiverse constructs are used indifferently by set-theorists as tools to study sets and determine their properties. The last two authors have presented their own version of the multiverse either to subsequently discard it (Woodin) or to suggest ways to reduce it to a universe view (Steel) and, through examining them, we also hope to receive some insight on how and why one may, at some point, get rid of the multiverse.

---

[5]Incidentally, it is not clear whether a realist in truth-value is best accommodated to the universe view. For instance, take Hauser, who seems to be only a realist in truth-value. He says: "At the outset mathematical propositions are treated as having determinate truth values, but no attempt is made to describe their truth by relying on a specific picture of mathematical objects. Instead one seeks to exhibit the truth or falsity of mathematical propositions by rational and reliable methods" [18, p. 266]. From this, it is far from clear that one single picture of sets would have to be found anyway, if not at the outset, at least in due course, presumably after the truth-value of such statements as CH has been reliably fixed. Hauser has also addressed truth-value realism and its conceptual emphasis on *objectivity* rather than on *objects* in Hauser [17]. See also Martin [29].

2.1 The Realist Multiverse View

2.1.1 The General Framework

Hamkins has made a full case for a realist multiverse view in his [16]. Before examining it in detail, let us briefly summarise it. Set theory deals with different model-theoretic constructions, wherein the truth-value of relevant set-theoretic statements may vary. The only way to make sense of this phenomenon is to acknowledge that there are different *set concepts*, each of which is instantiated by a specific model-theoretic construct, that is, a universe of sets. There is no a priori reason to ban any model-theoretic construction from the multiverse: any universe of sets is a legitimate member of the set-theoretic multiverse. This means that even such controversial models as ill-founded models are granted full citizenship in the multiverse.[6]

Furthermore, any universe in the multiverse describes an *existing* reality of sets. The latter thesis implies that any model-theoretic construct should also be taken to describe an existing reality in the platonistic sense.

This peculiar form of multiverse realism is the crux of Hamkins' conception and, thus, needs an extended commentary.

First of all, Hamkins leaves no doubt as to the platonistic character of his conception:

> The multiverse view is one of higher-order realism—Platonism about universes—and I defend it as a realist position asserting actual existence of the alternative set-theoretic universes into which our mathematical tools have allowed us to glimpse. [16, p. 417]

Now, as anticipated in Sect. 1, it may not have escaped other commentators that the conception Hamkins formulates in the passage above is connected to that peculiar version of platonism, due to Balaguer, known as full-blooded platonism (FBP). FBP is, ontologically, a richer version of platonism, as, on this conception, *any* theory of sets describes an existing realm of objects (that is, a universe of sets), and, consequently, FBP has to accommodate the platonistic conception of truth to this enlarged ontological realm. In Balaguer's words:

> It is worth noting that what FBP does *not* advocate is a *shift* in our conception of mathematical truth. Now, it *does* imply (when coupled with a corresponding theory of truth) that the consistency of a mathematical sentence is sufficient for its truth. [...] What mathematicians *ordinarily* mean when they say that some set-theoretic claim is true is that it is true of the *actual* universe of sets. Now, as we have seen, according to FBP, there is no *one* universe of sets. There are many, but nonetheless, a set-theoretic claim is true just in case it is true of *actual* sets. What FBP says is that there are so many different kinds of sets that every consistent theory is true of an *actual* universe of sets. [2, p. 315]

---

[6]Hamkins epitomises this conception through the adoption of the naturalistic maxim 'maximise', by virtue of which one should not place "undue limitations on what universes might exist in the multiverse. This is simply a higher-order analogue of the same motivation underlying set-theorists' ever more expansive vision of set theory. We want to imagine the multiverse as big as possible." [16, p. 437].

Balaguer gives one further neat exemplification of the situation described above, by explaining that:

> According to FBP, both ZFC and ZF+ not-C[7] truly describe parts of the mathematical realm; but there is nothing wrong with this, because they describe *different* parts of that realm. This might be expressed by saying that ZFC describes the universe of sets$_1$, while ZF+not-C describes sets$_2$, where sets$_1$ and sets$_2$ are different kinds of things. [2, p. 315]

Analogously, Hamkins sees universes of sets as being tightly related to specific *set concepts*, the latter being, presumably, embodied by an axiom or a collection of axioms (in Balaguer's example, sets$_1$ is the universe or region of the multiverse which instantiates the set concept expressed by AC and sets$_2$ that which instantiates the set concept expressed by the negation of AC):

> Often, the clearest way to refer to a set concept is to describe the universe of sets in which it is instantiated, and in this article I shall simply identify a set concept with the model of set theory to which it gives rise. [16, p. 417]

It is not entirely clear from what Hamkins says whether a set concept should be automatically and uniquely identified with the universe(s) that instantiate it, or whether concepts of sets have an independent (and *prioritary*) status, something which would presumably differentiate Hamkins' position from Balaguer's. What is sure is that the correspondence between set concepts and instantiating universes should be construed in terms of a correspondence between axioms and models, as demonstrated by the following general observation:

> The background idea of the multiverse, of course, is that there should be a large collection of universes, each a model of (some kind of) set theory. There seems to be no reason to restrict inclusion only to ZFC models, as we can include models of weaker theories ZF, ZF$^-$, KP and so on, perhaps even down to second order number theory, as this is set-theoretic in a sense. [16, p. 436]

So much for the ontology of the multiverse. As we have seen, its underlying philosophy does not seem to differ to a significant extent from that of Balaguer's FBP-ist. Where, on the contrary, Hamkins seems to supplement it, is on set-theoretic truth, and, in particular, with regard to the truth-value of the undecidable statements. An FBP-ist is supposed to be very liberal on this: the answer to a set-theoretic problem (and, possibly, also to a non-set-theoretic problem which has a strong dependency upon set theory) depends on the universe of sets one is talking about. Accordingly, CH may be true in some universes and false in others, but there is no a priori reason to consider one of the answers provided by a universe in the multiverse as more relevant or more strongly motivated than any other.

On this point, Hamkins seems to want to expand on FBP. Although, at the purely ontological level, all universes are equally legitimate, set-theoretic practice may still dictate which are more relevant in view of specific needs. We should still pay attention to specific versions of truth within universes, as presumably there are some which look more 'attractive' than others, as explained in the following quote :

---

[7] ZF+ the negation of the Axiom of Choice.

... there is no reason to consider all universes in the multiverse equally, and we may simply be more interested in the parts of the multiverse consisting of universes satisfying very strong theories, such as ZFC plus large cardinals. The point is that there is little need to draw sharp boundaries as to what counts as a set-theoretic universe, and we may easily regard some universes as more set-theoretic than others. [16, p. 436–437]

Elsewhere, he expresses such concerns even more neatly. For instance, when he talks about CH, he says that

On the multiverse view, consequently, the continuum hypothesis is a settled question; it is *incorrect* [*our italics*] to describe the CH as an open problem. The answer to CH consists of the expansive, detailed knowledge set theorists have gained about the extent to which it holds and fails in the multiverse, about how to achieve it or its negation in combination with other diverse set-theoretic properties. Of course, there are and will always remain questions about whether one can achieve CH or its negation with this or that hypothesis, but the point is that the most important and essential facts about CH are deeply understood, and these facts constitute the answer to the CH question. [16, p. 429]

The emphasis, here, is more on the fact that a shared solution to the Continuum Problem is represented by our 'knowledge' of how CH varies across the multiverse, rather than on its truth-value in specific universes. As a consequence, here Hamkins seems to conjure an epistemically 'active' role for his multiverse conception, as a study of the *relationships* among universes which may provide us with detailed knowledge of the alternative answers to mathematical problems. Such a view is also expressed in the quote below:

On the multiverse view, set theory remains a foundation for the classical mathematical enterprise. The difference is that when a mathematical issue is revealed to have a set-theoretic dependence, then the multiverse is a careful explanation that the mathematical fact of the matter depends on which concept of set is used, and this is almost always a very interesting situation, in which one may weigh the desirability of various set-theoretic hypotheses with their mathematical consequences. [16, p. 419]

2.1.2 Problems with the Realist Multiverse View

We now want to proceed to examine some potential difficulties with the realist multiverse view. In fact, in what follows, these are formulated as objections to FBP rather than to Hamkins' views. However, if one believes, as we do, that Hamkins' views are modelled upon (or, at least, connected to) the former, then the realist multiverse view supporter should take such objections to FBP very seriously.

One issue is that of whether we have sufficient grounds to assert that the existence of different models can be construed in terms of the existence of different universes of sets instantiating different concepts (and theories) of sets, as required by FBP (and presumably, as we have seen, also by the Hamkinsian multiverse view supporter).

A second, but parallel, issue is that of whether models (universes) are sufficiently characterised, conceptually, to be viewed as more than mere alternative character-isations of a unique existing universe, V. This is precisely the issue we mentioned at the beginning: in fact, each model can be seen to be living 'inside' V. This issue inevitably puts some pressure on the realist multiverse view supporter, who believes that there is no 'real' V and, most crucially, that knowledge about set-theoretic truth depends on knowledge of universes other than V.

As far as the first issue is concerned, first we wish to review an objection to FBP by Colyvan and Zalta. In [5], the authors argue that, by FBP, when we consider two different models M and N, we should think of them as two entirely different realms of sets. But this is hardly the case. For instance, all forcing extensions of a transitive model of the axioms of ZFC leave the truth-value of sentences at the arithmetical level unchanged, and, thus, it is hard to imagine that the finite numbers in M are different from the finite numbers in N.

If, on the other hand, M and N contain the same finite numbers, then these models will hardly be *entirely* different realms of sets. For instance, $\mathcal{P}(\omega)$ in M may be different from $\mathcal{P}(\omega)$ in N, but a set like 7 will be the same in both.

It is not clear whether Hamkins is aware of this criticism, when, at some point in his paper, he even challenges the fixedness of the concept of natural number. He argues that Zermelo-style categoricity arguments, which would give us grounds to believe the universe view, are unpersuasively based on an absolute concept of set. This may extend to categoricity arguments with regard to arithmetic, insofar as

> ...although it may seem that saying "1, 2, 3, ...and so on" has to do with a highly absolute concept of finite number, the fact that the structure of the finite numbers is uniquely determined depends on our much murkier understanding of which subsets of the natural numbers exist. [...] My long-term expectation is that technical developments will eventually arise that provide a forcing analogue for arithmetic. [16, p. 14]

We do not know on what grounds Hamkins makes such a prediction and whether the discovery of a forcing analogue for arithmetic would help him rebut the mentioned objection efficaciously. In any case, if we take FBP as asserting that, whenever we have two different models, then we have entirely different sets of objects in them, we inevitably fall back on the thesis that there are as many types of finite sets as models, something which seems to defeat our well-established, pre-theoretic assumption of the full determinacy of finite numbers. If, on the contrary, universes may share some, but not all sets, then it is less easy to recognise them as entirely alternative realms of sets (universes), although the latter case is less problematic.

To introduce the second issue, we will start with a quote from Potter's [30]. The author says:

> ...for a view to count as realist, [...], it must hold the truth of the sentences in question to be metaphysically constrained by their subject matter more substantially than Balaguer can allow. A realist conception of a domain is something we win through to when we have gained an understanding of the nature of the objects the domain contains and the relations that hold between them. For the view that bare consistency entails existence to count as realist, therefore, it would be necessary for us to have a quite general conception of the whole of logical space as a domain populated by objects. But it seems quite clear to me that we simply have no such conception. [30, p. 11]

Potter's criticism goes to the heart of the FBP-ist's conception, which essentially consists in the claim that 'consistency guarantees existence'. That this doctrine is tenable and conceptually cogent is crucial to Hamkins' purposes: if members of the multiverse do not exist on the grounds of consistency alone, then the entire multiverse construct might be seen as shaky. It should be clarified that the problem, here, is not whether the form of existence of mathematical entities set forth by FBP-ists is inadequate to hold that FBP is a realist conception, but rather that the form of existence set forth by FBP-ists is not sufficient to spell out a multiverse conception.

The crucial case study is the 'ontology of forcing'. The realist multiverse view supporter commits himself to asserting that the forcing extension of the universe V associated to a V-generic filter G, $V[G]$, really exists.[8] But how can he do that, if V is everything there is? The response is contained in the following remark:

> On the multiverse view, the use of the symbol V to mean "the universe" is something like an introduced constant that might refer to any of the universes in the multiverse, and for each of these the corresponding forcing extensions $V[G]$ are fully real. [16, p. 5]

Once he has ascertained that there is a consistent reading of such objects as $V[G]$, via FBP he can legitimately say that such objects exist. But that is precisely where we find Potter's objection strike at the roots. What kind of existence is that? Suppose Potter is right, and set-generic extensions cannot be said to be existing only by virtue of their consistency. That would put the multiverse view supporter in trouble, as he might face up with the objection that set-generic extensions are, in fact, illusory. This is precisely the strategy used by a universe view supporter. Hamkins acknowledges this fact himself:

> Of course, one might on the universe view simply use the naturalist account of forcing as the means to explain the illusion: the forcing extensions don't really exist, but the naturalist account merely makes it seem as though they do. [16, p. 10]

Yet, Hamkins thinks that FBP is more in line with our experience of these objects as expressed within the naturalist account:

> ... the philosophical position [higher-order realism, *our note*] makes sense of our experience – in a way that the universe view does not – simply by filling in the gaps, by positing as a philosophical claim the actual existence of the generic objects which forcing comes close to grasping, without actually grasping. [16, p. 11]

However, this position is, at least, controversial. Admittedly, forcing comes close to grasping generic objects, but actually never grasps them. How may all this ever help us believe that these objects are existent? Is consistency alone a sufficient reason?

Similar concerns have been expressed by Koellner, in his [24], where the author extends them to other model-theoretic constructions:

> In summary, on the face of it, all three methods provide us with models that are either sets in V or inner models (possibly non-standard) of V or class models that are not two-valued. In each case one sees by construction that (just as in the case of arithmetic) the model is non-standard. One can by an act of imagination treat the new model as the "real" universe. The broad multiverse position is a *consistent* position. But we have been given no reason for taking that imaginative leap. [24, p. 22]

---

[8]Hamkins defines this interpretation of forcing 'naturalist', as opposed to the 'original' interpretation of forcing, whereby one starts the construction with a countable transitive ground model M and extends it to an $M[G]$, by adding an M-generic filter G.

In fact, as we have seen, the reason we have been given for the 'imaginative leap' Koellner is referring to here, is the central ontological position of FBP, that 'consistency guarantees existence'. But if we doubt that such a position is at all tenable, then it is reasonable to assert that the existence of a multiverse, at least in the terms conjured by Hamkins, may be seen as not entirely unproblematic.

2.2 The Non-realist Multiverse View

None of these difficulties has to be addressed by a non-realist multiverse view supporter. This kind of *pluralist* does not believe in the existence of universes and, thus, is likely to dismiss all the ontological concerns we have previously reviewed. This person may see the set-theoretic multiverse not as a structured, or independent, for that matter, reality, but only as a phenomenon arising in practice. As a consequence, he may not attach any relevance either to 'real V' or to the 'multiverse': in this person's view, these are just labels one may use indifferently on the basis of one's personal needs and theoretical convenience. Given such presuppositions, this person may see the multiverse essentially as a tool to produce independence proofs.

As Koellner has painstakingly shown in one of his recent articles, one of the most committed proponents of this form of 'radical pluralism' has been Carnap.[9] The Carnapian pluralist typically believes that any theory, say ZFC+CH or ZFC+¬CH, has its own appeal and that adopting either is only a matter of *expedience*. Given that the meaning of the axioms is not dependent upon any prior knowledge of their content, the only theoretical concern a Carnapian pluralist has is that of finding what one can prove from those axioms. It follows that models are only needed indirectly, to understand which statements are provable from the axioms and, thus, to allow us carry out independence proofs.

The kind of realist we have scrutinised in the preceding section may also be viewed as a radical pluralist as far as truth is concerned. However, as we have seen, the Hamkinsian multiverse view supporter has a somewhat different view of truth, insofar as she may acknowledge that it is not only theoretical expedience which dictates our choices of the axioms and of the corresponding models.

The typical attempt to counteract pluralism consists in showing that (1) *expedience* is not sufficient to adopt a theory[10]; and/or (2) that issues of meaning cannot be entirely circumvented. An example of the latter strategy is given by Gödel's response to Errera's attacks in his 1964 version of his Cantor paper. Errera had stated that set theory is bifurcated by CH in the same way as geometry by Euclid's fifth postulate, and that one should feel free to see each of the two theories as equally justified. Gödel attempted to rebut this view, by pointing out that even the axioms of

---

[9]See Koellner [23], especially Sect. 2.

[10]This is, for instance, Koellner's line of attack in the aforementioned paper.

Euclidean geometry may have a fixed meaning: at least they refer to laws concerning bodies and, accordingly, a decision on their truth might depend on that given system of physical objects.

Within the community of set-theorists, it is Shelah who has voiced most vividly the non-realist multiverse conception. In his 'foundational' paper, he says:

> My mental picture is that we have many possible set theories, all conforming to ZFC. I do not feel "a universe of ZFC" is like "the Sun", it is rather like "a human being" or "a human being of some fixed nationality". [35, p. 211]

Shelah is, admittedly, a 'mild' formalist,[11] but he does not seem to adhere to the full-blown radical pluralistic point of view about truth. Instead, he talks about different degrees of 'typicality' of models of the axioms (in this case, ZFC). Always referring to models of ZFC as 'citizens', he specifies:

> ...a typical citizen will not satisfy $(\forall \alpha)[2^{\aleph_\alpha} = \aleph_{\alpha+\alpha+7}]$ but will probably satisfy $(\exists \alpha)[2^{\aleph_\alpha} = \aleph_{\alpha+\alpha+7}]$. However, some statements do not seem to me clearly classified as typical or atypical. You may think "Does CH, i.e., $2^{\aleph_0} = \aleph_1$ hold?" is like "Can a typical American be Catholic?" [35, p. 211]

In fact, it should be pointed out that radical, unmitigated pluralism is hardly the preferred choice among pluralists, as all of them seem to be keen on finding correctives (such as the aforementioned preference for 'typicality'). For instance, Field says:

> ...we can still advance aesthetic criteria for preferring certain values of the continuum over others; we must now view these not as *evidence that* the continuum has a certain value, but rather as *reason for refining our concepts so as to give* the continuum that value, [...]. [in 6, p. 300]

Analogously, Balaguer finds that:

> There are at least two ways in which the FBP-ist can salvage the objective bite of mathematical disputes. The first has to do with the notion of *inclusiveness* or *broadness*: the dispute over CH, for instance, might be construed as a dispute about whether ZF+CH or ZF+not-CH characterizes a broader notion of set. And a second way in which FBP-ists can salvage objective bite is by pointing out that certain mathematical disputes are disputes about whether some sentence is true in a *standard model*. [2, p. 317]

Some of these correctives may also be dictated by naturalistic concerns: there are certain axioms which solve problems very nicely, or give us better pictures of the realm of sets, or are more elegant, concise and with a stronger 'unificatory' power.

As we have seen, Shelah admits that there are models which are more 'typical' than others, insofar as they would satisfy specific set-theoretic statements which, in turn, are more typical than others. Such statements he proposes to call 'semi-axioms'. The label is very aptly chosen, as it is supposed to convey the idea that none

---

[11]We say 'mild', as he seems to want to deny to be a fully committed (an 'extreme', in his words) formalist. He says: "..I reject also the extreme formalistic attitude which says that we just scribble symbols on paper or all consistent set theories are equal." [35, p. 212].

of these set-theoretic statements might eventually be viewed as an axiom. Shelah says:

> Generally, I do not think that the fact that a statement solves everything really nicely, even deeply, even being the best semi-axiom (if there is such a thing, which I doubt), is a sufficient reason to say that it is a "true" axiom. In particular, I do not find it compelling at all to see it as true. [35, p. 212]

So we fittingly go back to 'unfettered' pluralism: although we may want to introduce a hierarchy of more or less convenient, of more or less typical set-theoretic statements, there is no hope to see any of these as more true of the set concept.

For the sake of completeness, we should mention that another way of being a non-realist multiverse view supporter might be that propounded by Feferman, that of a 'default', so to speak, pluralist. In Feferman's conception, the set concept is *vague*, and so is that of the 'linear' continuum. Consequently, the multiverse is a 'practically' inevitable construct, which will never be reduced to the simplicity of a single, definitive universe. The case for the solvability of such problems as CH is hopeless and, thus, any solution, in a sense, goes, insofar any solution is only a partial characterisation of both the set concept and of the continuum concept.

However, it should be noticed that Feferman is a 'default' pluralist only for truths concerning levels of the set-theoretic hierarchy beyond V_ω, and that, on the contrary, he attributes a full meaning to the whole of finite mathematics. Feferman's concerns, thus, are entirely different from the Carnapian's: the latter has no pre-existing theory of meaning, but only general criteria for adopting theories, whereas the Fefermanian pluralist is just a constructivist who is at a loss with the higher reaches of the set-theoretic hierarchy.[12]

2.3   The Set-Generic Multiverse

2.3.1   Woodin

The next two multiverse views we are going to discuss are, in fact, one and the same conception, namely the *set-generic multiverse* view. However, as we shall see, they differ, to a certain extent, in their ultimate goal and in some other features. For instance, Woodin's conception leaves it open whether the set-generic multiverse view is at all plausible, and, in fact, its author suggests that this may not be the case. On the other hand, Steel argues that there is some significant evidence that truth in the set-generic multiverse could, ultimately, be reduced to truth in some simpler fragment of the multiverse itself, that is, its *core*. In the end, it would probably be more correct to describe the authors as universe view supporters, although their positions on this point are not always transparent.

---

[12]These views have been stated by the author several times. See, in particular, Feferman [8, 10] and the more recent [9]. A careful response to Feferman's concerns is in Hauser [18].

In Woodin's case, our claim can be more strongly and convincingly substantiated. For instance, take the following crucial passage in [40], where the author sets forth his central epistemological view:

> It is a fairly common (informal) claim that the quest for truth about the universe of sets is analogous to the quest for truth about the physical universe. However, I am claiming an important distinction. While physicists would rejoice in the discovery that the conception of the physical universe reduces to the conception of some simple fragment or model, the Set Theorist rejects this possibility. I claim that by the very nature of its conception, the set of all truths of the transfinite universe (the universe of sets) cannot be reduced to the set of truths of some explicit fragment of the universe of sets. [...] The latter is the basic position on which I shall base my arguments. [41, p. 104]

As we shall see, this philosophical position, as announced by the author, has some bearings on the 'decease' of the multiverse and, thus, orientates its construction from the beginning.

In very rough terms, the set-generic multiverse is generated by picking up a universe M from an initial multiverse ('the collection of possible universes of sets') and taking all generic extensions and refinements (inner models) of M. Suppose one starts with a countable transitive model M satisfying ZFC and let $\mathbb{V}_M$ the smallest set such that: (1) $M \in \mathbb{V}_M$; (2) for any M_1 and M_2, if M_1 is a model of ZFC and M_2 is a generic extension of M_1 and if either M_1 or M_2 are in $\mathbb{V}_M$, then both are in $\mathbb{V}_M$. We say that $\mathbb{V}_M$ is the set-generic multiverse generated in V from M.

As far as truth is concerned, it is natural to expect it to vary throughout the multiverse. As a matter of fact, there are some truths which hold in all set-generic extensions, that is, in all $N \in \mathbb{V}_M$. Suppose ϕ is one such truth: ϕ is, then, said to be a *multiverse truth*.

> The *generic multiverse conception of truth* is the position that a sentence is true if and only if it holds in each universe of the generic multiverse generated by V. This can be formalized within V in the sense that for each sentence ϕ there is a sentence ϕ^* such that ϕ is true in each universe of the generic multiverse generated by V if and only if ϕ^* is true in V. [40, p. 103–4]

The generic multiverse conception of truth is entirely reasonable from the point of view of a universe view supporter: truth *in* the multiverse ought to be defined as truth *in all members of* the multiverse, as long as truths holding in only one or some universes may not be seen as 'real' truths. It is not clear, however, that this position spells out a plausible multiverse view. After all, the multiverse was articulated precisely to make sense of our 'abundance' of truth, and, possibly, to understand what the reason for such an abundance was (e.g., by studying relationships among universes), whereas, by the generic multiverse conception of truth, Woodin's preoccupation, on the contrary, seems to be more that of bolstering a pre-multiverse attitude.

In fact, presumably, Woodin's idea from the beginning is that such constructs as set-generic extensions should not be taken as 'separately existing' constructs in the same way as Hamkins held (and we have seen that Hamkins' position may also be problematic). Furthermore, it is the very construction of the multiverse which makes appeal to a 'meta-universe', from which the multiverse is supposed to be 'generated' and this fact may be viewed as lending support to the generic multiverse conception of truth.

Now, Woodin can prove that, under certain conditions, into which we cannot delve here,[13] the set of the Π_2-multiverse truths is recursive in the set of the truths of V_{δ_0+1}, where δ_0 is the least Woodin cardinal.[14]

This result, according to Woodin, violates the multiverse *laws*. These laws precisely prescribe that the set of Π_2-multiverse truths is not recursive in the set of the truths of V_{δ_0+1}. But then, just what is the rationale behind the multiverse laws? In plain terms, it is the aforementioned doctrine that no reduction of 'truth in V' to 'truth in a simpler fragment of V' is possible. Admittedly, the truths we are dealing with here are only a small fragment of the truths holding in V, so why should we ever bother formulating such a multiverse law? Because the multiverse analogue of 'truth in V', that is, of 'real' truths, as we said, is precisely 'truth in all members of the multiverse' and, among multiverse truths, Π_2-multiverse truths hold a special position, as they express that something is true at a certain V_α. In simpler terms, Π_2-multiverse truths would be the multiverse analogue of 'universe truths'.

Now, if the set of Π_2-multiverse truths is recursive in the set of truths of V_{δ_0+1}, then, there is a sense in which the set-generic multiverse fully captures 'truth in V'. But this, in turn, runs counter to Woodin's platonistic assumptions: set-theoretic truth cannot be reduced to truth in a certain fragment of V (that is, V_{δ_0+1}). This is, essentially, Woodin's argument for the rejection of the multiverse.

There may be legitimate reasons of concern about the philosophical tenability of the argument, but, even before that, it should be noticed that the argument rests upon some yet unverified mathematical hypotheses, and, therefore, until these hypotheses are not proved correct, in principle, it is not even known if such an argument can be produced. Some further, more general, concerns over the set-generic multiverse we will express in the next subsection, after examining Steel's framework.

2.3.2 Steel's Programme

Unlike Woodin, Steel aims to articulate a *formal* theory of the set-generic multiverse. This is a first-order theory (**MV**), with two kinds of variables, one for *sets* and one for *worlds*. Within the theory, worlds are treated as proper classes and contain sets. In particular, one of the axioms of **MV** prescribes that an object is a set if and only if it belongs to some world.

Therefore, **MV** is a first-order theory which expands on ZFC, by specifying what models of ZFC constitute the multiverse of ZFC. In particular, worlds are either 'initial' worlds or *set-generic* extensions of initial worlds. The reason why we need a formal theory of the multiverse immediately reveals Steel's intents: a multiverse

---

[13]Details on these can also be found in Woodin [40].

[14]For the definition of Woodin cardinals, see Kanamori [21, p. 360].

theory is construed first and foremost by Steel as a foundational theory, wherein one can develop both 'concrete' and set-theoretic mathematics. He says:

> ...we don't want everyone to have his own private mathematics. We want one framework theory, to be used by all, so that we can use each other's work. It's better for all our flowers to bloom in the same garden. If truly distinct frameworks emerged, the first order of business would be to unify them. [36, p. 11]

But Steel also has another foundational concern, related to Gödel's programme. As known, the ZFC axioms are insufficient to solve lots of set-theoretic problems, and, what is worse, it is not clear what axioms one should adopt to solve them. Steel's concern is precisely that of finding an 'optimal' set theory extending ZFC, and he settles on *large cardinal axioms* as being the most 'natural' extensions of the ZFC axioms.

According to Steel, one optimal requisite of a natural extension T of ZFC is that of being able to *maximise* interpretative power, that is, of including and, possibly, extending the set of provable sentences of ZFC or of any of its extensions. There is a 'tool' we can use to see how theories perform in this respect: the linearly ordered scale of consistency strengths associated to axiomatic theories of the form ZFC + large cardinals. It can be proved that, given any two examples of large cardinals, if H and T are two set theories containing them as axioms, then one invariably obtains that $H \leq_{Con} T$, $T \leq_{Con} H$ or $H \equiv_{Con} T$. In the first and second case, one says that T is stronger, consistency-wise, than H or viceversa, whereas, in the latter, H and T have the same consistency strength.

This gives us a linear arrangement of theories, for which a general interesting fact may hold: that is, if $H \leq_{Con} T$, then a fragment, and possibly all of, $Th(H)$ may be included in $Th(T)$. This holds, for instance, for statements at the level of arithmetic.[15] Now, given two theories H and T whose consistency strength is that of "infinitely many Woodin cardinals", then one may extend this result to second-order arithmetic. The hope is that, by strengthening the large cardinal assumptions associated to extensions of ZFC, the set of provable statements 'widens' accordingly. At the moment, this conjecture only holds in some specific mathematical structures.[16]

At any rate, by now it should be clear what Steel means by 'maximising interpretative power': a 'master' set theory should be one which, in a linear scale of theories, maximises over the set of provable statements. If the linear scale is given by the consistency strengths of 'natural' theories of the form ZFC+ large cardinals, such a maximisation is simply a function of the consistency strength of such theories. To recapitulate, using Steel's words:

> Maximizing interpretative power entails maximizing consistency strength, but it requires more, in that we want to be able to translate other theories/languages into our framework theory/language in such a way as to preserve their meaning. The way we interpret set

---

[15]In fact, this result, as Steel clarifies, only holds for 'natural' theories, that is theories with 'natural' mathematical axioms, not quite like, for instance, the 'Rosser sentence'.

[16]For instance, Steel's result on p. 7 only holds in $L(\mathbb{R})$.

theories today is to think of them as theories of inner models of generic extensions of models satisfying some large cardinal hypothesis, and this method has had amazing success. [...] It is natural then to build on this approach. [36, p. 11]

But Steel has to confront a crucial problem: even within **MV** there is no hope to solve problems like CH. This is because, as known from the late 1960s, no large cardinal axiom solves CH. Therefore, Steel makes a step forward: we may want to reduce the complexity of **MV** by identifying a 'core' world within the framework we have set up. Steel preliminarily discusses two conceptions about the multiverse: *absolutism* and *relativism*, each coming with two different degrees of strength ('weak' and 'strong'). Relativism implies that there is no preferred world within **MV**, whereas absolutism identifies **MV** as only a transitory framework. Weak absolutism is what appeals most to Steel: there is a multiverse, but the multiverse has a core.

Weak absolutism aims to avail to itself two mathematical results. The first says that, if a multiverse has a definable world, then that world is unique and is included in all the others. The second is the existence of an axiom (Axiom **H**), which may be in line with the goal of maximising interpretative power in the way indicated and implies that the multiverse has a core.

If the axiom is true, then the multiverse of *V* has a core, which is, more or less, the HOD of any *M* which satisfies AD, the Axiom of Determinacy.[17] The axiom has many consequences and, in particular, implies CH.

2.3.3 Problems with the Set-Generic Multiverse

We have already raised some concerns about Woodin's conception. We now want to review further potential objections, directed, this time, at both Woodin's and Steel's accounts.

First of all, the set-generic multiverse fosters a restrictive conception of the multiverse. As we have seen, Hamkins' radical viewpoint is consistent with an FBP-ist's presuppositions and ultimately depends on them for its justification. As far as the set-generic multiverse is concerned, the only grounds to accept it seem to be pre-eminently 'practical'. Woodin seems to be fully aware of this. In a revealing remark in his [40], he says:

> Arguably, the generic-multiverse view of truth is only viable for Π_2 sentences and not, in general, for Σ_2 sentences [...]. This is because of the restriction to *set forcing* in the definition of the generic multiverse. At present there is no reasonable candidate for the definition of an expanded version of the generic multiverse that allows for *class forcing* extensions and yet preserves the existence of large cardinals across the multiverse. [40, p. 104]

---

[17]For more accurate mathematical details, we refer the reader to Steel's cited paper.

Analogously, Steel motivates his ban on models other than set-generic extensions in **MV** in the following manner:

> Our multiverse is an equivalence class of worlds under "has the same information". Definable inner models and sets may lose information, and we do not wish to obscure the original information level. For the same reason, our multiverse does not include class-generic extensions of the worlds. There seems to be no way to do this without losing track of the information in what we are now regarding as the multiverse, no expanded multiverse whose theory might serve as a foundation. We seem to lose interpretative power. [36, p. 13]

From both quotations, it seems clear that the authors' main concern is about losing large cardinals and their associated interpretative power, as class forcing does not, in general, *preserve* large cardinals. In particular, Steel's reasons to cling on large cardinals seemed to be well-motivated, in the light of his foundational programme. However, restrictiveness remains a patent weakness of the set-generic multiverse view, and one which cannot be easily repaired: in our view, the authors propose it essentially because, otherwise, they could not obtain the results they are most interested in, that is 'unification' via the core hypothesis (Steel) or the retreat to a universe view (Woodin).

It should be noticed that Steel himself has expressed some significant criticisms of Woodin's conceptual framework. In a footnote, he says:

> ...the decision to stay within the multiverse language does not commit one to a view as to what the multiverse looks like. The "multiverse laws" do not follow from the weak relativist thesis. The argument that they do is based on truncating worlds at their least Woodin cardinal. However, this leaves one with nothing, an unstructured collection of sets with no theory. [36, p. 16]

He continues:

> The Ω-conjecture does not imply a paradoxical reduction of **MV** or its language to something simpler, because there is no simpler language or theory describing a "reduced multiverse". [36, p. 16]

As a matter of fact, Steel's criticism may well apply to his own framework: the preference for large cardinals as natural axioms also yields a 'reduced' multiverse, one where, presumably, there is no universe which does not contain large cardinals.

One further reason of concern with Steel's account of the multiverse may be raised by the staunch multiverse view supporter with respect to the purpose of 'unification'. Steel owes us a more coherent explanation of why **MV** with a core would be more suitable to our purposes than **MV** with no core. Steel's response would, presumably, be that in the former case we increase our level of information, by enlarging the set of provable statements. But at what price? For instance, one may still hold the legitimate view that CH is false, in the face of its being true in the core. Steel's way out of this is to ultimately appeal to naturalistic concerns, which would override all other sorts of concerns about truth. He says:

> The strong absolutist who believes that V does not satisfy CH must still face the question whether the multiverse has a core satisfying Axiom **H**. If he agrees that it does, then the argument between him and someone who accepts Axiom **H** as a strong absolutist seems to have little practical importance. [36, p. 17]

What Steel seems to suggest here is that even the strong absolutist who believes that CH is false must acknowledge that the multiverse has a core and, as a consequence of this, he might come to hold the truth of CH. Now, it may well be that the strong absolutist might come to accept the truth of CH via the acceptance of the Axiom **H**, that is through accepting **MV** with a core for 'extrinsic' reasons, but then, in turn, it would be the Axiom **H** which would be in strong need of that kind of justification the strong absolutist would be more naturally inclined to accept. In other terms, it is far from obvious that all strong absolutists would come to accept the Axiom **H** on purely naturalistic grounds. They might want to have some stronger justification, probably stronger than the one Steel can, at present, offer them.

3 The 'Vertical' Multiverse

In this section, we first discuss the actualist and potentialist conceptions of V and then examine Zermelo's account of the universe of sets, which contains features of both conceptions. Our aim is to argue that Zermelo's account can be viewed as one further multiverse conception (featuring a 'vertical' multiverse), which allows V to be heightened while keeping its width fixed. As said in the beginning (Sect. 1.1), this multiverse conception is preferable for the implementation of the hyperuniverse programme, whose general features and goals we briefly review in Sect. 3.3.1.

Finally, in the last subsection, we shall show how an infinitary logic (V-logic) can be used to express the horizontal maximality of the 'vertical' multiverse.

3.1 Actualism and Potentialism

We take actualism as a position which construes V as an actual object, a fully actualised domain of all sets, as something given which, accordingly, cannot be modified. According to actualists, there is no way to 'stretch' V: any model-theoretic construct which seems to do this produces, in fact, a construction which is within V. We saw that Koellner's criticism of the Hamkinsian multiversist, ultimately, seemed to advocate such a view: all the universes Hamkins thought to exist separately did not require of one to conceive anything more than V and, thus, an actualist can still accommodate them to her conception.

A potentialist, on the other hand, sees V as an indefinite object, which can never be thought of as a 'fixed' entity. The potentialist may well believe that there are some fixed features of V, but she believes that these are not sufficient to fully make sense of an 'unmodifiable' V: the potentialist believes that V is indeed 'modifiable' in some sense.

The actualist/potentialist dichotomy seems to recapitulate a great part of the current philosophy of set theory, insofar as one of the latter's primary concerns is to make sense of the nature of V. Actualists are more naturally grouped with realists, whereas potentialists with non-realists. However, this does not have to be necessarily the case. For instance, one may be a potentialist about classes (like V itself) and nonetheless be a realist. That V is not a set we know from the early emergence of the paradoxes, but just what else it should be is unclear and, in fact, early set theory dealt with this issue from diverse angles.[18]

It seems natural to distinguish the following four types of actualism and potentialism[19]:

HEIGHT ACTUALISM: the height of V is fixed, that is, no new ordinals can be added,

WIDTH ACTUALISM: the width of V is fixed, that is no new subsets can be added.

HEIGHT POTENTIALISM: the height of V is not fixed, new ordinals can always be added,

WIDTH POTENTIALISM: the width of V is not fixed, new subsets can always be added.

One could also hold a 'mixed' position, that is, one could be a height potentialist and a width actualist or a width potentialist and a height actualist. While a priori a mixed position might seem less tenable, we will argue in the next section that the Zermelian concept of set naturally leads to such a position.

Another crucial factor which bears on the distinction between actualism and potentialism and the preference for one over the other is a concern for the 'maximality of V'. The 'maximality of V' seems to adumbrate the possibility that V be conceived as one among many objects, as a 'picture' among other 'pictures' of the universe. If one is a potentialist, then one can more easily make sense of different 'pictures' of V. In particular, one can make sense of the stretching of V, that is, its being 'extended' in height and width, in ways which are seen to be maximal in some respect. We sometimes call such extensions 'lengthenings' (when new ordinals are added) and 'thickenings' (when new subsets are added).

---

[18] A crucial reading on this is Hallett's book on the emergence of the *limitation of size* doctrine [15]. The distinction between actualists and potentialists may be construed as the result of different interpretations of Cantor's *absolute infinite*. One of the most exhaustive articles on Cantor's conception of absoluteness and of its inherent tension between actualism and potentialism is Jané [20]. For a discussion of actualism and potentialism, with reference to the justification of *reflection principles*, see Koellner [22]. For an accurate overview of several potentialist positions, see Linnebo [26].

[19] The distinction between 'height' and 'width' of the universe is firmly based on the iterative concept of set: the length of the ordinal sequence determines the height of the universe, while the width of the universe is given by the powerset operation.

3.2 Zermelo's Account: the 'Vertical' Multiverse

Our examination of Zermelo's views is based on his [42]. In this paper, Zermelo shows that the axioms of second-order set theory Z_2 are *quasi-categorical* it the sense that every model M of Z_2 is of the form $(V_\kappa, \in)$ where κ is a strongly inaccessible cardinal; V_κ is called a *natural domain*.

Zermelo construes the sequence of natural domains 'dynamically', as the unfolding of a temporary, endless actualisation of the universe. However, if one looks closer, within this process, there is no longer a single universe present. The universe V, in this construction, becomes just a collection of different V_α's whose width is fixed, and whose height can be extended.

Zermelo vividly recapitulates his approach in the following manner:

> To the unbounded series of Cantor ordinals there corresponds a similarly unbounded double-series of essentially different set-theoretic models, in each of which the whole classical theory is expressed. The two polar opposite tendencies of the thinking spirit, the idea of creative advance and that of collection and completion [*Abschluss*], ideas which also lie behind the Kantian 'antinomies', find their symbolic representation and their symbolic reconciliation in the transfinite number series based on the concept of well-ordering. This series reaches no true completion in its unrestricted advance, but possesses only relative stopping-points, just those 'boundary numbers' [*Grenzzahlen*] which separate the higher model types from the lower. Thus the set-theoretic 'antinomies', when correctly understood, do not lead to a cramping and mutilation of mathematical science, but rather to an, as yet, unsurveyable unfolding and enriching of that science. ([42], in [7], p. 1233)

Zermelo's natural models can be viewed as a tower-like multiverse, where each universe is indexed by an inaccessible cardinal. If we have a proper class of inaccessible cardinals, then every natural model can be extended to a higher natural model: in our current terminology, the Zermelian concept of set theory is an example of *potentialism in height*.

On the other hand, as said, Zermelo's account is second-order, insofar as it seems to adumbrate the availability of a collection of 'definite' properties of sets (over which the axioms and, in particular, the Axiom of Separation, quantify) and this, in turn, implies the fixedness of the power-set operation. So, Zermelo's account also constitutes an example of *actualism in width*.

To sum up, Zermelo's conception of V can be seen as a multiverse conception that features a 'vertical' multiverse which embraces height potentialism and width actualism.

Now, Zermelo's conception seems preferable for the hyperuniverse programme because it entirely befits its goals, that is the search for optimal mathematical principles expressing the maximality of V, and we now explain why.

There are two main forms of maximality: maximality in height (*vertical maximality*) and maximality in width (*horizontal maximality*). Vertical maximality can be formulated in many ways, but perhaps its ultimate form was introduced and studied in [11] by two of the present authors. Such a principle is in line with the height potentialism inherent in Zermelo's account.[20]

As far as horizontal maximality is concerned, it would seem that width potentialism would best suit the goals of the hyperuniverse programme. However, we feel there is no need to drop Zermelo's width actualism, but the reasons for our choice are different from Zermelo's. As said above, Zermelo committed to a second-order version of the axioms, which implied the determinacy of the power-set operation. We do not hold this position. In our view, height potentialism is natural in light of the clear and coherent way one can add new V-levels by extending the ordinal numbers through iteration, whereas width potentialism is not, as there is no analogous clear and coherent iteration process for enlarging power-sets. In other terms, the addition of ordinals can be carried out in an orderly, stage-like manner, whereas the addition of subsets cannot. This is the reason why we accept Zermelo's width actualism in our account of V, and, as we shall show in Sect. 3.3.2, there is a way of exploring horizontal maximality which makes Zermelo's account fully compatible with the hyperuniverse programme. On the other hand, as explained above, height potentialism is entirely natural, and, moreover, within the hyperuniverse programme, height actualism puts severe restrictions on formulations of the maximality of V.

We mentioned the programme several times. It is now time to provide the reader with more details about the programme. In Sect. 3.3 we briefly review it and then describe recent results which make sense of a theory of horizontal maximality.

3.3 The 'Vertical' Multiverse Within the Hyperuniverse Programme

3.3.1 A Brief Review of the Hyperuniverse Programme

The programme was introduced by the second author and Arrigoni in [1]. In that paper, the word *hyperuniverse* was introduced to denote the collection of all transitive countable models of ZFC. Within the programme, such models are viewed as a technical tool allowing set-theorists to use the standard model-theoretic and forcing techniques (the Omitting Types Theorem and the existence of generic extensions, respectively). The underlying idea is that the study of the members of the hyperuniverse allows one to indirectly examine properties of the real universe V (that we construe, by now, as the 'vertical' multiverse discussed above).

The hyperuniverse programme is essentially concerned with the notion of the *maximality* of the universe. As already mentioned above, we construe the 'maximality of V' as implying that V should be maximal among its different

---

[20] Also Reinhardt's theory of *legitimate candidates* (see Reinhardt [32]) seems to follow Zermelo's account. Finally, Hellman (in [19]), develops a structuralist account of the universe in line with Zermelo's concerns, but based on modal assumptions.

'pictures', that is, candidates for universes. It is precisely here that the hyperuniverse is helpful, because it provides the context, i.e., the collection of candidates, where we can look for a maximal universe.

Thus the hyperuniverse programme analyses the maximality of V through the study of non-first-order maximality properties of members of the hyperuniverse. Among the earliest examples of such properties is the IMH, the Inner Model Hypothesis:

Definition (IMH) Let M be a member of the hyperuniverse. For any φ, whenever there is an outer model W of M where φ holds, there is a definable inner model $M' \subseteq M$ where φ also holds.

This principle clearly postulates that M is 'maximal' in *width*, in some sense, among the members of the hyperuniverse. Using large cardinals, one can prove the consistency of IMH.[21]

The programme is based on the conviction—still to be verified—that the different formalisations of the notion of maximality will lead to an optimal such formalisation, and that the first-order sentences which hold in all universes which exhibit that optimal form of maximality will be regarded as first-order consequences of the notion of the maximality of V and, therefore, as good candidates for new axioms of set theory. The programme also aims to make a case for the 'intrinsicness' of such axiom candidates, insofar as their selection, in the end, would only depend upon a thorough analysis of the concept of the 'maximality of V'.

Now, just what is the connection between the hyperuniverse and the 'vertical' multiverse described above? We view the hyperuniverse only as a technical tool. However, one may also view it as one further 'auxiliary' multiverse, more suited to the kind of mathematical investigations to be carried out within the programme. In this sense, one could say that the 'vertical' multiverse is supplemented—for strictly mathematical reasons—by such an auxiliary multiverse.

3.3.2 Width Actualism and Infinitary Logic

We now proceed to present mathematical results which show that one can address horizontal maximality within the 'vertical' multiverse, that is, in a V which is potential in height but actual in width, by using the hyperuniverse as a mathematical tool.

Let us, first, briefly summarise in conceptual terms what we will, then, be showing in a rigorous mathematical fashion. In the hyperuniverse programme we refer to both 'lengthenings' and 'thickenings' of V and, in particular, the IMH contains a reference to 'thickenings'. 'Lengthenings' are entirely unproblematic for a height potentialist, whereas 'thickenings' would appear to collide with the width actualism inherent in Zermelo's account we chose to adopt. However, we argue

---

[21]The proof is in Friedman et al. [12].

that, using V-logic, we can express properties of 'thickenings' of V without actually requiring the existence of such 'thickenings', and, moreover, these properties are first-order over what we call $\mathrm{Hyp}(V)$, a modest 'lengthening' of V. Therefore, with the mathematical tools we will be employing, there is no violation of the Zermelian conception to which we commit ourselves.

We start by noticing that the Löwenheim-Skolem theorem allows one to argue that any first-order property of V reflects to a countable transitive model (that is, a member of the hyperuniverse). However, on a closer look, one needs to deal with the problem that not all relevant properties of V are first-order over V; in particular, the property of V 'having an outer model (a 'thickening') with some first-order property' is a higher-order property. We show now that, with a little care, all reasonable properties of V formulated with reference to outer models are actually first-order over a slight extension ('lengthening') of V.[22]

We first have to recall some basic notions regarding the infinitary logic $L_{\kappa,\omega}$, where κ is a regular cardinal.[23] For our purposes, the language is composed of κ-many variables, up to κ-many constants, symbols $\{=, \in\}$, and auxiliary symbols. Formulas in $L_{\kappa,\omega}$ are defined by induction: (1) All first-order formulas are in $L_{\kappa,\omega}$; (2) whenever $\{\varphi\}_{i<\mu}$, $\mu < \kappa$ is a system of formulas in $L_{\kappa,\omega}$ such that there are only finitely many free variables in these formulas taken together, then the infinite conjunction $\bigwedge_{i<\mu} \varphi_i$ and the infinite disjunction $\bigvee_{i<\mu} \varphi_i$ are formulas in $L_{\kappa,\omega}$; (3) if φ is in $L_{\kappa,\omega}$, then its negation and its universal closure are in $L_{\kappa,\omega}$. Barwise developed the notion of proof for $L_{\kappa,\omega}$ and showed that this syntax is complete, when $\kappa = \omega_1$, with respect to the semantics (see discussion below and Theorem 3.3.2).

A special case of $L_{\kappa,\omega}$ is the so-called V-logic. Suppose V is a transitive set of size κ. Consider the logic $L_{\kappa^+,\omega}$, augmented by κ-many constants $\{\bar{a}_i\}_{i<\kappa}$ for all the elements a_i in V. In this logic, one can write a single infinitary sentence which ensures that if M is a model of this sentence (which is set up to ensure some desirable property of M), then M is an outer model of V (satisfying that desirable property). Now, the crucial point is the following: if V is countable, and this sentence is consistent in the sense of Barwise, then such an M really exists in the ambient universe.[24] However, if V is uncountable, the model itself may not exist

---

[22]The use of the Löwenheim-Skolem theorem, while completely legitimate, is actually optional: if one wishes to analyse the outer models of V without 'going countable', one can do it by using the V-logic introduced below. However, there is a price to pay: instead of having the elegant clarity of countable models, one will just have to refer to different theories. This has analogies in forcing: to have an actual generic extension, one needs to start with a countable model; if the initial model is larger, one can still deal with forcing syntactically, but a generic extension may not exist (see, for instance, Kunen [25]). A more relevant analogy in our case is that the Omitting Types Theorem (which is behind V-logic) works for countable theories, but not necessarily for larger cardinalities.

[23]Full mathematical details are in Barwise [4]. We wish to stress that the infinitary logic discussed in this section appears only at the level of theory as a tool for discussing outer models. The ambient axioms of ZFC are still formulated in the usual first-order language.

[24]Again, for more details we refer the reader to Barwise [4].

in the ambient universe, but, in that case, we still have the option of staying with the syntactical notion of a consistent sentence.

We have to introduce one further ingredient, that of an *admissible set*. M is an admissible set if it models some very weak fragment of ZFC, called Kripke-Platek set theory, KP. What is important for us here is that for any set N, there is a smallest admissible set M which contains N as an element – M is of the form $L_\alpha(N)$ for the least α such that M satisfies KP. We denote this M as Hyp(N).

And now for the following crucial result:

Theorem (Barwise) *Let V be a transitive set model of ZFC. Let $T \in V$ be a first-order theory extending ZFC. Then there is an infinitary sentence $\varphi_{T,V}$ in V-logic such that following are equivalent:*

1. $\varphi_{T,V}$ is consistent.
2. Hyp(V) $\models$ "$\varphi_{T,V}$ is consistent."
3. If V is countable, then there is an outer model M of V which satisfies T.

By the theorem above, if we wish to talk about outer models of V ('thickenings'), we can do it in Hyp(V)—a slight lengthening of V—by means of theories, without the need to really thicken our V (and indeed, we cannot thicken it if we are width actualists). However, if we wish to have models of the resulting consistent theories, then, using the Löwenheim-Skolem theorem, we can shift to countable transitive models. And this is precisely where the hyperuniverse comes into play.[25]

Now, we also want to make sure that members of the hyperuniverse really witness statements expressing the horizontal maximality of V. One such statement was the mentioned IMH.

Recall that V satisfies IMH if for every first-order sentence ψ, if ψ is satisfied in some outer model W of V, then there is a definable inner model $V' \subseteq V$ satisfying ψ. Ostensibly, the formulation of IMH requires the reference to all outer models of V, but with the use of infinitary logic, we can formulate IMH syntactically in Hyp(V) as follows: V satisfies IMH if for every $T = \text{ZFC}+\psi$, if $\varphi_{T,V}$ from Theorem 3.3.2 above is consistent in Hyp(V), then there is an inner model of V which satisfies T. Finally, with an application of the Löwenheim-Skolem theorem to Hyp(V), this becomes a statement about elements of the hyperuniverse.

4   Concluding Summary

We have seen that the multiverse construct can be spelt out in different ways. We proposed a unified way to classify different positions, centered on the realism/non-realism conceptual dichotomy.

---

[25]This is in clear analogy to the treatment of set-forcing, see footnote 22. However, note that unlike in set-forcing, where the syntactical treatment can be formulated inside V, to capture arbitrary outer models, we need a bit more, i.e. Hyp(V).

The realist multiverse view was represented by the Balaguer-Hamkins multiverse, wherein different universes instantiated different concepts of set (or, alternatively, different models instantiated different collections of axioms), whereas the non-realist multiverse view (whose main representative was Shelah) was construed as a form of radical pluralism with no explicit commitment to an ontological position. The other two conceptions we reviewed (Woodin's and Steel's) contribute to the discussion about the multiverse in the following manner: they explore a limited, but mathematically rich, concept of the set-theoretic multiverse, that consisting of a collection of set-generic extensions.

Finally, in the last section, building on the Zermelian account of V, with its conceptual reliance on height potentialism and width actualism, we have described one further multiverse conception that features a 'vertical' multiverse.

We believe that this conception—along with the use of infinitary logic and the hyperuniverse as 'auxiliary' multiverse—is the preferable multiverse conception for the hyperuniverse programme, insofar as: (1) it allows one to formulate maximality principles addressing 'lengthenings' and 'thickenings' of the universe (for 'thickenings', though, we also need to use V-logic), and, at the same time, (2) it does not compel us to embrace the potentiality of the power-set operation, for which, as discussed earlier, we do not have a clear and conceptually satisfying framework. In our view, the fact that the hyperuniverse programme exhibits the potential for generating new axioms of set theory through the study of maximality can be used as an argument in favour of this multiverse conception.

References

1. T. Arrigoni, S. Friedman, The hyperuniverse program. Bull. Symb. Log. **19**(1), 77–96 (2013)
2. M. Balaguer, A platonist epistemology. Synthese **103**, 303–25 (1995)
3. M. Balaguer, *Platonism and Anti-Platonism in Mathematics* (Oxford University Press, Oxford, 1998)
4. J. Barwise, *Admissible Sets and Structures* (Springer, Berlin, 1975)
5. M. Colyvan, E. Zalta, Review of Balaguer's 'Platonism and Anti-Platonism in Mathematics'. Philos. Math. **7**, 336–349 (1999)
6. H.G. Dales, G. Oliveri (eds.), *Truth in Mathematics* (Oxford University Press, Oxford, 1998)
7. W. Ewald (ed.), *From Kant to Hilbert: A Source Book in the Foundations of Mathematics*, vol. II (Oxford University Press, Oxford, 1996)
8. S. Feferman, Does mathematics need new axioms? Am. Math. Mon. **106**, 99–111 (1999)
9. S. Feferman, The Continuum Hypothesis is neither a definite mathematical problem or a definite logical problem, preprint (2014)
10. S. Feferman, P. Maddy, H. Friedman, J.R. Steel, Does mathematics need new axioms? Bull. Symb. Log. **6**(4), 401–446 (2000)
11. S. Friedman, R. Honzik, On strong forms of reflection in set theory. Math. Log. Quart. **62**(1–2), 52–58 (2016)
12. S. Friedman, P. Welch, W.H. Woodin, On the consistency strength of the Inner Model Hypothesis. J. Symb. Log. **73**(2), 391–400 (2008)
13. K. Gödel, What is Cantor's continuum problem? Am. Math. Mon. **54**, 515–525 (1947)

14. K. Gödel, *Collected Works, II: Publications 1938–1974* (Oxford University Press, Oxford, 1990)
15. M. Hallett, *Cantorian Set Theory and Limitation of Size* (Clarendon Press, Oxford, 1984)
16. J.D. Hamkins, The set-theoretic multiverse. Rev. Symb. Log. **5**(3), 416–449 (2012)
17. K. Hauser. Objectivity over objects: a case study in theory formation. Synthèse **128**(3), 245–285 (2001)
18. K. Hauser, Is the continuum problem inherently vague? Philos. Math. **10**, 257–285 (2002)
19. G. Hellman, *Mathematics without Numbers. Towards a Modal-Structural Interpretation* (Clarendon Press, Oxford, 1989)
20. I. Jané, The role of the absolute infinite in Cantor's conception of set. Erkenntnis **42**, 375–402 (1995)
21. A. Kanamori, *The Higher Infinite* (Springer, Berlin, 2003)
22. P. Koellner, On reflection principles. Ann. Pure Appl. Log. **157**(2–3), 206–219 (2009)
23. P. Koellner, Truth in mathematics: the question of pluralism, in *New Waves in the Philosophy of Mathematics*, ed. by O. Bueno, Ø. Linnebo (Palgrave Macmillan, London/New York, 2009), pp. 80–116
24. P. Koellner, Hamkins on the Multiverse, preprint (May 2013)
25. K. Kunen, *Set Theory. An Introduction to Independence Proofs* (College Publications, London, 2011)
26. Ø. Linnebo, The Potential Hierarchy of Sets. Rev. Symb. Log. **6**(2), 205–228 (2013)
27. P. Maddy, *Naturalism in Mathematics* (Oxford University Press, Oxford, 1997)
28. P. Maddy, *Defending the Axioms* (Oxford University Press, Oxford, 2011)
29. D.A. Martin, Mathematical evidence, in *Truth in Mathematics*, ed. by H.G. Dales, G. Oliveri (Clarendon Press, Oxford, 1998), pp. 215–231
30. M. Potter, *Set Theory and its Philosophy* (Oxford University Press, Oxford, 2004)
31. H. Putnam, Models and reality. J. Symb. Log. **45**(3), 464–482 (1979)
32. W.N. Reinhardt, Remarks on reflection principles, large cardinals and elementary embeddings, in *Proceedings of Symposia in Pure Mathematics*, vol. XIII, 2, ed. by T. Jech (American Mathematical Society, Providence (Rhode Island), 1974)
33. S. Shapiro, *Thinking About Mathematics* (Oxford University Press, Oxford, 2000)
34. S. Shapiro (ed.), *Oxford Handbook of Philosophy of Mathematics* (Oxford University Press, Oxford, 2005)
35. S. Shelah, Logical dreams. Bull. Am. Math. Soc. **40**(2), 203–228 (2003)
36. J.R. Steel, Gödel's program, in *Interpreting Gödel. Critical Essays*, ed. by J. Kennedy (Cambridge University Press, Cambridge, 2014), pp. 153–179
37. M. van Atten, J. Kennedy, On the philosophical development of Kurt Gödel. Bull. Symb. Log. **9**(4), 425–476 (2003)
38. H. Wang, *From Mathematics to Philosophy* (Routledge & Kegan Paul, London, 1974)
39. H. Wang, *A Logical Journey* (MIT Press, Cambridge, MA, 1996)
40. W.H. Woodin, The realm of the infinite, in *Infinity. New Research Frontiers*, ed. by W.H. Woodin, M. Heller (Cambridge University Press, Cambridge, 2011), pp. 89–118
41. W.H. Woodin, The transfinite universe, in *Horizons of Truth. Kurt Gödel and the Foundations of Mathematics*, ed. by M. Baaz, C.H. Papadimitriou, D.S. Scott, H. Putnam (Cambridge University Press, Cambridge, 2011), pp. 449–474
42. E. Zermelo, Über Grenzzahlen und Mengenbereiche: neue Untersuchungen über die Grundlagen der Mengenlehre. Fundam. Math. **16**, 29–47 (1930)

Evidence for Set-Theoretic Truth and the Hyperuniverse Programme

Sy-David Friedman

Abstract I discuss three potential sources of evidence for truth in set theory, coming from set theory's roles as a branch of mathematics and as a foundation for mathematics as well as from the intrinsic maximality feature of the set concept. I predict that new *non first-order* axioms will be discovered for which there is evidence of all three types, and that these axioms will have significant first-order consequences which will be regarded as true statements of set theory. The bulk of the paper is concerned with the *Hyperuniverse Programme*, whose aim is to discover an optimal mathematical principle for expressing the maximality of the set-theoretic universe in height and width.

1 Introduction

The truth of the axioms of ZFC is commonly accepted for at least two reasons. One reason is *foundational*, as they endow set theory with the ability to serve as a remarkably good foundation for mathematics as a whole, and another is *intrinsic*, as (with the possible exception of AC, the axiom of choice) they can be seen to be derivable from the concept of set as embodied by the maximal iterative conception.

In fact a little bit more than ZFC is justifiable on intrinsic and perhaps also foundational grounds. I refer here to *reflection principles* and their related *small large cardinals*, which are also derivable from the maximal iterative conception through *height (ordinal) maximality* and, at least in the case of inaccessible cardinals, are occasionally useful for the development of certain kinds of highly abstract mathematics (such as *Grothendieck universes*). These extensions of ZFC are *mild* in the sense that they are compatible with the *powerset-minimality* principle $V = L$.

Originally published in S. Friedman, Evidence for set-theoretic truth and the Hyperuniverse Programme. IfCoLog J. Log. Appl. "Proof, Truth, Computation" **3**(4), 517–555 (2016).

S.-D. Friedman (✉)
Kurt Gödel Research Center, University of Vienna, Wien, Austria
e-mail: sdf@logic.univie.ac.at

But finding strong evidence for the truth of axioms that contradict $V = L$ has been exceedingly difficult. There are a number of reasons for this. One is the fact that mild extensions of ZFC have been in a sense *too good*, in that they alone have until recently been sufficient to serve the needs of set theory as a foundation for mathematics. Another is the difficulty of squeezing more out of the maximal iterative conception through a *width (powerset) maximality* analogue of the height maximality principles that give rise to reflection. And the development of set theory as a branch of mathematics has been so dramatic, diverse and ever-changing that it has been impossible to select those perspectives on the subject whose choices of new axioms can be regarded as "the most true".

My aim in this article is to provide evidence for the following three predictions.

The Richness of Set-Theoretic Practice The development of set theory as a branch of mathematics is so rich that there will never be a consensus about which first-order axioms (beyond ZFC plus small large cardinals) best serve this development.

A Foundational Need Just as AC is now accepted due to its essential role for mathematical practice, a systematic study of independence results across mathematics will uncover first-order statements contradicting CH (and hence also $V = L$) which are best for resolving such independence.

An Optimal Maximality Criterion Through the *Hyperuniverse Programme* it will be possible to arrive at an optimal *non first-order* axiom expressing the maximality of the set-theoretic universe in height and width; this axiom will have first-order consequences contradicting CH (and hence also $V = L$).

And as a synthesis of these three predictions I propose the following optimistic scenario for making progress in the study of set-theoretic truth.

Thesis of Set-Theoretic Truth There will be first-order statements of set theory that well serve the needs of set-theoretic practice and of resolving independence across mathematics, and which are derivable[1] from the maximality of the set-theoretic universe in height and width. Such statements will come to be regarded as *true statements of set theory*.

This *Thesis* has a converse: In order for a first-order statement contradicting $V = L$ to be regarded as *true*, in my view it must well serve the needs of set-theoretic practice and of resolving independence in mathematics, and it must at least be compatible with the maximality of the set-theoretic universe as expressed by the optimal maximality criterion. Indeed the strength of the evidence for such a statement's truth is in my view measured by the extent to which it fulfills these three requirements.

---

[1] For a discussion of this notion of *derivability* see the final Sect. 4.13.

An important consequence of the *Thesis* is the failure of CH. Thus part of my prediction is that CH will be regarded as false.

Note that in the *Thesis* I do not refer to *true first-order axioms* but only to *true first-order statements*. The reason is the following additional claim.

Beyond First-Order There will never be a consensus about the truth of proposed first-order axioms that contradict $V = L$; instead true first-order statements will arise solely as consequences of true *non first-order* axioms.

One reason for this claim is the inadequacy of first-order statements to capture the maximality of the set-theoretic universe.

The plan of this paper is as follows. First I'll review some of the popular first-order axioms that well serve the needs of set-theoretic practice and argue for the *Richness* prediction above. Second I'll discuss what little is known about independence across mathematics, discussing the role of forcing axioms as evidence for the *Foundational* prediction above. And by far the bulk and central aim of the paper is the third part, in which I present the *Hyperuniverse Programme*, including its philosophical foundation and most recent mathematical developments.

2 Set-Theoretic Practice

Set theory is a burgeoning subject, rife with new ideas and new developments, constantly leading to new perspectives. Naturally certain of these perspectives stand out among the chaotic mass of new results being proved, and it is worth focusing on a few of these to expose the difficulty of settling on particular new axioms as being "the true ones".

I have emphasized the need to find evidence for the truth of axioms that contradict $V = L$, but purely in terms of the value of an axiom for the development of good set theory, what I will refer to as *Type 1 evidence*, this is not possible. Jensen's deep work unlocking the power of this axiom reveals the power of $V = L$, indeed it appears to give us, when combined with small large cardinals, a theory that is complete for all natural set-theoretic statements! That is a remarkable achievement and speaks volumes in favour of declaring $V = L$ to be *true* based on Type 1 evidence.

A natural Type 1 objection to $V = L$ is that it doesn't take forcing into account, a fundamental method for building new models of set theory. Admittedly, even in L one has forcing extensions of countable models, but it is more natural to force over the full L and not just over some small piece of it. So now we contradict $V = L$ in favour of "V contains many generic extensions of L" or something similar.

Having lots of forcing extensions of L sounds good, but then what is our canonical universe now? Shouldn't we also have a sentence that is true only in V, and not in any of its proper inner models, while at the same time having many generic extensions of L? Indeed this is possible with class forcing (see [11]). So now we have a nice Type 1 axiom: V is a canonical universe which is class-generic over L, containing many set-generic extensions of L. This is an excellent context for doing set theory, as the forcing method is now available.

In fact we can do even better and take V to be $L[0^\#]$. Not only does this model contain many generic extensions of L, it is also a canonical universe and we recover all of the powerful methods that Jensen developed under $V = L$, relativised now to the real $0^\#$. So our Type 1 evidence leads us to the superb axiom $V = L[0^\#]$.

Objection! What about measurable cardinals? Recall the important hierarchy of consistency strengths: Natural theories are wellordered (up to bi-interpretability) by their consistency strengths and the consistency strengths of large cardinal axioms provide a nice collection of consistency strengths which is cofinal in a large initial segment of (if not all of) this hierarchy. This does not mean that large cardinals must exist but at the very least there should be inner models having them. So now based on Type 1 evidence we get some version of "There are inner models with large cardinals", an attractive environment in which to do good set theory.

Moreover, notice that if we have inner models for large cardinals we haven't lost the option of looking at L or its generic extensions, they are still available as inner models. So we seem to have reached the best Type 1 axiom yet.

But we could ask for even more. Recall that L has a nice internal structure, very powerful for deriving consequences of $V = L$. Can V not only have inner models for large cardinals but also an L-like internal structure? Of course the answer is positive, as we can adopt the axiom "There are inner models with large cardinals and $V = L[x]$ for some real x". A better answer is provided in [14], where it is shown that V can be L-like together with arbitrary large cardinals, not only in inner models but in V itself. However, as attractive as this may sound, it fails to address a key problem, and this is where we see the multiple perspectives of set theory, with no single perspective having a claim to being "the best".

Even if we produce a nice axiom[2] of the form "There are large cardinals and V is a canonical generalisation of L", doing so commits us to an L-like environment in which to do set theory. Indeed there are other compelling perspectives on set theory which lead us to non L-like environments and correspondingly to entirely different Type 1 axioms. I will mention two of them. (Further information about the notions mentioned below is available in [22].)

Forcing axioms have a long history, dating back to Martin's axiom (MA), a special case of which asserts the existence of generics for ccc partial orders (i.e. partial orders with only countable antichains) over models of size $\aleph_1$. This simple axiom can be used to establish in one blow the relative consistency of a huge range of set-theoretic statements. Naturally there has been interest in strengthenings of MA, and a popular one is the Proper Forcing Axiom (PFA), which strengthens this[3] to the wider class of *proper* partial orders.

---

[2]Woodin has in fact proposed such an axiom which he calls *Ultimate L*.

[3]For the experts, to get PFA one must allow non-transitive models of size $\aleph_1$.

Now with regard to Type 1 evidence the point is that PFA has even more striking consequences than MA, qualifying it as a central and important tool for solving combinatorial problems in set theory. A powerful case can be made for its truth based on Type 1 evidence. But of course PFA conflicts with any axiom which asserts that V is L-like, as it implies the negation of CH. In fact PFA implies that the size of the continuum is $\aleph_2$.

The diversity of Type 1 evidence goes beyond just L-likeness and forcing axioms; there are also cardinal characteristics. These are natural and heavily-investigated cardinal numbers that arise when studying definability-theoretic and combinatorial properties of sets of real numbers. Each of these cardinal characteristics is an uncountable cardinal number of size at most the continuum. Now given the variety of such characteristics together with the fact that they can consistently differ from each other, isn't it compelling to adopt the axiom that cardinal characteristics provide a large spectrum of distinct uncountable cardinals below the size of the continuum and therefore the continuum is indeed quite large, in contradiction to both L-likeness and forcing axioms?[4]

Thus we have three distinct types of axioms with excellent Type 1 evidence: L-likeness with large cardinals, forcing axioms and cardinal characteristic axioms. They contradict each other yet each is consistent with the existence of inner models for the others. In my view, this makes a clear case that Type 1 evidence is insufficient to establish the truth of axioms of set theory; it is also insufficient to decide whether or not CH is true.

3   Set Theory as a Foundation for Mathematics

Of course axiomatic set theory can be heartily congratulated for its success in providing a foundation for mathematics. An overwhelming case can be made that when theorems are proved in mathematics they can be regarded as theorems of a mild extension of ZFC (compatible with $V = L$). In particular, we routinely expect questions in mathematics to be answerable (perhaps with great difficulty!) in a mild extension of ZFC.

A consequence is that an independence result for such mild extensions is indeed an independence result for mathematics as a whole. This is of course of minor importance if the independence result in question is a statement of set theory, as set theory is just a small part of mathematics. But this is of considerable importance when independence arises with questions of mathematics outside of set theory, as is the case with the Borel, Kaplansky and Whitehead Conjectures of measure theory, functional analysis and group theory, respectively. Let us not forget the

---

[4]As a specific example, let $\mathfrak{a}$ denote the least size of an infinite almost disjoint family of subsets of ω, and $\mathfrak{b}$ ($\mathfrak{d}$) the least size of an unbounded (dominating) family of functions from ω to ω ordered by eventual domination. Then $\mathfrak{b} < \mathfrak{a} < \mathfrak{d}$ is consistent; shouldn't it in fact be true?

great mathematician David Hilbert's thesis that the questions of mathematics can be resolved using the powerful tools of the subject. An understanding of how to deal with independence is needed to restore the status of mathematics as the complete and definitive field of study that Hilbert envisaged.

The time is ripe for set-theorists to focus on this problem. The central question is:

Foundational or Type 2 Evidence Are there particular axioms of set theory which best serve the needs of resolving independence in other areas of mathematics?

Recently there are signs that a positive answer to this question is emerging, as new applications of set theory to functional analysis, topology, abstract algebra and model theory (a field of logic, but still outside of set theory) are being found. The *Foundational Need* that I expressed earlier is precisely the prediction that a pattern will emerge from these applications to reveal that particular axioms of set theory are best for bringing set theory closer to the complete foundation that Hilbert was hoping for.

Now where are these *foundationally advantageous* axioms of set theory to be found? Consider the following list of candidates with good Type 1 evidence:

$V = L$

V is a canonical and rich class-generic extension of L

Large Cardinal Axioms (like supercompacts)

Forcing Axioms like MA, PFA

Determinacy Axioms like AD in $L(R)$

Cardinal Characteristic Axioms like $\mathfrak{b} < \mathfrak{a} < \mathfrak{d}$

As already said, each of these axioms is important for the development of set theory, providing a unique perspective on the subject. But perhaps it is surprising to discover that only two of them, $V = L$ and Forcing Axioms, have had any significant impact on mathematics outside of set theory! The impact of Large Cardinal Axioms (like supercompacts) and Cardinal Characteristic Axioms has been minimal and that of Determinacy Axioms non-existent so far.

To give a bit more detail, both $V = L$ and Forcing Axioms can be used to answer the following questions (in different ways):

Functional Analysis Must every homomorphism from $C(X)$, X compact Hausdorff, into another Banach algebra be continuous (the Kaplansky Problem)? Is the ideal of compact operators on a separable Hilbert space in the ring of all bounded operators the sum of two smaller ideals?; Are all automorphisms of the Calkin Algebra inner?

Topology and Measure Theory Is every normal Moore Space metrizable? Are there S-spaces (regular, hereditarily separable spaces where some open cover has no countable subcover)? Is every strong measure 0 set of reals countable (the Borel conjecture)?

Abstract Algebra Is every Whitehead group free (the Whitehead Problem)? What is the homological dimension of $R(x, y, z)$ as an $R[x, y, z]$-module where R is the field of real numbers? Does the direct product of countably many fields have global dimension 2?

One could also mention the field of Model Theory (part of Logic, but not part of Set Theory), where new axioms of Set Theory may play an important role in the study of Morley's theorem for Abstract Elementary Classes or perhaps even in the resolution of Vaught's Conjecture.

My prediction is that $V = L$ and Forcing Axioms will be the definite winners among choices of axioms of Set Theory that resolve independence across mathematics as a whole. But as $V = L$ is in conflict with the maximality of the set-theoretic universe in width, it is not suitable as a realization of the *Thesis of Set-Theoretic Truth*, leaving Forcing Axioms as the current leading candidate for that.

4 The Maximality of the Set-Theoretic Universe and the HP

The letters HP stand for the *Hyperuniverse Programme*, which I now discuss in detail. This programme had its origins in [2, 12], was introduced in [3] and was further discussed in [1, 17, 32].

4.1 The Iterative Conception of Set

As Gödel put it, the *iterative conception* of set expresses the idea that a set is something obtainable from well-defined objects by iterated application of the powerset operation. In more detail (following Boolos [7]; also see [27]): Sets are formed in *stages*, where only the empty set is formed at stage 0 and at any stage greater than 0, one forms collections of sets formed at earlier stages. (Said this way, a set is re-formed at every stage past where it is first formed, but that is OK.) Any set is formed at some least stage, after its elements have been formed. This conception excludes anomalies: We can't have $x \in x$, there is no set of all sets, there are no cycles $x_0 \in x_1 \in \cdots \in x_n \in x_0$ and there are no infinite sequences $\cdots \in x_n \in x_{n-1} \in x_{n-2} \in \cdots \in x_1 \in x_0$, as there must be a least stage at which one of the x_n's is formed. We'll assume that there are infinite sets,[5] so the iteration process leads to a limit stage ω, which is not 0 and is not a successor stage.

The iterative conception yields that the universe of sets is a model of the axioms of Zermelo Set Theory, i.e. ZFC without Replacement and without the Axiom of Choice. The standard model for this theory is $V_{\omega+\omega}$.

---

[5]This is derivable once we add *maximality* to the iterative conception, but is convenient to assume already as part of the iterative conception.

Nevertheless, Replacement and AC (the Axiom of Choice) are included as part of the standard axioms of Set Theory, for very different reasons. The case for AC is typically made on *extrinsic grounds*, citing its *fruitfulness* for the development of mathematics and its corresponding necessity for Set Theory as a foundation for mathematics (a case of what I have called *Type 2 evidence*). It is not clear to me that Choice is derivable from the iterative conception, nor from its necessity for doing good Set Theory (*Type 1 evidence*).

Replacement, on the other hand, is derivable from the concept of set. To see this, we need to extend the iterative conception to the stronger *maximal iterative conception*, also implicit in the set-concept.

4.2   Maximality and the Iterative Conception

The term *maximal* is used in many different senses in Set Theory, what I have in mind here is a very specific use associated to the iterative conception (IC). Recall that according to the IC, sets appear inside levels indexed by the ordinal numbers, where each successor level $V_{\alpha+1}$ is the powerset of the previous. As Boolos explained, the IC alone takes no stand on how many levels there are (the *height* of the universe V) or on how fat the individual levels are (the *width* of V). However it is generally regarded as implicit in the set-concept that both of these should be *maximal*:

Height (or Ordinal) Maximality The universe V is *as tall as possible*, i.e., the sequence of ordinals is *as long as possible*.

Width (or Powerset) Maximality The universe V is *as wide (or thick) as possible*, i.e., the powerset of each set is *as large as possible*.

If we conjunct the IC with maximality we arrive at the MIC, the *maximal iterative conception*, also part of the set-concept but more of a challenge to explain than the simple IC.

It is natural to see a *comparative* aspect to maximality, as to be *as large as possible* suggests *as large as possible within the realm of possibilities*. Thus a natural way to explain height and width maximality would be to compare V to other possible universes.

But now we face a serious problem. If V is the fixed universe of *all* sets, then there are no universes other than those already included in V. In other words V is maximal by default, as no other universe can threaten its maximality, and therefore we are limited in what we can say about this concept.

I will postpone this problem for now, and instead discuss an easier one: Let M denote a countable transitive model of ZFC (ctm). What could it mean to say that M is maximal?

Now we have a different problem. The natural way to express the maximality of M is to say that M cannot be expanded to a larger universe. Let us call this *structural maximality*. But under a very mild assumption (there is a set-model of

ZFC containing all of the reals) this is impossible: Any ctm M is an element (and therefore proper subset) of a larger ctm.

So instead we move to a milder form of maximality, called *syntactic maximality*, expressed as follows.

In the case of (syntactic-) height maximality, we consider *lengthenings* of M, i.e. ctm's M^* of which M is a rank-initial segment (the ordinals of M form an initial segment of the ordinals of M^* and the powerset operations of these two universes agree on the sets in M).

In the case of width maximality, we consider *thickenings* of M, i.e. ctm's M^* of which M is an inner model (M and M^* have the same ordinals and M is included in M^*).

In this way we can produce forms of *height maximality* and *width maximality* for ctm's as follows.

If M is *height maximal* then a property of M also holds of some rank-initial segment of M. This is the typical formulation of *reflection*. (However we will see that *height maximality* is stronger than *reflection*.) Of course specific realizations of height maximality must specify which properties are to be taken into account.

If M is *width maximal* then a property of a thickening of M also holds of some inner model of M. In the case of first-order properties this is called the *Inner Model Hypothesis*, or *IMH* (introduced in [12]).

The above discussion of maximality for ctm's, although brief, will suffice for establishing the strategy of the HP.

We return now to the problem of maximality for V. Can the above discussion for ctm's also be applied to V? Does it make sense to talk about *lengthenings* and *thickenings* of V in the way we talk about them for ctm's? There are differences of opinion about this, which I'll take up next.

4.3 Actualism and Potentialism

Recall that in the *IC* we describe V, the universe of sets, via a process of iteration of the powerset operation. Does this process come to an end, or is it indefinite, always extendible further to a longer iteration? The former possibility, that there is a "limit" to the iteration process is referred to as *height actualism* and the latter view is called *height potentialism*. Analogously there is a question of the definiteness of the powerset operation: For a given set, is its powerset determined or is it always possible to extend it further by adding more subsets? The former is called *width actualism* and latter *width potentialism*.

There is a vast literature on this topic [4, 19–21, 23–26, 29, 31, 33]. However as the Hyperuniverse Programme is very flexible on the choice of ontology, we will not engage here in a lengthy discussion of the actualism/potentialism debate, but only mention some points in favour of a *Zermelian view*, combining height potentialism with width actualism, the view which we choose to adopt for our analysis of maximality via the HP.

We can summarize the situation as follows. Without difficulty, height potentialism facilitates an analysis of height maximality. Surprisingly, we will show that even with width actualism, it also facilitates an analysis of width maximality, using the method of *V-logic*. A further benefit of height potentialism is that we can reduce the study of maximality for V to the study of maximality for ctm's.[6] Our arguments also show that height actualism is viable for our analysis of width maximality, provided it is enhanced with a strong enough fragment of MK (Morse-Kelley class theory; one only needs Σ_1^1 comprehension). Thus the only problematic ontology for the HP is height actualism supported by only a weak class theory; otherwise the choice of ontology is not critical for the HP (although the programme develops slightly differently with width potentialism than it does with width actualism).[7]

I will now present some arguments due to Geoffrey Hellman [32][8] in favour of height potentialism and width actualism, *the Zermelian view*. Hellman says:

> The idea that any universe of sets can be properly extended (in height, not width) is extremely natural, endorsed by many mathematicians (e.g. MacLane, seemingly by Gödel et al.) ... As Maddy and others say, if it's possible that sets beyond some (putatively maximal) level exist, then they do exist ... Thus, if 'imaginable' (end) extensions of V are not incoherent, then they are possible, and then, on an actualist, platonist reading, they are actual, and V wasn't really maximal after all. ... such extensions are always possible, so that the notion of a single fixed, absolutely maximal universe V of sets is really an incoherent notion.

And again:

> I have no earthly or heavenly idea what 'as high as possible' could mean, since the notion of a set domain that absolutely could not in logic be extended seems to me incoherent (or at any rate empty). As Putnam put it in his controversial paper, 'Mathematics without Foundations' (1967), 'Even God couldn't make a universe for Zermelo set theory that it would be impossible to extend.' And I agree, theology aside."

Regarding width potentialism, Hellman says [32]:

> I have a good idea, I think, about 'as thick as possible', since the notion of full power set of a given set makes perfect sense to me ... Granted that forcing extensions can be viewed as 'thickenings' of the cumulative hierarchy, as usually described, when we assert the standard Power Sets axiom, we implicitly build in bivalence, i.e. that either x belongs to y or it doesn't, i.e. we are in effect ruling forcing extensions or Boolean-valued generalizations as *non-standard* [my italics], i.e. 'full power set' is to be understood only in the standard way.

---

[6]The set of ctm's is called the *Hyperuniverse*; hence we arrive at the *Hyperuniverse Programme*.

[7]Height actualism with just GB (Gödel-Bernays) appears inadequate for a fruitful analysis of maximality. A referee has informed us about *agnostic Platonism*, the view that there is a well-determined universe V of all sets but without taking a position on whether ZFC holds in it. But as this perspective allows for the possibility of height actualism with just GB, it is problematic for the HP.

[8]These comments were made during a lively e-mail exchange among numerous set-theorists and philosophers of set theory from August until November 2014, triggered by my response to Sol Feferman's preprint *The Continuum Hypothesis is neither a definite mathematical problem nor a definite logical problem*. Some of this discussion is documented at <http://logic.harvard.edu/blog/?cat=2>, but regrettably Hellman's comments do not appear there.

And further:

> Thus, to my way of thinking, there is an important disanalogy between 'all ordinals' ... and 'all subsets of a given set'. The latter is 'already relativized'; there is nothing implicit in the notion of 'subset' that allows for indefinite extensions, so long as we are speaking of 'subsets of a fixed, given set' ... In contrast, 'all ordinals' cries out for relativization (a point I find in Zermelo's [1930]); without it, it does allow for indefinite extensibility, by the very operations that we use to describe ordinals

I do appreciate Hellman's point here, and indeed will (for the most part) adopt the *Zermelian perspective*, height potentialism with width actualism, in this paper. Another strong point in favour of this view is that although we have a clear and coherent way of generating the ordinals through a process of iteration, there is currently no analogous iteration process for generating increasingly rich power sets.[9]

In light of this adoption of potentialism in height, I will now use the symbol V ambiguously, not to denote the fixed universe of all sets (which does not exist) but as a variable to range over universes within the *Zermelian multiverse* in which each universe is a rank initial segment of the next.

Despite my adoption of the Zermelian view, I will for expository purposes also consider a form of potentialism in both height and width which I will call *radical potentialism*. The HP can be run with either point of view. Although it is simpler with radical potentialism, there are interesting issues (both mathematical and philosophical) which arise when employing the Zermelian view which are worth exploring.

To describe radical potentialism, let me begin with something less radical, *width potentialism*. First as motivation, consider a Platonist view, so that V is the fixed universe of all sets, and consider the method of forcing for producing generic sets. If M is a ctm we can easily build a generic extension $M[G]$ of M using the countability of M. But of course generic extensions $V[G]$ of V do not exist, as our "real V" has all the sets. Despite this we can talk definably in V about what can be true in such a *generic extension* without actually having such extensions in V, by constructing the Boolean universe V^B within V and taking *true in a generic extension of V* to just mean *of nonzero Boolean truth value in V^B*. Thus the Platonist view is in fact dualistic: It allows for the possibility of making sense of truth in universes (generic extensions) without allowing these universes to actually exist.

Width potentialism is a view in which any universe can be thickened, keeping the same ordinals, even to the extent of making ordinals countable. Thus for example it allows for the existence of the generic extensions of V (now a variable ranging over the multiverse of all possible universes) that are prohibited by the Platonist. So for any ordinal α of V we can thicken V to a universe where α is countable; i.e., any ordinal is *potentially countable*. But that does not mean that every ordinal of V is

---

[9]But I am not 100% sure that there could not be such an analogous iteration process, perhaps provided by a wildly successful theory of inner models for large cardinals.

countable in V, it is only countable in a larger universe. So this *potential countability* does not threaten the truth of the powerset axiom in V.

Now *radical potentialism* is in effect a unification of width and height potentialism. It entails that any V (in the multiverse of possible universes) looks countable inside a larger universe: We allow V to be lengthened and thickened simultaneously. Note that even just width potentialism (allowing universes to be thickened) forces us also into height potentialism: If we were to keep thickening to make every ordinal of V countable then after $\mathrm{Ord}(V)$ steps we are forced to also lengthen to reach a universe that satisfies the powerset axiom. In that universe, the original V looks countable. But then we could repeat the process with this new universe until it too is seen to be countable. The height potentialist aspect is that we cannot end this process by taking the union of all of our universes, as this would not be a model of ZFC (the powerset axiom will fail) and therefore would have to be lengthened. Note that once again, the *potential countability of V* does not threaten the truth of the axioms of ZFC in V.

4.4 Maximality in Height and #-Generation

The analysis of height maximality is the first major success of the HP. The programme has produced a robust principle expressing the maximality of V in height which appears to encompass all prior height maximality principles, including reflection, and to constitute the definitive expression of the height maximality of V in mathematical terms. This principle shares some features with ideas discussed in [28] .

For our discussion of height maximality, height potentialism will suffice (radical potentialism is not needed). Thus we allow ourselves the option of lengthening V to universes V^* which have V as a rank-initial segment. Of course we can also consider shortenings of V, replacing V by one of its own rank-initial segments. Let us now make use of lengthenings and shortenings to formulate a height maximality principle for V, expressing the idea that the sequence of ordinals is *as long as possible*.

But before embarking on our analysis of height maximality we should take note of the following: No *first-order* statement φ can be adequate to fully capture height maximality. This is simply because a first-order statement true in V will reflect to one of its rank initial segments and we are then naturally led from φ to the stronger first-order statement "φ holds both in V and in some transitive set model of ZFC". We will also see that no first-order statement is adequate to capture width maximality. This is an instance of the *Beyond First-Order* claim of the introduction: True first-order statements contradicting $V = L$ only arise as consequences of true *non first-order* axioms.

But how do we capture height maximality with a non first-order axiom? We do this via a detailed analysis of the relationship between V and its lengthenings and shortenings.

Standard Lévy reflection tells us that a single first-order property of V with parameters will hold in some V_κ which contains those parameters. It is natural to strengthen this to the simultaneous reflection of all first-order properties of V to some V_κ, allowing arbitrary parameters from V_κ. Thus we have reflected V to a V_κ which is an elementary submodel of V.

Repeating this process leads us to an increasing, continuous sequence of ordinals $(\kappa_i \mid i < \infty)$, where ∞ denotes the ordinal height of V, such that the models $(V_{\kappa_i} \mid i < \infty)$ form a continuous chain $V_{\kappa_0} \prec V_{\kappa_1} \prec \cdots$ of elementary submodels of V whose union is all of V.

Let C be the proper class consisting of the κ_i's. We can apply reflection to V with C as an additional predicate to infer that properties of (V, C) also hold of some $(V_\kappa, C \cap \kappa)$. But the unboundedness of C is a property of (V, C) so we get some $(V_\kappa, C \cap \kappa)$ where $C \cap \kappa$ is unbounded in κ and therefore κ belongs to C. As a corollary, properties of V in fact hold in some V_κ where κ belongs to C. It is convenient to formulate this in its contrapositive form: If a property holds of V_κ for all κ in C then it also holds of V.

Now note that for all κ in C, V_κ can be *lengthened* to an elementary extension (namely V) of which it is a rank-initial segment. By the contrapositive form of reflection of the previous paragraph, V itself also has such a lengthening V^*.

But this is clearly not the end of the story. For the same reason we can also infer that there is a continuous increasing sequence of such lengthenings $V = V_{\kappa_\infty} \prec V^*_{\kappa_\infty+1} \prec V^*_{\kappa_\infty+2} \prec \cdots$ of length the ordinals. For ease of notation, let us drop the *'s and write W_{κ_i} instead of $V^*_{\kappa_i}$ for $\infty < i$ and instead of V_{κ_i} for $i \leq \infty$. Thus V equals W_∞.

But which tower $V = W_{\kappa_\infty} \prec W_{\kappa_\infty+1} \prec W_{\kappa_\infty+2} \prec \cdots$ of lengthenings of V should we consider? Can we make the choice of this tower *canonical*?

Consider the entire sequence $W_{\kappa_0} \prec W_{\kappa_1} \prec \cdots \prec V = W_{\kappa_\infty} \prec W_{\kappa_\infty+1} \prec W_{\kappa_\infty+2} \prec \cdots$. The intuition is that all of these models resemble each other in the sense that they share the same first-order properties. Indeed by virtue of the fact that they form an elementary chain, these models all satisfy the same first-order sentences. But again in the spirit of "resemblance", the following should hold: For $i_0 < i_1$ regard $(W_{\kappa_{i_1}}, W_{\kappa_{i_0}})$ as the structure $(W_{\kappa_{i_1}}, \in)$ together with $W_{\kappa_{i_0}}$ as a unary predicate. Then it should be the case that any two such pairs $(W_{\kappa_{i_1}}, W_{\kappa_{i_0}})$, $(W_{\kappa_{j_1}}, W_{\kappa_{j_0}})$ (with $i_0 < i_1$ and $j_0 < j_1$) satisfy the same first-order sentences, even allowing parameters which belong to both $W_{\kappa_{i_0}}$ and $W_{\kappa_{j_0}}$. Generalising this to triples, quadruples and n-tuples in general we arrive at the following situation:

(∗)V occurs in a continuous elementary chain $W_{\kappa_0} \prec W_{\kappa_1} \prec \cdots \prec V = W_{\kappa_\infty} \prec W_{\kappa_\infty+1} \prec W_{\kappa_\infty+2} \prec \cdots$ of length $\infty + \infty$, where the models W_{κ_i} form a *strongly-indiscernible chain* in the sense that for any n and any two increasing n-tuples $\vec{i} = i_0 < i_1 < \cdots < i_{n-1}, \vec{j} = j_0 < j_1 < \cdots < j_{n-1}$, the structures $W_{\vec{i}} = (W_{\kappa_{i_{n-1}}}, W_{\kappa_{i_{n-2}}}, \cdots, W_{\kappa_{i_0}})$ and $W_{\vec{j}}$ (defined analagously) satisfy the same first-order sentences, allowing parameters from $W_{\kappa_{i_0}} \cap W_{\kappa_{j_0}}$.

We are getting closer to the desired axiom of #-generation. Surely we can impose higher-order indiscernibility on our chain of models. For example, consider the pair of models $W_{\kappa_0} = V_{\kappa_0}$, $W_{\kappa_1} = V_{\kappa_1}$. We can require that these models satisfy the same second-order sentences; equivalently, we require that $H(\kappa_0^+)^V$ and $H(\kappa_1^+)^V$ satisfy the same first-order sentences. But as with the pair $H(\kappa_0)^V$, $H(\kappa_1)^V$ we would want $H(\kappa_0^+)^V$, $H(\kappa_1^+)^V$ to satisfy the same first-order sentences *with parameters*. How can we formulate this? For example, consider κ_0, a parameter in $H(\kappa_0^+)^V$ that is second-order with respect to $H(\kappa_0)^V$; we cannot simply require $H(\kappa_0^+)^V \vDash \varphi(\kappa_0)$ iff $H(\kappa_1^+)^V \vDash \varphi(\kappa_0)$, as κ_0 is the largest cardinal in $H(\kappa_0^+)^V$ but not in $H(\kappa_1^+)^V$. Instead we need to replace the occurrence of κ_0 on the left side with a "corresponding" parameter on the right side, namely κ_1, resulting in the natural requirement $H(\kappa_0^+)^V \vDash \varphi(\kappa_0)$ iff $H(\kappa_1^+)^V \vDash \varphi(\kappa_1)$. More generally, we should be able to replace each parameter in $H(\kappa_0^+)^V$ by a "corresponding" element of $H(\kappa_1^+)^V$. It is natural to solve this parameter problem using embeddings.

Definition 1 (See [10]) A structure $N = (N, U)$ is called a # *with critical point* κ, or just a #, if the following hold:

(a) N is a model of ZFC$^-$ (ZFC minus powerset) in which κ is both the largest cardinal and strongly inaccessible.
(b) (N, U) is amenable (i.e. $x \cap U \in N$ for any $x \in N$).
(c) U is a normal measure on κ in (N, U).
(d) N is iterable, i.e., all of the successive iterated ultrapowers starting with (N, U) are well-founded, yielding iterates (N_i, U_i) and Σ_1 elementary iteration maps $\pi_{ij} : N_i \to N_j$ where $(N, U) = (N_0, U_0)$.

We let κ_i denote the largest cardinal of the ith iterate N_i.

If N is a # and λ is a limit ordinal then $LP(N_\lambda)$ denotes the union of the $(V_{\kappa_i})^{N_i}$'s for $i < \lambda$. (LP stands for *lower part*.) $LP(N_\infty)$ is a model of ZFC.

Definition 2 We say that a transitive model V of ZFC is #-*generated* iff there is $N = (N, U)$, a # with iteration $N = N_0 \to N_1 \to \cdots$, such that V equals $LP(N_\infty)$ where ∞ denotes the ordinal height of V.

#-generation fulfills our requirements for vertical maximality, with powerful consequences for reflection. L is #-generated iff $0^\#$ exists, so this principle is compatible with $V = L$. If V is #-generated via (N, U) then there are elementary embeddings from V to V which are canonically-definable through iteration of (N, U): In the above notation, any order-preserving map from the κ_i's to the κ_i's extends to such an elementary embedding. If $\pi : V \to V$ is any such embedding then we obtain not only the indiscernibility of the structures $H(\kappa_i^+)$, for all i but also of the structures $H(\kappa_i^{+\alpha})$ for any $\alpha < \kappa_0$ and more. Moreover, #-generation evidently provides the maximum amount of vertical reflection: If V is generated by (N, U) as $LP(N_\infty)$ where ∞ is the ordinal height of V, and x is any parameter in a further iterate $V^* = N_{\infty^*}$ of (N, U), then any first-order property $\varphi(V, x)$ that holds in V^* reflects to $\varphi(V_{\kappa_i}, \bar{x})$ in N_j for all sufficiently large $i < j < \infty$, where $\pi_{j,\infty^*}(\bar{x}) = x$. This implies any known form of vertical reflection and summarizes the amount of reflection one has in L under the assumption that $0^\#$ exists, the maximum amount of reflection in L. This is reinforced by a Jensen's #-*generated coding theorem* (Theorem 9.1. of [6]) which states that if V is #-generated then V can be coded into a #-generated model $L[x]$ for a real x where the given # which generates V extends to the natural generator $x^\#$ for the model $L[x]$.

From this we can conclude that #-generated models have the same large cardinal and reflection properties as does L when $0^\#$ exists.

#-generation also answers our question about which *canonical* tower of lengthenings of V to look at in reflection, namely the further lower parts of iterates of any # that generates V. This tower of lengthenings is independent of the choice of generating # for V and is therefore entirely *canonical*. And #-generation fully realizes the idea that V should look exactly like closed unboundedly many of its rank initial segments as well as its *canonical* lengthenings of arbitrary ordinal height.

In summary, #-generation stands out as the correct formalization of the principle of *height maximality*, and we shall refer to #-generated models as being *maximal in height*. It is not first-order (we have argued that no optimal height maximality principle can be), however it is second-order in a very restricted way: For a countable V, the property of being a # that generates V is expressible by quantifying universally over the models $L_\alpha(V)$ as α ranges over the countable ordinals.

4.5 Maximality in Width and the IMH

Whereas in the case of *maximality in height* we can use height potentialism (i.e., the option of lengthening V to taller universes) to arrive at an optimal principle, the case of *maximality in width* is of a very different nature. Unlike in the case of height maximality, we will see that there are many distinct criteria for width maximality and will not easily arrive at an optimal criterion. Moreover, to get a fair picture of maximality in both height and width, it is necessary to *synthesise* or *unify* width maximality criteria with #-generation, the optimal height maximality criterion.

A thorough analysis of the different possible width maximality criteria and their synthesis with #-generation, with an aim towards arriving at an optimal criterion, is the principal aim of the *Hyperuniverse Programme*.

I'll begin with a discussion of width maximality in the context of radical potentialism, as this offers a simpler theory than that provided by the Zermelian view. Thus we use the symbol V to be a variable ranging not over the Zermelian multiverse (in which universes are ordered by the relation of rank-initial segment) but over elements of the rich multiverse provided by radical potentialism, in which each universe is potentially countable. We begin with the fundamental:

Inner Model Hypothesis (IMH [12]) If a first-order sentence holds in some outer model of V then it holds in some inner model of V.

For the current presentation, we may take *outer model* to mean a transitive set V^* containing V, with the same ordinals as V, which satisfies ZFC. An *inner model* in this presentation is a V-definable subclass of V with the same ordinals as V which satisfies ZFC. By radical potentialism, any transitive model of ZFC is countable in a larger such model and from this we can infer the existence of a rich collection of outer models of V.

The consistency of #-generation follows from the existence of $0^{\#}$. But the consistency of the IMH, i.e. the assertion that there are universes V satisfying the IMH, requires more.

Consistency of the IMH

Theorem 3 ([18]) *Assuming large cardinals there exists a countable transitive model M of ZFC such that if a first-order sentence φ holds in an outer model N of M then it also holds in an inner model of M.*

Proof For any real R let $M(R)$ denote the least transitive model of ZFC containing R. We are assuming large cardinals so indeed such an $M(R)$ exists (the existence of just an inaccessible is sufficient for this). We will need the following consequence of large cardinals:

($*$) There is a real R such that for any real S in which R is recursive, the (first-order) theory of $M(R)$ is the same as the theory of $M(S)$.

One can derive ($*$) from large cardinals as follows. Large cardinals yield Projective Determinacy (PD). A theorem of Martin is that PD implies the following *Cone Theorem*: If X is a projective set of reals closed under Turing-equivalence then for some real R, either S belongs to X for all reals S in which R is recursive or S belongs to the complement of X for all reals S in which R is recursive.

Now for each sentence φ consider the set $X(\varphi)$ consisting of those reals R such that $M(R)$ satisfies φ. This set is projective and closed under Turing-equivalence. By the cone theorem we can choose a real $R(\varphi)$ so that either φ is true in $M(S)$ for all reals S in which $R(\varphi)$ is recursive or this holds for $\sim \varphi$. Now let R be any real in which every $R(\varphi)$ is recursive; as there are only countably-many φ's this is possible. Then R witnesses the property ($*$).

We claim that if N is an outer model of $M(R)$ satisfying ZFC and φ is a sentence true in N then φ is true in an inner model of $M(R)$. For this we need the following deep theorem of Jensen.

Coding Theorem (see [6]) Let α be the ordinal height of N. Then N has an outer model of the form $L_\alpha[S]$ for some real S which satisfies ZFC and in which N is Δ_2-definable with parameters.

As R belongs to $M(R)$ it also belongs to N and hence to $L_\alpha[S]$ where S codes N as above. Also note that since α is least so that $M(R) = L_\alpha[R]$ models ZFC, it is also least so that $L_\alpha[S]$ satisfies ZFC and therefore $L_\alpha[S]$ equals $M(S)$.

Clearly we can choose S to be Turing above R (simply replace S by its join with R). But now by the special property of R, the theories of $M(R)$ and $M(S)$ are the same. As N is a definable inner model of $M(S)$, part of the theory of $M(S)$ is the statement "There is an inner model of φ which is Δ_2-definable with parameters" and therefore there is an inner model of $M(R)$ satisfying φ, as desired. $\square$

Note that the model that we produce above for the IMH, $M(R)$ for some real R, is the minimal model containing the real R and therefore satisfies "there are no inaccessible cardinals". This is no accident:

Theorem 4 ([12]) *Suppose that M satisfies the IMH. Then in M: There are no inaccessible cardinals and in fact there is a real R such that there is no transitive model of ZFC containing R.*

Proof A theorem of Beller and David (also in [6]) extends Jensen's Coding Theorem to say that any model M has an outer model of the form $M(R)$ for some real R, where as above $M(R)$ is the minimal transitive model of ZFC containing R. Now suppose that M satisfies the IMH and consider the sentence "There is no inaccessible cardinal". This is true in an outer model $M(R)$ of M and therefore in an inner model of M. It follows that there are no inaccessibles in M. The same argument with the sentence "There is a real R such that there is no transitive model of ZFC containing R" gives an inner model M_0 of M with this property for some real R; but then also M has this property as any transitive model of ZFC containing R in M would also give such a model in the $L[R]$ of M and therefore in M_0, as M_0 contains the $L[R]$ of M.   □

It follows that if M satisfies the IMH then some real in M has no # and therefore boldface Π_1^1 determinacy fails in M (although $0^{\#}$ does exist and lightface Π_1^1 determinacy does hold).

Width Actualism

So far I have presented the IMH in the context of radical potentialism, which allows us to talk freely about *outer models (thickenings)* of the universe V. This is of course unacceptable to the width actualist, who sees a fixed meaning to V_α for each ordinal α (although possibly an unfixed, potentialist view of what the ordinals are). Is it possible to nevertheless talk about the *maximality of V in width* from a width actualist perspective (where V is now a variable ranging over the Zermelian multiverse)? Can we express the idea that V is *as thick as possible* without actually comparing V to thicker universes (which do not exist)?

A positive answer to the latter question emerges through a study of V-logic, to which I turn next. A useful reference for this material is Barwise's book [5].

V-Logic

Let's start with something simpler, V_ω-logic. In V_ω-logic we have constant symbols $\bar{a}$ for $a \in V_\omega$ as well as a constant symbol $\bar{V}_\omega$ for V_ω itself (in addition to $\in$ and the other symbols of first-order logic). Then to the usual logical axioms and the rule of *Modus Ponens* we add the rules:

For $a \in V_\omega$: From $\varphi(\bar{b})$ for each $b \in a$ infer $\forall x \in \bar{a}\, \varphi(x)$.
From $\varphi(\bar{a})$ for each $a \in V_\omega$ infer $\forall x \in \bar{V}_\omega\, \varphi(x)$.

Introducing the second of these rules generates new provable statements via proofs which are now infinite. The idea of V_ω-logic is to capture the idea of a model *in which V_ω is standard*. By the *ω-completeness theorem*, the logically provable sentences of V_ω-logic are exactly those which hold in every model in which $\bar{a}$ is interpreted as a for $a \in V_\omega$ and $\bar{V}_\omega$ is interpreted as the (real, standard) V_ω. Thus a theory T in V_ω-logic is consistent in V_ω-logic iff it has a model in which V_ω is the real, standard V_ω.

Now the set of logically-provable formulas (i.e. validities) in V_ω-logic, unlike in first-order logic, is not arithmetical, i.e. it is not definable over the model V_ω. Instead it is definable over a larger structure, a *lengthening of* V_ω. Let me explain.

As proofs in V_ω-logic are no longer finite, they do not naturally belong to V_ω. Instead they belong to *the least admissible set* $(V_\omega)^+$ containing V_ω as an element, this is known to higher recursion-theorists as $L_{\omega_1^{ck}}$, where ω_1^{ck} is the least non-recursive ordinal. Something very nice happens: Whereas proofs in first-order logic belong to V_ω and therefore provability is Σ_1 definable over V_ω (*there exists a proof* is Σ_1), proofs in V_ω-logic belong to $(V_\omega)^+$ and provability is Σ_1 definable over $(V_\omega)^+$.

For our present purposes the point is that $(V_\omega)^+$ is a lengthening, not a thickening of V_ω and in this lengthening we can formulate theories which describe arbitrary models in which V_ω is standard. For example the existence of a real R such that (V_ω, R) satisfies a first-order property can be formulated as the consistency of a theory in V_ω-logic. As the structure (V_ω, R) can be regarded as a "thickening" of V_ω, we have described what can happen in "thickenings" of V_ω by a theory in $(V_\omega)^+$, a lengthening of V_ω. This is even more dramatic if we start not with V_ω but with $(V_\omega)^+ = L_{\omega_1^{ck}}$ and introduce $L_{\omega_1^{ck}}$-*logic*, a logic for ensuring that the recursive ordinals are standard. Then in the lengthening $(L_{\omega_1^{ck}})^+$ of $L_{\omega_1^{ck}}$, the least admissible set containing $L_{\omega_1^{ck}}$ as an element, we can express the existence of a thickening of $L_{\omega_1^{ck}}$ in which a first-order statement holds, and such thickenings can contain new reals and more as elements.

V-logic is analogous to the above. It has the following constant symbols:

1. A constant symbol $\bar{a}$ for each set a in V.
2. A constant symbol $\bar{V}$ to denote the universe V.

Formulas are formed in the usual way, as in any first-order logic. To the usual axioms and rules of first-order logic we add the new rules:

($*$) From $\varphi(\bar{b})$ for all $b \in a$ infer $\forall x \in \bar{a}\, \varphi(x)$.
($**$) From $\varphi(\bar{a})$ for all $a \in V$ infer $\forall x \in \bar{V}\, \varphi(x)$.

This is the logic to describe models in which V is standard. The proofs of this logic appear in V^+, the least admissible set containing V as an element; this structure V^+ is a special lengthening of V of the form $L_\alpha(V)$, the αth level of Gödel's L-hierarchy built over V. We refer to such lengthenings as *Gödel lengthenings*. Recall that with our height potentialist perspective, we can lengthen V to models V^* with V as a rank-initial segment, and therefore surely lengthen V to the Gödel lengthening V^+. (This is also the case with a height actualist perspective, provided we allow our classes to satisfy MK (Morse-Kelley), as in MK we can construct a class coding V^+.)

The Inner Model Hypothesis for a Width Actualist

As width actualists we cannot talk directly about outer models or even about sets that do not belong to V. However using V-logic we can talk about them indirectly, as I'll now illustrate. Consider the theory in V-logic where we not only have constant symbols $\bar{a}$ for the elements of V and a constant symbol $\bar{V}$ for V itself, but also a constant symbol $\bar{W}$ to denote an "outer model" of V. We add the new axioms:

1. The universe is a model of ZFC (or at least the weaker KP, admissibility theory).
2. $\bar{W}$ is a transitive model of ZFC containing $\bar{V}$ as a subset and with the same ordinals as V.

So now when we take a model of our axioms which obeys the rules of V-logic, we get a universe modelling ZFC (or at least KP) in which $\bar{V}$ is interpreted correctly as V and $\bar{W}$ is interpreted as an outer model of V. Note that this theory in V-logic has been formulated without "thickening" V, indeed it is defined inside V^+, the least admissible set containing V, a Gödel lengthening of V. Again the latter makes sense thanks to our adoption of height (not width) potentialism.

So what does the *IMH* really say for a width actualist? It says the following:

IMH Suppose that φ is a first-order sentence and the above theory, together the axiom "$\bar{W}$ satisfies φ" is consistent in V-logic. Then φ holds in an inner model of V.

In other words, instead of talking directly about "thickenings" of V (i.e. "outer models") we instead talk about the consistency of a theory formulated in V-logic and defined in V^+, a (mild) Gödel lengthening of V.

Note that this also provides a powerful extension of the Definability Lemma for set-forcing. The latter says that definably in V we can express the fact that a sentence with parameters holds in a "set-generic extension" (for sentences of bounded complexity, such as Σ_n sentences for a fixed n). The above shows that we can do the same for arbitrary "thickenings" of V, but where the definability takes place not in V but in V^+. (In the case of *omniscient* universes V, we can in fact obtain definability in V, and under mild large cardinal assumptions, V will be omniscient. See Sect. 4.12 for a discussion of this.)

So far we have worked with V, its lengthenings and its "thickenings" (via theories expressed in its lengthenings). We next come to an important step, which is to reduce this discussion to the study of certain properties of countable transitive models of ZFC, i.e., to the *Hyperuniverse* (the set of countable transitive models of ZFC). The net effect of this reduction is to show that our width actualist discussion of maximality is in fact equivalent to a radical potentialist discussion in which all models under consideration belong to the Hyperuniverse.

4.6   The Reduction to the Hyperuniverse

Of course it would be much more comfortable to remove the quotes in "thickenings" of V, as we could then dispense with the need to reformulate our intuitions about outer models via theories in V-logic. Indeed, if we were to have this discussion not about V but about a countable transitive ZFC model *little-V*, then our worries evaporate, as genuine thickenings become available. For example, if P is a forcing notion in little-V then we can surely build a P-generic extension to get a little-$V[G]$. Of course we can't do this for V itself as in general we cannot construct generic sets for partial orders with uncountably many maximal antichains.

But the way we have analysed things with V-logic allows us to *reduce* our study of maximality criteria for V to a study of countable transitive models. As the collection of countable transitive models carries the name *Hyperuniverse*, we are then led to what is known as the *Hyperuniverse Programme*.

I'll illustrate the reduction to the Hyperuniverse with the specific example of the IMH. Suppose that we formulate the IMH as above, using V-logic, and want to know what first-order consequences it has.

Lemma 5 *Suppose that a first-order sentence φ holds in all countable models of the IMH. Then it holds in all models of the IMH.*

Proof Suppose that φ fails in some model V of the IMH, where V may be uncountable. Now notice that the IMH is first-order expressible in V^+, a lengthening of V. But then apply the downward Löwenheim-Skolem theorem to obtain a countable little-V which satisfies the IMH, as verified in its associated little-V^+, yet fails to satisfy φ. But this is a contradiction, as by hypothesis φ must hold in all *countable* models of the IMH. $\qquad\qquad\Box$

So *without loss of generality*, when looking at first-order consequences of maximality criteria as formulated in V-logic, we can restrict ourselves to countable little-V's. The advantage of this is then we can dispense with the little-V-logic and the quotes in "thickenings" altogether, as by the Completeness Theorem for little-V-logic, consistent theories in little-V-logic do have models, thanks to the countability of little-V. Thus for a countable little-V, we can simply say:

IMH for little-V's Suppose that a first-order sentence holds in an outer model of little-V. Then it holds in an inner model of little-V.

This is exactly the radical potentialist version of the IMH with which we began. Thus the width actualist and radical potentialist versions of the IMH coincide on countable models.

#-Generation Revisited

The reduction of maximality principles to the Hyperuniverse is however not always so obvious, as we will now see in the case of #-generation. This reveals a difference in the development of the HP form a Zermelian perspective versus a radical potentialist perspective.

First consider the following encouraging analogue for #-generation of our earlier reduction claim for the IMH.

Lemma 6 *Suppose that a first-order sentence φ holds in all countable models which are #-generated. Then it holds in all models which are #-generated.*

Proof Suppose that φ fails in some #-generated model V, where V may be uncountable. Let (N, U) be a generating # for V and place both V and (N, U) inside some transitive model of ZFC minus powerset T. Now apply Löwenhiem-Skolem to T to produce a countable transitive $\bar{T}$ in which there is a $\bar{V}$ which $\bar{T}$ believes to be generated by $(\bar{N}, \bar{U})$ with an elementary embedding of $\bar{T}$ into T, sending $\bar{V}$ to V and $(\bar{N}, \bar{U})$ to (N, U). But the fact that (N, U) is iterable and $(\bar{N}, \bar{U})$ is embedded into (N, U) is enough to conclude that also $(\bar{N}, \bar{U})$ is iterable. So we now have a countable $\bar{V}$ which is #-generated (via $(\bar{N}, \bar{U})$) in which φ fails, contrary to hypothesis.   □

However the difficulty is this: How do we express #-generation from a width actualist perspective? Recall that to produce a generating # for V we have to produce a set of rank less than Ord(V) which does not belong to V, in violation of width actualism.

And recall that a # is a structure (N, U) meeting certain first-order conditions which is in addition *iterable*: For any ordinal α if we iterate (N, U) for α steps then it remains wellfounded. V is #-generated if there is a # which generates it. But notice that to express the iterability of a generating # for V we are forced to consider theories T_α formulated in $L_\alpha(V)$-logic for *arbitrary* Gödel lengthenings $L_\alpha(V)$ of V: T_α asserts that V is generated by a *pre-#* (i.e. by a structure that looks like a # but may not be fully iterable) which is α-*iterable*, i.e. iterable for α-steps. Thus we have no fixed theory that captures #-generation but only a tower of theories T_α (as α ranges over ordinals past the height of V) which capture closer and closer approximations to it.

Definition 7 V is *weakly #-generated* if for each ordinal α past the height of V, the theory T_α which expresses the existence of an α-iterable pre-# which generates V is consistent.

Weak #-generation is meaningful for a width actualist (who accepts enough height potentialism to obtain Gödel lengthenings) as it is expressed entirely in terms of theories internal to Gödel lengthenings of V.

For a countable little-V, weak #-generation can be expressed semantically. First a useful definition:

Definition 8 Let little-V be a countable transitive model of ZFC and α an ordinal. Then little-V is α-*generated* if there is an α-iterable pre-# which generates little-V (as the union of the lower parts of its first γ iterates, where γ is the ordinal height of little-V).

Then a countable little-V is weakly #-generated if it is α-generated for each countable ordinal α (where the witness to this may depend on α). Little-V is #-generated iff it is α-generated when $\alpha = \omega_1$ iff it is α-generated for all ordinals α.

Just as a syntactic approach is needed for a width actualist formulation of #-generation, the reduction of this weakened form of #-generation to the Hyper-universe takes a syntactic form:

Lemma 9 *Suppose that a first-order sentence φ holds in all countable little-V which are weakly #-generated, and this is provable in ZFC. Then φ holds in all models which are weakly #-generated.*

Proof Let W be a weakly #-generated model (which may be uncountable). Thus for each ordinal α above the height of W, the theory $T_\alpha + \sim \varphi$ expressing that φ fails in W and W is generated by an α-iterable pre-# is consistent. If we choose α so that $L_\alpha(W)$ is a model of ZFC (or enough of ZFC where the truth of φ in countable #-generated models provable) then $L_\alpha(W)$ is a model of (enough of) ZFC in which W is weakly #-generated. Apply Löwenheim-Skolem to obtain a countable $\bar{W}$ and $\bar{\alpha}$ such that $L_{\bar{\alpha}}(\bar{W})$ embeds elementarily into $L_\alpha(W)$ and therefore satisfies (enough of) ZFC plus "$\bar{W}$ is weakly #-generated". Now let g be generic over $L_{\bar{\alpha}}(\bar{W})$ for the Lévy collapse of (the height of) $\bar{W}$ to ω; then $L_{\bar{\alpha}}(\bar{W})[g]$ is a model of (enough of) ZFC in which $\bar{W}$ is both countable and weakly #-generated. By hypothesis $L_{\bar{\alpha}}(W)[g]$ satisfies "$\bar{W}$ satisfies φ" and therefore $\bar{W}$ really does satisfy φ. Finally, by elementarity W satisfies φ as well, as desired. $\square$

   To summarise: As radical potentialists we can comfortably work with full #-generation as our principle of height maximality. But as width actualists we instead work with weak #-generation, expressed in terms of theories inside Gödel lengthenings $L_\alpha(V)$ of V. Weak #-generation is sufficient to maximise the height of the universe. And properly formulated, the reduction to the Hyperuniverse applies to weak #-generation: To infer that a first-order statement follows from weak #-generation it suffices to show that in ZFC one can prove that it holds in all weakly #-generated countable models.

   Weak #-generation is indeed strictly weaker than #-generation for countable models: Suppose that $0^\#$ exists and choose α to be least so that α is the αth Silver indiscernible (α is countable). Now let g be generic over L for Lévy collapsing α to ω. Then by Lévy absoluteness, L_α is weakly #-generated in $L[g]$, but it cannot be #-generated in $L[g]$ as $0^\#$ does not belong to a generic extension of L.

   In what follows I will primarily work with #-generation, as at present the mathematics of weak #-generation is poorly understood. Indeed, as we'll see in the next section, a synthesis of #-generation with the IMH is consistent, but this remains an open problem for weak #-generation.

4.7   Synthesis

We introduced the IMH as a criterion for width maximality and #-generation as a criterion for height maximality. It is natural to see how these can be combined into a single criterion which recognises both forms of maximality. We achieve this in this section through *synthesis*. Note that the IMH implies that there are no inaccessibles yet #-generation implies that there are. So we cannot simply take the conjunction of these two criteria.

A #-generated model M satisfies the IMH# iff whenever a sentence holds in a #-generated outer model of M it also holds in an inner model of M.

Note that IMH# differs from the IMH by demanding that both M and M^*, the outer model, are #-generated (while the outer models considered in IMH are arbitrary). The motivation behind this requirement is to impose width maximality only with respect to those models which are height maximal.

Theorem 10 ([15]) *Assuming that every real has a # there is a real R such that any #-generated model containing R satisfies the IMH#.*

Proof (Woodin) Let R be a real with the following property: Whenever X is a lightface and nonempty Π_2^1 set of reals, then X has an element recursive in R. We claim that any #-generated model M containing R as an element satisfies the IMH#.

Suppose that φ holds in M^*, a #-generated outer model of M. Let (m^*, U^*) be a generating # for M^*. Then the set X of reals S such that S codes such an (m^*, U^*) (generating a model of φ) is a lightface Π_2^1 set. So there is such a real recursive in R and therefore in M. But then M has an inner model satisfying φ, namely any model generated by a # coded by an element of X in M.   □

The argument of the previous theorem is special to the weakest form of IMH#. The original argument from [15], used *#-generated Jensen coding* to prove the consistency of a stronger principle, SIMH#(ω_1); see Theorem 15.

Corollary 11 *Suppose that φ is a sentence that holds in some V_κ with κ measurable. Then there is a transitive model which satisfies both the IMH# and the sentence φ.*

Proof Let R be as in the proof of Theorem 10 and let U be a normal measure on κ. The structure $N = (H(\kappa^+), U)$ is a #; iterate N through a large enough ordinal ∞ so that $M = LP(N_\infty)$, the lower part model generated by N, has ordinal height ∞. Then M is #-generated and contains the real R. It follows that M is a model of the IMH#. Moreover, as M is the union of an elementary chain $V_\kappa = V_\kappa^N \prec V_{\kappa_1}^{N_1} \prec \cdots$ where φ is true in V_κ, it follows that φ is also true in M.   □

Note that in Corollary 11, if we take φ to be any large cardinal property which holds in some V_κ with κ measurable, then we obtain models of the IMH# which also satisfy this large cardinal property. This implies the compatibility of the IMH# with arbitrarily strong large cardinal properties.

Question 12 Reformulate IMH# using weak #-generation, as follows: V is weakly #-generated and for each sentence φ, if the theories expressing that V has an outer model satisfying φ with an α-iterable generating pre-# are consistent for each α, then φ holds in an inner model of V. Is this consistent?

The above formulation of IMH# for weak #-generation takes the following form for a countable V: V is α-generated for each countable α and for all φ, if φ holds in an α-generated outer model of V for each countable α then φ holds in an inner model of V. It is not known if this is consistent.

Remark An even weaker form of #-generation asserts that V is just $\text{Ord}(V) + \text{Ord}(V)$-generated, a sufficient amount of iterability to obtain ordinal maximality.

However a synthesis of the IMH with this very weak #-generation yields a consistent principle that contradicts large cardinals (indeed the existence of #'s for arbitrary reals). These different forms of #-generation, and of their synthesis with the IMH, are in need of further philosophical discussion.

We have now laid the foundations for the HP and discussed the two most basic maximality principles, #-generation and the IMH. Most of the mathematical work in the HP remains to be done. Therefore what I will do in the remainder of this article is simply present a range of maximality criteria which are yet to be fully analysed and which give the flavour of how the HP is intended to proceed. These criteria are also referred to as *H-axioms*, formulated as properties of elements of the Hyperuniverse H, expressible as maximality properties within H.

4.8   The Strong IMH

Our discussion of the IMH has been always with regard to sentences, without parameters. Stronger forms result if we introduce parameters.

First note the difficulties with introducing parameters into the IMH. For example the statement

> If a sentence with parameter ω_1^V holds in an outer model of V then it holds in an inner model

is inconsistent, as the parameter ω_1^V could become countable in an outer model and therefore the above cannot hold for the sentence "ω_1^V is countable". If we however require that ω_1 is preserved then we get a consistent principle.

Theorem 13 *Let SIMH(ω_1) be the following principle: If a sentence with parameter ω_1 holds in an ω_1-preserving outer model then it holds in an inner model. Then the SIMH(ω_1) is consistent (assuming large cardinals).*

Proof Again use PD to get a real R such that the theory of $M(S)$, the least transitive ZFC model containing S, is fixed for all S Turing above R. Now suppose that $\varphi(\omega_1)$ is a sentence true in an ω_1-preserving outer model N of $M(R)$, where ω_1 denotes the ω_1 of $M(R)$. Then as in the proof of consistency of the IMH, we can code N into $M(S)$ for some real S Turing above R, and moreover this coding is ω_1-preserving. As $\varphi(\omega_1)$ holds in a definable inner model of $M(S)$ and ω_1 is the same in $M(R)$ and $M(S)$, it follows that $M(R)$ also has an inner model satisfying $\varphi(\omega_1)$.   □

The above argument uses the fact that Jensen-coding is ω_1-preserving. It is however not ω_2-preserving unless CH holds, and therefore we have the following open question:

Question 14 Let SIMH(ω_1, ω_2) be the following principle: If a sentence with parameters ω_1, ω_2 holds in an ω_1-preserving and ω_2-preserving outer model then it holds in an inner model. Then is the SIMH(ω_1, ω_2) consistent (assuming large cardinals)?

The $SIMH(\omega_1, \omega_2)$ implies that CH fails, as any model has a cardinal-preserving outer model in which there is an injection from ω_2 into the reals. Is there an analogue $M^*(R)$ of the minimal model $M(R)$ which does not satisfy CH? Is there a coding theorem which says that any outer model of $M^*(R)$ which preserves ω_1 and ω_2 has a further outer model of the form $M^*(S)$, also with the same ω_1 and ω_2? If so, then one could establish the consistency of the $SIMH(\omega_1, \omega_2)$.

The most general from of the SIMH makes use of *absolute parameters*. A parameter p is *absolute* if some formula defines it in all outer models which preserve cardinals up to and including the hereditary cardinality of p, i.e. the cardinality of the transitive closure of p. Then $SIMH(p)$ for an absolute parameter p states that if a sentence with parameter p holds in an outer model which preserves cardinals up to the hereditary cardinality of p then it holds in an inner model. The full SIMH (Strong Inner Model Hypothesis) states that this holds for every absolute parameter p.

The SIMH is closely related to strengthenings of Lévy absoluteness. For example, define Lévy(ω_1) to be the statement that Σ_1 formulas with parameter ω_1 are absolute for ω_1-preserving outer models; this follows from the $SIMH(\omega_1)$ and is therefore consistent. But the consistency of Lévy(ω_1, ω_2), i.e. Σ_1 absoluteness with parameters ω_1, ω_2 for outer models which preserve these cardinals, is open.

The SIMH#

A synthesis of the SIMH with #-generation can be formulated as follows: V satisfies the SIMH# if V is #-generated and whenever a sentence φ with absolute parameters holds in a #-generated outer model having the same cardinals as V up to the hereditary cardinality of those parameters, φ also holds in an inner model of V. A special case is SIMH#(ω_1), where the only parameter involved is ω_1 and we are concerned only with ω_1-preserving outer models.

Theorem 15 ([15]) *Assuming large cardinals, the SIMH#(ω_1) is consistent.*

Proof Assume there is a Woodin cardinal with an inaccessible above. For each real R let $M^\#(R)$ be $L_\alpha[R]$ where α is least so that $L_\alpha[R]$ is #-generated. The Woodin cardinal with an inaccessible above implies enough projective determinacy to enable us to use Martin's Lemma to find a real R such that the theory of $M^\#(S)$ is constant for S Turing-above R. We claim that $M^\#(R)$ satisfies SIMH#(ω_1): Indeed, let M be a #-generated ω_1-preserving outer model of $M^\#(R)$ satisfying some sentence $\varphi(\omega_1)$. Let α be the ordinal height of $M^\#(R)$ (= the ordinal height of M). By the result of Jensen quoted before (Theorem 9.1 of [6]), M has a #-*generated* ω_1-preserving outer model W of the form $L_\alpha[S]$ for some real S with $R \leq_T S$. Of course α is least so that $L_\alpha[S]$ is #-generated. So W equals $M^\#(S)$ and the ω_1 of W equals the ω_1 of $M^\#(R)$. By the choice of R, $M^\#(R)$ also has a definable inner model satisfying $\varphi(\omega_1)$. $\square$

However as with the $SIMH(\omega_1, \omega_2)$, the consistency of SIMH#(ω_1, ω_2) is open.

4.9   A Maximality Protocol

This protocol aims to organise the study of height and width maximality into three stages.

Stage 1.   Maximise the ordinals (height maximality).
Stage 2.   Having maximised the ordinals, maximise the cardinals.
Stage 3.   Having maximised the ordinals and cardinals, maximise powerset (width maximality).

Stage 1 is taken care of by #-generation. So we focus now on Stage 2, cardinal-maximisation.

In light of Stage 1, we assume now that V is #-generated and when discussing outer models of V we only consider those which are also #-generated.

We would like a criterion which says that for each cardinal κ, κ^+ is *as large as possible*. To get started let's consider the case $\kappa = \omega$, so we want to maximise ω_1. The basic problem of course is the following. As set-generic extensions of #-generated models are also #-generated:

Fact V has a #-generated outer model in which ω_1^V is countable.

But surely we would want something like: $\omega_1^{L[x]}$ is countable for each real x. The reason for this is that $\omega_1^{L[x]}$, unlike ω_1^V in general, is *absolute* between V and all of its outer models.

Definition 16 Let p be a parameter in V and P a set of parameters in V. Then p is *strongly absolute relative to P* if there is a formula φ with parameters from P that defines p in V and all #-generated outer models of V which preserve cardinals up to and including the hereditary cardinality of the parameters mentioned in φ.[10]

Typically we will take P to consist of all subsets of some infinite cardinal κ, in which case the cardinal-preservation in the above definition refers to cardinals up to and including κ.

4.10   CardMax(κ^+) (for κ an Infinite Cardinal)

Suppose that the ordinal α is strongly absolute relative to subsets of κ. Then α has cardinality at most κ.

It is possible to show that if κ is regular then there is a set-forcing extension in which CardMax(κ^+) holds.

---

[10]We thank one of the referees for pointing out that an earlier version of cardinal-maximality with a weaker parameter-absoluteness assumption is inconsistent. A similar phenomenon with weakly absolute parameters occurs in Theorem 10 of [18].

Question 17 Is CardMax consistent, where CardMax denotes CardMax(κ^+) for all infinite cardinals κ, both regular and singular?

Internal Cardinal Maximality

Another approach to cardinal maximality is to relate the cardinals of V to those of its inner models. Two large inner models are HOD, the class of hereditarily ordinal-definable sets, and the smaller inner model $\mathbb{S}$, the *Stable Core* of [13]. V is class-generic over each of these models.

Let M denote an inner model.

M-cardinal Violation For each infinite cardinal κ, κ^+ is greater than the κ^+ of M.

In [8] it is shown that HOD-cardinal violation is consistent. Can we strengthen this?

Question 18 Is it consistent that for each infinite cardinal κ, κ^+ is inaccessible, measurable or even supercompact in HOD? Is this consistent with HOD replaced by the Stable Core $\mathbb{S}$?

A result of Shelah states that all subsets of κ belong to HOD_x for some fixed subset x of κ when κ is a singular strong limit cardinal of uncountable cofinality. By Cummings et al. [9] this need not be true at countable cofinalities.

Question 19 Is it consistent that for each infinite cardinal κ, κ^+ is greater than κ^+ of $\mathbb{S}_x$ (the Stable Core relativised to x) for each subset x of κ?

A major difference between HOD and $\mathbb{S}$ is that while any set is set-generic over HOD, this is not the case for $\mathbb{S}$.

Question 20 Is it consistent that for each infinite cardinal κ, some subset of κ^+ is not set-generic over $\mathbb{S}_x$ for any subset x of κ?

A positive answer to any of these three questions would yield a strong internal cardinal-maximality principle for V.

Stage 3: Having maximised the ordinals and cardinals, maximise powerset.

This is where we revisit the SIMH, but only in the context of #-generation and cardinal-preservation. Again assume that V is #-generated.

A parameter p in V is *cardinal-absolute* if there is a parameter-free formula which defines p in all #-generated outer models of V which have the same cardinals as V.

SIMH#(CP) (Cardinal-Preserving SIMH#) Suppose that p is a cardinal-absolute parameter, V^* is a #-generated outer model of V with the same cardinals as V and φ is a sentence with parameter p which holds in V^*. Then φ holds in an inner model of V.

Question 21 Is the SIMH#(CP) consistent?

Note that SIMH#(CP) implies a strong failure of CH.

4.11   Width Indiscernibility

An alternative to the Maximality Protocol (which ideally should be synthesised with it) is *Width Indiscernibility*. The motivation is to provide a description of V in width analogous to its description in height provided by #-generation.

Recall that with #-generation we arrive at the following:

$$V_0 \prec V_1 \prec \cdots \prec V = V_\infty \prec V_{\infty+1} \prec \cdots$$

where for $i < j$, V_i is a *rank-initial segment* of V_j. Moreover the models V_i form a collection of *indiscernible models* in a strong sense. This picture was the result of an analysis which began with *height reflection*, starting with the idea that V must have unboundedly many rank-initial segments V_i which are elementary in V.

Analogously, we introduce *width reflection*. We would like to say that V has proper inner models which are "elementary in V". Of course this cannot literally be true, as if V_0 is an elementary submodel of V with the same ordinals as V then it is easy to see that V_0 equals V. Instead, we use elementary embeddings.

Width Reflection   For each ordinal α, there is a proper elementary submodel H of V such that $V_\alpha \subseteq H$ and H is *amenable*, i.e. $H \cap V_\beta$ belongs to V for each ordinal β.

Equivalently:

Width Reflection   For each ordinal α, there is a nontrivial elementary embedding $j : V_0 \to V$ with critical point at least α such that j is *amenable*, i.e. $j \restriction (V_\beta)^{V_0}$ belongs to V for each ordinal β.

Let's write $V_0 < V$ if there is a nontrivial amenable $j : V_0 \to V$, as in the second formulation of width reflection. This relation is transitive.

Proposition 22

(a) If $V_0 < V$ then V_0 is a proper inner model of V.
(b) Width Reflection is consistent relative to the existence of a Ramsey cardinal.

Proof

(a) This follows from Kunen's Theorem that there can be no nontrivial elementary embedding from V to V.
(b) Suppose that κ is Ramsey. Then it follows that any structure of the form $\mathcal{M} = (V_\kappa, \in, \ldots)$ has an unbounded set of indiscernibles, i.e. an unbounded subset I of κ such that for each n, any two increasing n-tuples from I satisfy the same formulas in $\mathcal{M}$. Now apply this to $\mathcal{M} = (V_\kappa, \in, <)$ where $<$ is a wellorder of V_κ of length κ. Let J be any unbounded subset of I such that $I \setminus J$ is unbounded and for any $\alpha < \kappa$, let $H(J \cup \alpha)$ denote the Skolem hull of $J \cup \alpha$ in $\mathcal{M}$. Then $H(J \cup \alpha)$ is an elementary submodel of V_κ and is not equal to V_κ because no element of $I \setminus J$ greater than α belongs to it. As V_κ contains all bounded subsets of κ it follows that $H(J \cup \alpha)$ is amenable. $\square$

A variant of the argument in (b) above yields the consistency of arbitrarily long finite chains $V_0 < V_1 < \cdots < V_n$. But obtaining infinite such chains seems more difficult, and even more ambitiously we can ask:

Question 23 Is it consistent to have $V_0 < V_1 < \cdots < V$ of length Ord $+ 1$ such that the union of the V_i's equals V?

The latter would be a good start on the formulation of a consistent criterion of *Width Indiscernibility*, as an analogue for maximality in width to the criterion of maximality in height provided by #-generation.

4.12   Omniscience

By OMT(V), the *outer model theory of V*, we mean the class of sentences with arbitrary parameters from V which hold in all outer models of V. We have seen using V-logic that OMT(V) is definable over V^+. However for many universes V, OMT(V) is in fact first-order definable over V. These universes are said to be *omniscient*.

Recall the following version of Tarski's result on the undefinability of truth:

Proposition 24 *The set of sentences with parameters from V which hold in V is not (first-order) definable in V with parameters.*

Surprisingly, Mack Stanley showed however that OMT(V) can indeed be V-definable.

Theorem 25 (Stanley [30]) *Suppose that in V there is a proper class of measurable cardinals, and indeed this class is V^+-stationary, i.e. Ord(V) is regular with respect to V^+-definable functions and this class intersects every club in Ord(V) which is V^+-definable. Then OMT(V) is V-definable.*

Proof Using V-logic we can translate the statement that a first-order sentence φ (with parameters from V) holds in all outer models of V to the validity of a sentence φ^* in V-logic, a fact expressible over V^+ by a Σ_1 sentence. Using this we show that the set of φ which hold in all outer models of V is V-definable.

As Ord(V) is regular with respect to V^+-definable functions we can form a club C in Ord(V) such that for κ in C there is a Σ_1-elementary embedding from Hyp(V_κ) into V^+ (with critical point κ, sending κ to Ord(V)). Indeed C can be chosen to be V^+-definable.

For any κ in C let φ_κ^* be the sentence of V_κ-logic such that φ holds in all outer models of V_κ iff φ_κ^* is valid (a Σ_1 property of Hyp(V_κ)). By elementarity, φ_κ^* is valid iff φ^* is valid.

Now suppose that φ holds in all outer models of V, i.e. φ^* is valid. Then φ_κ^* is valid for all κ in C and since the measurables form a V^+-stationary class, there is a measurable κ such that φ_κ^* is valid.

Conversely, suppose that φ_κ^* is valid for some measurable κ. Now choose a normal measure U on κ and iterate $(H(\kappa^+), U)$ for Ord(V) steps to obtain a wellfounded structure $(H^*, U*)$. (This structure is wellfounded, as for any admissible set A, any measure in A can be iterated without losing wellfoundedness for α steps, for any ordinal α in A.) Then H^* equals $\mathrm{Hyp}(V^*)$ for some $V^* \subseteq V$. By elementarity, the sentence φ^*_{V*} which asserts that φ holds in all outer models of V^* is valid. But as V^* is an inner model of V, φ also holds in all outer models of V.

Thus φ belongs to $\mathrm{OMT}(V)$ exactly if it belongs to $\mathrm{OMT}(V_\kappa)$ for some measurable κ, and this is first-order expressible.   □

Are measurable cardinals needed for omniscience? Actually, Stanley was able to use just Ramsey cardinals, but as far as the consistency of omniscience we have the following:

Theorem 26 ([16]) *Suppose that κ is inaccessible and GCH holds. Then there is an omniscient model of the form $V_\kappa[G]$ where G is generic over V. Moreover, $V_\kappa[G]$ carries a definable wellorder.*

Omniscience demonstrates that it is possible to treat truth in arbitrary outer models internally in a way similar to how truth in set-generic extensions can be handled using the standard definability and truth lemmas of set-forcing. In fact, the situation is even better in that the entire outer model theory is first-order definable, not just the restriction of this theory to sentences of bounded complexity, as is the case for set-forcing. (The key difference is that in the case of set-forcing, the ground model V is uniformly definable in its set-generic extensions and therefore the full $\mathrm{OMT}(V)$ cannot be first-order definable in V by Proposition 24. An omniscient V cannot be uniformly definable in its arbitrary outer models for the same reason.)

Note also that by Theorem 25, omniscience synthesises well with #-generation: We need only work with models that have sufficiently many measurable cardinals.

4.13   The Future of the HP

We have discussed evidence of Type 1, coming from set theory's role as a branch of mathematics, and evidence of Type 2, coming from set theory's role as a foundation for mathematics. In the first case, evidence is judged by its value for the mathematical development of set theory and in the second case it is judged by its value for resolving independence in (and providing tools for) other areas of mathematics. In both cases the weight of the evidence is measured by a consensus of researchers working in the field.

Type 3 evidence is also measured by a consensus of researchers working in set theory (and its philosophy) but emanates instead from an analysis of the intrinsic maximality feature of the set concept as expressed by the maximal iterative conception. The Hyperuniverse Programme provides a strategy for *deriving* mathematical consequences from this conception.

To illustrate more clearly how the HP derives consequences of the maximality of V I'll discuss the case of #-generation and the search for an *optimal maximality criterion*.

#-generation is a major success of the HP. It provides a powerful mathematical criterion for height maximality which implies all prior known height maximality principles and provides an elegant description of how the height of V is maximised in a way analogous to the way L is maximised in height by the existence of large cardinals (or equivalently, by the existence of $0^{\#}$). There are good reasons to believe that #-generation will be accepted by the community of set-theorists and philosophers of set theory as the definitive expression of height maximality.

Width maximality is of course much more difficult than height maximality and the formulation, analysis and synthesis of the various possible width maximality criteria is at its early stages. The basic IMH is a good start, but must be synthesised with #-generation. The biggest challenge at the moment is dealing with formulations of width maximality which make use of parameters. The *maximality protocol* is a promising approach. But it is important to emphasize that the mathematical analysis of width maximality principles is challenging and there are sure to be some false turns in the development of the programme, leading to inconsistent principles (this has already happened several times). Such false turns are not damaging to the programme, but rather provide valuable further understanding of the nature of maximality.

The aim of the HP is to arrive after extensive mathematical work at an *optimal criterion* of maximality for the height and width of the universe of sets, providing a full mathematical analysis of the maximal iterative conception. As already said, the validation of such a criterion as optimal depends on a consensus of researchers working in set theory and its philosophy. *Derivability* from the maximal iterative conceptions refers to formal derivability form this sought-after optimal criterion. Of greatest interest are the first-order statements derivable from maximality, but it is already clear that the criteria being developed in the programme, such as the ones mentioned in this paper, are almost exclusively *non first-order*. My prediction is that the optimal criterion will include some form of the SIMH and therefore imply the (first-order) failure of CH.

I remain optimistic that when the discoveries of this programme are combined with further work in set theory and its application to resolving problems of independence in other areas of mathematics, the prediction expressed by the *Thesis of Set-Theoretic Truth* will be satisfyingly realized. But there is first a lot of work to be done.

References

1. C. Antos, S. Friedman, R. Honzik, C. Ternullo, Multiverse conceptions in set theory. Synthèse **192**(8), 2463–2488 (2015)
2. T. Arrigoni, S. Friedman, Foundational implications of the inner model hypothesis. Ann. Pure Appl. Logic **163**, 1360–1366 (2012)
3. T. Arrigoni, S. Friedman, The hyperuniverse program. Bull. Symb. Log. **19**(1), 77–96 (2013)

4. N. Barton, Multiversism and concepts of set: how much relativism is acceptable? in *Objectivity, Realism, and Proof. FilMat Studies in the Philosophy of Mathematics*, ed. by F. Boccuni, A. Sereni. Boston Studies in the Philosophy and History of Science (Springer, New York, 2016), pp. 189–209.
5. J. Barwise, *Admissible Sets and Structures* (Springer, Berlin, 1975)
6. A. Beller, R. Jensen, P.Welch, *Coding the Universe* (Cambridge University Press, Cambridge, 1982)
7. G.Boolos, The iterative conception of set. J. Philos. **68**(8), 215–231 (1971)
8. J. Cummings, S. Friedman, M. Golshani, Collapsing the cardinals of HOD. J. Math. Log. **15**(02), 1550007 (2015)
9. J. Cummings, S. Friedman, M. Magidor, A. Rinot, D. Sinapova, Definable subsets of singular cardinals. Israel J. Math. (to appear)
10. A. Dodd, *The Core Model* (Cambridge University Press, Cambridge, 1982)
11. S. Friedman, *Fine Structure and Class Forcing* (de Gruyter, Berlin, 2000)
12. S. Friedman, Internal consistency and the inner model hypothesis. Bull. Symb. Log. **12**(4), 591–600 (2006)
13. S. Friedman, The stable core. Bull. Symb. Log. **18**(2), 261–267 (2012)
14. S. Friedman, P. Holy, A quasi-lower bound on the consistency strength of PFA. Trans. AMS **366**, 4021–4065 (2014)
15. S. Friedman, R. Honzik, On strong forms of reflection in set theory. Math. Log. Q. **62**(1–2), 52–58 (2016)
16. S. Friedman, R. Honzik, Definability of satisfaction in outer models. J. Symb. Log. **81**(03), 1047–1068 (2016)
17. S. Friedman, C. Ternullo, The search for new axioms in the Hyperuniverse Programme, in *Objectivity, Realism, and Proof. FilMat Studies in the Philosophy of Mathematics*, ed. by F. Boccuni, A. Sereni, Boston Studies in the Philosophy and History of Science (Springer, New York, 2016), pp. 165–188
18. S. Friedman, P. Welch, H. Woodin, On the consistency strength of the inner model hypothesis. J. Symb. Log. **73**(2), 391–400 (2008)
19. J.D. Hamkins, A multiverse perspective on the axiom of constructibility, in *Infinity and Truth*, vol. 25 (World Scientific Publishing, Hackensack, NJ, 2014), pp. 25–45
20. G. Hellman, *Mathematics Without Numbers: Towards a Modal-Structural Interpretation* (Oxford University Press, Oxford, 1989)
21. D. Isaacson, The reality of mathematics and the case of set theory, in *Truth, Reference and Realism*, ed. by Z. Novak, A. Simonyi (Central European University Press, Budapest, 2011), pp. 1–76
22. T. Jech, *Set Theory* (Springer, Berlin, 2003)
23. P. Koellner, On reflection principles. Ann. Pure Appl. Logic **157**(2), 206–219 (2009)
24. Ø. Linnebo, The potential hierarchy of sets. Rev. Symb. Log. **6**(2), 205–228 (2013)
25. P. Maddy, *Defending the Axioms: On the Philosophical Foundations of Set Theory* (Oxford University Press, Oxford, 2011)
26. T. Meadows, Naive infinitism: the case for an inconsistency approach to infinite collections. Notre Dame J. Formal Logic **56**(1), 191–212 (2015)
27. C. Parsons, What is the iterative conception of set? *Logic, Foundations of Mathematics, and Computability Theory*, The University of Western Ontario Series in Philosophy of Science, vol. 9 (University of Western Ontario, London, 1977), pp. 335–367
28. W. Reinhardt, Remarks on reflection principles, large cardinals, and elementary embeddings, in *Proceedings of Symposia in Pure Mathematics*, vol. 13 (1974), pp. 189–205
29. I. Rumfitt, Determinacy and bivalence, in *The Oxford Handbook of Truth*, ed. by M. Glanzberg (Clarendon Press, Oxford, to appear)
30. M. Stanley, *Outer Model Satisfiability*, preprint
31. J. Steel, Gödel's program, in *Interpreting Gödel*, ed. by J. Kennedy (Cambridge University Press, Cambridge, 2014)

32. *The Thread*, an e-mail discussion during June-November 2014 (with extensive contributions by S.Feferman, H.Friedman, S.Friedman, G.Hellman, P.Koellner, P.Maddy, R.Solovay and H.Woodin).
33. E. Zermelo, (1930) On boundary numbers and domains of sets, in *From Kant to Hilbert: A Source Book in the Foundations of Mathematics*, ed. by W.B. Ewald (Oxford University Press, Oxford, 1996), pp. 1208–1233

On the Set-Generic Multiverse

Sy-David Friedman, Sakaé Fuchino, and Hiroshi Sakai

Abstract The forcing method is a powerful tool to prove the consistency of set-theoretic assertions relative to the consistency of the axioms of set theory. Laver's theorem and Bukovský's theorem assert that set-generic extensions of a given ground model constitute a quite reasonable and sufficiently general class of standard models of set-theory.

In Sects. 2 and 3 of this note, we give a proof of Bukovský's theorem in a modern setting (for another proof of this theorem see Bukovský (Generic Extensions of Models of ZFC, a lecture note of a talk at the Novi Sad Conference in Set Theory and General Topology, 2014)). In Sect. 4 we check that the multiverse of set-generic extensions can be treated as a collection of countable transitive models in a conservative extension of ZFC. The last section then deals with the problem of the existence of infinitely-many independent buttons, which arose in the modal-theoretic approach to the set-generic multiverse by Hamkins and Loewe (Trans. Am. Math. Soc. 360(4):1793–1817, 2008).

1 The Category of Forcing Extensions as the Set-Theoretic Multiverse

The forcing method is a powerful tool to prove the consistency of set-theoretic (i.e., mathematical) assertions relative to (the consistency of) the axioms of set theory. If a sentence σ in the language $\mathcal{L}_{ZF}$ of set theory is proved to be relatively consistent with the axioms of set theory (ZFC) by some forcing argument then it is so in the

Originally published in S. Friedman, S. Fuchino, H. Sakai, On the set-generic multiverse, in *Sets and Computations*. IMS Lecture Notes Series, vol. 33 (Institute of Mathematical Sciences, National University of Singapore, Singapore, 2017), pp. 25–44.

S.-D. Friedman
Kurt Gödel Research Center for Mathematical Logic, University of Vienna, Vienna, Austria
e-mail: sdf@logic.univie.ac.at

S. Fuchino (✉) • H. Sakai
Graduate School of System Informatics, Kobe University, Kobe, Japan
e-mail: fuchino@diamond.kobe-u.ac.jp; hsakai@people.kobe-u.ac.jp

© Springer International Publishing AG 2018
C. Antos et al. (eds.), *The Hyperuniverse Project and Maximality*,
https://doi.org/10.1007/978-3-319-62935-3_5

sense of the strictly finitist standpoint of Hilbert: the forcing proof can be recast into an algorithm $\mathcal{A}$ such that, if a formal proof $\mathcal{P}$ of a contradiction from ZFC $+ \sigma$ is ever given, then we can transform $\mathcal{P}$ with the help of $\mathcal{A}$ to another proof of a contradiction from ZFC or even ZF alone.

The "working set-theorists" however prefer to see their forcing arguments not as mere discussions concerning manipulations of formulas in a formal system but rather concerning the "real" mathematical universe in which they "live". Forcing for them is thus a method of extending the universe of set theory where they originally "live" (the ground model, usually denoted as "V") to many (actually more than class many in the sense of V) different models of set theory called generic extensions of V. Actually, a family of generic extensions is constructed for certain V-definable partial orderings $\mathbb{P}$. Each such generic extension is obtained first by fixing a so-called generic filter G which is a filter over $\mathbb{P}$, sitting outside V with a "generic" sort of transcendence over V, and then by adding G to V to generate a new structure—the generic extension $V[G]$ of V—which is also a model of ZFC. Often this process of taking generic extensions over some model of set theory is even repeated transfinitely-many times. As a result, a set-theorist performing forcing constructions is seen to live in many different models of set theory simultaneously. This is manifested in many technical expositions of forcing where the reader very often finds narratives beginning with phrases like: "Working in $V[G]$, …", "Let $\alpha < \kappa$ be such that x is in the αth intermediate model $V[G_\alpha]$ and …", "Now returning to V, …", etc., etc.

Although this "multiverse" view of forcing is in a sense merely a modus loquendi, it is worthwhile to study the possible pictures of this multiverse per se. Some initial moves in this direction have been taken e.g. in [1, 2, 5–8, 11, 12, 21, 24] etc. The term "multiverse" probably originated in work of Woodin in which he considered the "set-generic multiverse", the "class" of set-theoretic universes which forms the closure of the given initial universe V under set-generic extension and set-generic ground models. Sometimes we also have to consider the constellations of the set-generic multiverse where V cannot be reconstructed as a set-generic extension of some of or even any of the proper inner models of V. To deal with such cases it is more convenient to consider the expanded generic multiverse where we also assume that the multiverse is also closed under the construction of definable inner models.

The set-generic universe should be distinguished from the "class-generic multiverse", defined in the same way but with respect to class-forcing extensions and ground models, as well as inner models of class-generic extensions that are not themselves class-generic (see [5]). It is even possible to go beyond class-forcing by considering forcings whose conditions are classes, so-called hyperclass forcings (see [6]). The broadest point of view with regard to the multiverse is expressed in [7], where the "hyperuniverse" is taken to consist of *all* universes which share the same ordinals as the initial universe (which is taken to be countable to facilitate the construction of new universes). The hyperuniverse is closed under all notions of forcing.

In this article we restrict our attention to the set-generic multiverse. The well-posedness of questions regarding the set-generic multiverse is established by the theorems of Laver and Bukovský which we discuss in Sect. 2. These theorems show that the set-generic extensions and set-generic ground models of a given universe represent a "class" of models with a natural characterization.

The straightforward formulation of the set-generic multiverse requires the notion of "class" of classes which cannot be treated in the usual framework of ZF set theory, but, as emphasized at the beginning, theorems about the set-generic multiverse are actually meta-theorems about ZFC. However we can also consider a theory which is a conservative extension of ZFC in which set-generic extensions and set-generic ground models are real objects in the theory and the set-generic multiverse a definable class. In Sect. 4, we consider such a system and show that it is a conservative extension of ZFC.

The multiverse view sometimes highlights problems which would never have been asked in the conventional context of forcing constructions (see [11]). As one such example we consider in Sect. 5 the problem of the existence of infinitely many independent buttons (in the sense of [12]).

2   Laver's Theorem and Bukovský's Theorem

In the forcing language, we often have to express that a certain set is already in the ground model, e.g. in a statement like: $p \Vdash_{\mathbb{P}}$ "... $\dot{x}$ is in V and ...". In such situations we can always find a large enough ordinal ξ such that the set in question should be found in that level of the cumulative hierarchy in the ground model. So we can reformulate a statement like the one above into something like $p \Vdash_{\mathbb{P}}$ "... $\dot{x} \in \check{V}_\xi$ and ..." which is a legitimate expression in the forcing language.

This might be one of the reasons why it is proved only quite recently that the ground model is always definable in an arbitrary set-generic extension:

Theorem 2.1 (Laver [17], Woodin [23])   *There is a formula $\varphi^*(x, y)$ in $\mathcal{L}_{ZF}$ such that, for any transitive model V of ZFC and set-generic extension $V[G]$ of V there is $a \in V$ such that, for any $b \in V[G]$*

$$b \in V \iff V[G] \models \varphi^*(a, b). \qquad \qquad \square$$

An important corollary of Laver's theorem is that a countable transitive model of ZFC can have at most countably many ground models for set forcing.

Bukovský's theorem gives a natural characterization of inner models M of V such that V is a set-generic extension of M.[1] Note that, by Laver's theorem Theorem 2.1, such an M is then definable in V. However the inner model M of V may be introduced as a class in the sense of von Neumann-Bernays-Gödel class theory

---

[1] In the terminology of [8], M is a ground of V.

(NBG) and in such a situation the definability of M in V may not be immediately clear.

Let us begin with the following observation concerning κ-c.c. generic extensions. We shall call a partial ordering *atomless* if each element of it has at least two extensions which are incompatible with each other.

Lemma 2.2 *Let κ be a regular uncountable cardinal. If $\mathbb{P}$ is a κ-c.c. atomless partial ordering, then $\mathbb{P}$ adds a new subset of $2^{<\kappa}$.*

Proof Without loss of generality, we may assume that $\mathbb{P}$ consists of the positive elements of a κ-c.c. atomless complete Boolean algebra. Note that $\mathbb{P}$ adds new subsets of On since $\mathbb{P}$ adds a new set (e.g. the $(V, \mathbb{P})$-generic set). Suppose that $\dot{S}$ is a $\mathbb{P}$-name of a new subsfet of On. Let θ be a sufficiently large regular cardinal and let $M \prec \mathcal{H}(\theta)$ be such that

(2.1) $|M| \leq 2^{<\kappa}$;
(2.2) $^{<\kappa}M \subseteq M$ and
(2.3) $\mathbb{P}, \dot{S}, \kappa \in M$.

Let $\dot{T}$ be a $\mathbb{P}$-name such that $\Vdash_{\mathbb{P}} \text{"} \dot{T} = \dot{S} \cap M \text{"}$. By (2.1), it is enough to show the following, where V denotes the ground model:

Claim 2.2.1 $\Vdash_{\mathbb{P}} \text{"} \dot{T} \notin V \text{"}$.
$\vdash$ Otherwise there would be $p \in \mathbb{P}$ and $T \in V$, $T \subseteq$ On such that

(2.4) $p \Vdash_{\mathbb{P}} \text{"} \dot{T} = \check{T} \text{"}$.

We show in the following that then we can construct a strictly decreasing sequence $\langle q_\alpha : \alpha < \kappa \rangle$ in $\mathbb{P} \cap M$ such that

(2.5) $p \leq_{\mathbb{P}} q_\alpha$ for all $\alpha < \kappa$.

But since $\{q_\alpha \cdot -q_{\alpha+1} : \alpha < \kappa\}$ is then a pairwise disjoint subset of $\mathbb{P}$, this contradicts the κ-c.c. of $\mathbb{P}$.

Suppose that $\langle q_\alpha : \alpha < \delta \rangle$ for some $\delta < \kappa$ has been constructed. If δ is a limit, let $q_\delta = \prod_{\alpha < \delta} q_\alpha$. Then we have $p \leq_{\mathbb{P}} q_\delta$ and $q_\delta \leq_{\mathbb{P}} q_\alpha$ for all $\alpha < \delta$. Since $\langle q_\alpha : \alpha < \delta \rangle \in M$ by (2.2), we also have $q_\delta \in M$.

If $\delta = \beta + 1$, then, since $M \models \text{"} q_\beta$ does not decide $\dot{S}$" by the elementarity of M, there are $\xi \in$ On $\cap M$ and $q, q' \in \mathbb{P} \cap M$ with $q, q' \leq_{\mathbb{P}} q_\beta$ such that $q \Vdash_{\mathbb{P}} \text{"} \xi \in \dot{S} \text{"}$ and $q' \Vdash_{\mathbb{P}} \text{"} \xi \notin \dot{S} \text{"}$. At least one of them, say q, must be incompatible with p. Then $q_\delta = q_\beta \cdot -q$ is as desired. $\dashv$ (Claim 2.2.1)
$\square$ (Lemma 2.2)

Note that, translated into the language of complete Boolean algebras, the lemma above just asserts that no κ-c.c. atomless Boolean algebra $\mathbb{B}$ is $(2^{<\kappa}, 2)$-distributive.

Suppose now that we work in NBG, V is a transitive model of ZF and M an inner model of ZF in V (that is M is a transitive class $\subseteq V$ with $(M, \in) \models$ ZF). For a regular uncountable cardinal κ in M, we say that M κ-*globally covers* V if for every function f (in V) with dom(f) $\in M$ and rng(f) $\subseteq M$, there is a function $g \in M$ with dom(g) = dom(f) such that $f(i) \in g(i)$ and $M \models |g(i)| < \kappa$ for all $i \in$ dom(f).

Theorem 2.3 (Bukovský [3] and [4]) *Suppose that V is a transitive model of ZFC, M ⊆ V an inner model of ZFC and κ is a regular uncountable cardinal in M. Then M κ-globally covers V if and only if V is a κ-c.c. set-generic extension of M.*[2]

As the referee of the paper points out, this theorem can be formulated more naturally in the von Neumann-Bernays-Gödel class theory (NBG) since in the framework of ZFC this theorem can only be formulated as a meta-theorem, that is, as a collection of theorems consisting corresponding statements for each formula which might define an inner model M.

Proof of Theorem 2.3: If V is a κ-c.c. set-generic extension of M, say by a partial ordering $\mathbb{P} \in M$ with $M \models$"$\mathbb{P}$ has the κ-c.c.", then it is clear that M κ-globally covers V (for f as above, let $\dot{f} \in M$ be a $\mathbb{P}$-name of f and g be defined by letting $g(\alpha)$ to be the set of all possible values $\dot{f}(\alpha)$ may take).

The proof of the converse is done via the following Lemma 2.4. Note that, by Grigorieff's theorem (see Corollary 2.6 below), the statement of this Lemma is a consequence of Bukovský's theorem:

Lemma 2.4 *Suppose that M is an inner model of a transitive model V of ZFC such that M κ-globally covers V for some κ regular uncountable in M. Then for any $A \in V$, $A \subseteq On$, M[A] is*[3] *a κ-c.c. set-generic extension of M.*

Note that it can happen easily that $M[A]$ is not a set generic extension of M. For example, $0^\#$ exists and $M = L$, then $M[0^\#]$ is not a set-generic extension of M.

We first show that Theorem 2.3 follows from Lemma 2.4. Assume that M κ-globally covers V. We have to show that V is a κ-c.c. set-generic extension of M. In V, let λ be a regular cardinal such that $\lambda^{<\kappa} = \lambda$ and $A \subseteq On$ be a set such that

$$(2.6) \quad (\mathcal{P}(\lambda))^{M[A]} = (\mathcal{P}(\lambda))^V.$$

Then, by Lemma 2.4, $M[A]$ is a κ-c.c. generic extension of M and hence we have $M[A] \models$ "κ is a regular cardinal". Actually we have $M[A] = V$. Otherwise there would be a $B \in V \setminus M[A]$ with $B \subseteq On$. Since $M[A]$ κ-globally covers $M[A][B]$, we may apply Lemma 2.4 on this pair and conclude that $M[A][B]$ is a (non trivial) κ-c.c. generic extension of $M[A]$. By Lemma 2.2, there is a new element of $\mathcal{P}((2^{<\kappa})^{M[A]}) \subseteq \mathcal{P}(\lambda)$ in $M[A][B]$. But this is a contradiction to (2.6). □ (Theorem 2.3)

Proof of Lemma 2.4: We work in M and construct a κ-c.c. partial ordering $\mathbb{P}$ such that $M[A]$ is a $\mathbb{P}$-generic extension over M.

Let $\mu \in On$ be such that $A \subseteq \mu$ and let $\mathcal{L}_\infty(\mu)$ be the infinitary sentential logic with atomic sentences

$$(2.7) \quad \text{"}\alpha \in \dot{A}\text{" for } \alpha \in \mu$$

---

[2]Tadatoshi Miyamoto told us that James Baumgartner independently proved this theorem in an unpublished note using infinitary logic.

[3] $M[A]$ may be defined by $M[A] = \bigcup_{\alpha \in On} L(V_\alpha^M \cup \{A\})$. $M[A]$ is a model of ZF: this can be seen easily e.g. by applying Theorem 13.9 in [13]. If M also satisfies AC then $M[A]$ satisfies AC as well since, in this case, it is easy to see that a well-ordering of $(V_\alpha)^M \cup \{A\}$ belongs to $M[A]$ for all $\alpha \in On$.

and the class of sentences closed under $\neg$ and $\bigvee$ where $\neg$ is to be applied to a formula and $\bigvee$ to an arbitrary set of formulas. To be specific let us assume that the atomic sentences "$\alpha \in \dot{A}$" for $\alpha \in \mu$ are coded by the sets $\langle \alpha, 0 \rangle$ for $\alpha \in \mu$, the negation $\neg \varphi$ by $\langle \varphi, 1 \rangle$ and the infinitary disjunction $\bigvee \Phi$ by $\langle \Phi, 2 \rangle$. We regard the usual disjunction $\vee$ of two formulas as a special case of $\bigvee$ and other logical connectives like "$\bigwedge$", "$\wedge$", "$\rightarrow$" as being introduced as abbreviations of usual combinations of $\neg$ and $\bigvee$. For a sentence $\varphi \in \mathcal{L}_\infty(\mu)$ and $B \subseteq \mu$, we write $B \models \varphi$ when φ holds if each atomic sentence of the form "$\alpha \in \dot{A}$" in φ is interpreted by "$\alpha \in B$" and logical connectives in φ are interpreted in canonical way. For a set Γ of sentences, we write $B \models \Gamma$ if $B \models \psi$ for all $\psi \in \Gamma$. For $\Gamma \subseteq \mathcal{L}_\infty(\mu)$ and φ, we write $\Gamma \models \varphi$ if $B \models \Gamma$ implies $B \models \varphi$ for all $B \subseteq \mu$ (in V).

Let $\vdash$ be a notion of provability for $\mathcal{L}_\infty(\mu)$ in some logical system which is correct (i.e. $\Gamma \vdash \varphi$ always implies $\Gamma \models \varphi$),[4] upward absolute (i.e. $M \subseteq N$ and $M \models$ "$\Gamma \vdash \varphi$" always imply $N \models$ "$\Gamma \vdash \varphi$" for any transitive models M, N of ZF) and sufficiently strong (so that all the arguments used below work for this $\vdash$). In Sect. 3 we introduce one such deductive system (as well as an alternative approach without using such a deduction system, based on Lévy Absoluteness).

Let $\lambda = \max\{\kappa, \mu^+\}$ and $\mathcal{L}_\lambda(\mu) = \mathcal{L}_\infty(\mu) \cap (V_\lambda)^M$. Let $f \in V$ be a mapping $f : \big(\mathcal{P}(\mathcal{L}_\lambda(\mu))\big)^M \setminus \{\emptyset\} \to \big(\mathcal{L}_\lambda(\mu)\big)^M$ such that, for any $\Gamma \in \big(\mathcal{P}(\mathcal{L}_\lambda(\mu))\big)^M \setminus \{\emptyset\}$, we have $f(\Gamma) \in \Gamma$ and $A \models f(\Gamma)$ if $A \models \bigvee \Gamma$. Since M κ-globally covers V, there is a $g \in M$ with $g : \big(\mathcal{P}(\mathcal{L}_\lambda(\mu))\big)^M \setminus \{\emptyset\} \to \mathcal{P}_{<\kappa}\big(\mathcal{L}_\lambda(\mu)\big)^M$ such that $f(\Gamma) \in g(\Gamma) \subseteq \Gamma$ for all $\Gamma \in (\mathcal{P}(\mathcal{L}_\lambda(\mu)))^M \setminus \{\emptyset\}$.

In M, let

$$(2.8) \quad T = \{\bigvee \Gamma \to \bigvee g(\Gamma) : \Gamma \in \mathcal{P}(\mathcal{L}_\lambda(\mu)) \setminus \{\emptyset\}\}.$$

Note that $M[A] \models$ "$A \models T$". It follows that T is consistent with respect to our deduction system (in V). In M, let

$$(2.9) \quad \mathbb{P} = \{\varphi \in \mathcal{L}_\lambda(\mu) : T \not\vdash \neg\varphi\}$$

and for $\varphi, \psi \in \mathbb{P}$, let

$$(2.10) \quad \varphi \leq_\mathbb{P} \psi \Leftrightarrow T \vdash \varphi \to \psi.$$

Claim 2.4.1 For $\varphi \in \mathcal{L}_\lambda(\mu)$, if $A \models \varphi$ then we have $\varphi \in \mathbb{P}$. In particular, "$\alpha \in \dot{A}$" $\in \mathbb{P}$ for all $\alpha \in A$ and "$\neg(\alpha \in \dot{A})$" $\in \mathbb{P}$ for all $\alpha \in \mu \setminus A$.

$\vdash$ Suppose $A \models \varphi$. We have to show $T \not\vdash \neg\varphi$: If $T \vdash \neg\varphi$ in M, then we would have $V \models$ "$T \vdash \neg\varphi$". Since $A \models T$ in V, it follows that $A \models \neg\varphi$. This is a contradiction. $\dashv$ (Claim 2.4.1)

Claim 2.4.2 For $\varphi, \psi \in \mathbb{P}$, φ and ψ are compatible if and only if

$$(2.11) \quad T \not\vdash \neg(\varphi \wedge \psi).$$

---

[4]More precisely, we assume that ZF proves the correctness of $\vdash$.

Note that (2.11) is equivalent to

(2.12) $T \nvdash \neg\varphi \vee \neg\psi$ ($\Leftrightarrow$ $T \nvdash \varphi \rightarrow \neg\psi$).

$\vdash$ Suppose that $\varphi, \psi \in \mathbb{P}$ are compatible. By the definition of $\leq_{\mathbb{P}}$ this means that there is $\eta \in \mathbb{P}$ such that $T \vdash \eta \rightarrow \varphi$ and $T \vdash \eta \rightarrow \psi$. For this η we have $T \vdash \eta \rightarrow (\varphi \wedge \psi)$. Since $T \nvdash \neg\eta$ by the consistency of T, it follows that $T \nvdash \neg(\varphi \wedge \psi)$.

Conversely if $T \nvdash \neg(\varphi \wedge \psi)$. Then $(\varphi \wedge \psi) \in \mathbb{P}$. Since $T \vdash (\varphi \wedge \psi) \rightarrow \varphi$ and $T \vdash (\varphi \wedge \psi) \rightarrow \psi$, we have $(\varphi \wedge \psi) \leq_{\mathbb{P}} \varphi$ and $(\varphi \wedge \psi) \leq_{\mathbb{P}} \psi$. Thus φ and ψ are compatible with respect to $\leq_{\mathbb{P}}$. $\dashv$ (Claim 2.4.2)

Claim 2.4.3 $\mathbb{P}$ has the κ-c.c.

$\vdash$ Suppose that $\Gamma \subseteq \mathbb{P}$ is an antichain. Since $|g(\Gamma)| < \kappa$, it is enough to show that $g(\Gamma) = \Gamma$. Suppose otherwise and let $\varphi_0 \in \Gamma \setminus g(\Gamma)$. Since "$\bigvee\Gamma \rightarrow \bigvee g(\Gamma)$" $\in T$ and $\vdash \varphi_0 \rightarrow \bigvee\Gamma$, we have

(2.13) $T \vdash \varphi_0 \rightarrow \bigvee g(\Gamma)$.

It follows that there is $\varphi \in g(\Gamma)$ such that φ_0 and φ are compatible. This is because otherwise we would have $T \vdash \varphi_0 \rightarrow \neg\varphi$ for all $\varphi \in g(\Gamma)$ by Claim 2.4.2. Hence $T \vdash \varphi_0 \rightarrow \bigwedge\{\neg\varphi : \varphi \in g(\Gamma)\}$ which is equivalent to $T \vdash \varphi_0 \rightarrow \neg\bigvee g(\Gamma)$. From this and (2.13), it follows that $T \vdash \neg\varphi_0$. But this is a contradiction to the assumption that $\varphi_0 \in \mathbb{P}$.

Now, since Γ is pairwise incompatible, it follows that $\varphi_0 = \varphi \in g(\Gamma)$. This is a contradiction to the choice of φ_0. $\dashv$ (Claim 2.4.3)

In V, let $G(A) = \{\varphi \in \mathbb{P} : A \models \varphi\}$. By Claim 2.4.1, we have $G(A) = \{\varphi \in \mathcal{L}_\lambda(\mu) : A \models \varphi\}$ and A is definable from $G(A)$ over M as $\{\alpha \in \mu : "\alpha \in \dot{A}" \in G(A)\}$. Thus we have $M[G(A)] = M[A]$.

Hence the following two Claims prove our Lemma:

Claim 2.4.4 $G(A)$ is a filter in $\mathbb{P}$.

$\vdash$ Suppose that $\varphi \in G(A)$ and $\varphi \leq_{\mathbb{P}} \psi$. Since this means that $A \models \varphi$ and $T \vdash \varphi \rightarrow \psi$, it follows that $A \models \psi$. That is, $\psi \in G(A)$.

Suppose now that $\varphi, \psi \in G(A)$. This means that

(2.14) $A \models \varphi$ and $A \models \psi$.

Hence we have $A \models \varphi \wedge \psi$. By Claim 2.4.1, it follows that $(\varphi \wedge \psi) \in \mathbb{P}$, that is, $T \nvdash \neg(\varphi \wedge \psi)$. Thus φ and ψ are compatible by Claim 2.4.2. $\dashv$ (Claim 2.4.4)

Claim 2.4.5 $G(A)$ is $\mathbb{P}$-generic.

$\vdash$ Working in M, suppose that Γ is a maximal antichain in $\mathbb{P}$. By Claim 2.4.3, we have $|\Gamma| < \kappa$ and hence we have $\bigvee\Gamma \in \mathcal{L}_\lambda(\mu)$ and hence $\bigvee\Gamma \in \mathbb{P}$: For $\varphi \in \Gamma$, since $\varphi \in \mathbb{P}$ we have $T \nvdash \neg\varphi$ and $\vdash \varphi \rightarrow \bigvee\Gamma$. It follows $T \nvdash \bigvee\Gamma$.

Moreover we have $T \vdash \bigvee\Gamma$: Otherwise $\neg\bigvee\Gamma$ would be an element of $\mathbb{P}$ incompatible with every $\varphi \in \Gamma$. A contradiction to the maximality of Γ.

Hence $A \models \bigvee\Gamma$ and thus there is $\varphi \in \Gamma$ such that $A \models \varphi$. That is, $\varphi \in G(A)$.
   $\dashv$ (Claim 2.4.5)
   $\square$ (Lemma 2.4)

The proof of Theorem 2.3 from Lemma 2.4 relies on Lemma 2.2 and the Axiom of Choice is involved both in the statement and the proof of Lemma 2.2.

On the other hand, Lemma 2.4 can be proved without assuming the Axiom of Choice in M: It suffices to eliminate choice from the proof of Claim 2.4.5.

Proof of Claim 2.4.5 Without the Axiom of Choice in M Working in M, suppose that D is a dense subset of $\mathbb{P}$. Then $A \models \bigvee D$: Otherwise we would have $T \nvdash \bigvee D$. Since

(2.15) $T \vdash \bigvee D \leftrightarrow \bigvee g(D),$

it follows that $T \nvdash \bigvee g(D)$. Since $\bigvee g(D) \in \mathcal{L}_\lambda(\mu)$, this implies $\neg \bigvee g(D) \in \mathbb{P}$. Since D is dense in $\mathbb{P}$ there is $\varphi_0 \in D$ such that $T \vdash \varphi_0 \rightarrow \neg \bigvee g(D)$. By (2.15), it follows that $T \vdash \varphi_0 \rightarrow \neg \bigvee D$. On the other hand, since $\varphi_0 \in D$ we have $T \vdash \varphi_0 \rightarrow \bigvee D$. Hence we have $T \vdash \neg \varphi_0$ which is a contradiction to $\varphi_0 \in \mathbb{P}$.

Thus there is $\varphi_1 \in D$ such that $A \models \varphi_1$, that is, $\varphi_1 \in G(A)$.

$\quad\quad\quad\quad\quad\quad\quad\quad\quad\quad\quad\quad\quad\quad\quad\quad\quad\quad\quad\quad\quad\quad\quad$ □ (Claim 2.4.5 without AC in M)

The next corollary follows immediately from this remark:

Corollary 2.5 *Work in NBG. Suppose that V is a model of ZFC and M is an inner model of V (of ZF) such that M κ-globally covers V. If $V = M[A]$ for some set $A \subseteq On$ then V is a κ-c.c. set-generic extension of M.* □

We do not know if Corollary 2.5 is false without the added assumption that V is $M[A]$ for a set of ordinals A.

More generally, it seems to be open if there is a characterisation of the set-generic extensions of an arbitrary model of ZF; or at least of such extensions given by partial orders which are well-ordered in the ground model.

Grigorieff's theorem can be also obtained by a modification of the proof of Theorem 2.3.

Corollary 2.6 (Grigorieff [10]) *Suppose that M is an inner model of a model V of ZFC and V is a set-generic extension of M. Then any inner model N of V (of ZFC) with $M \subseteq N$ is a set-generic extension of M and hence definable in V. Also, for such N, V is a set-generic extension of N.*

If V is κ-c.c. set-generic extension of M in addition, then N is a κ-c.c. set-generic extension of M and V is a κ-c.c. set-generic extension of N. □

Similarly to Theorem 2.3, we can also characterize generic extensions obtained via a partial ordering of cardinality $\leq \kappa$.

For M and V as above, we say that V is *κ-decomposable* into M if for any $a \in V$ with $a \subseteq M$, there are $a_i \in M$, $i \in \kappa$ such that $a = \bigcup_{i<\kappa} a_i$.

Theorem 2.7 *Suppose that V is a transitive model of ZFC and M an inner model of ZFC definable in V and κ is a cardinal in M. Then V is a generic extension of M by a partial ordering in M of size $\leq \kappa$ (in M) if and only if M κ^+-globally covers V and V is κ-decomposable into M.*

Proof If V is a generic extension of M by a generic filter G over a partial ordering $\mathbb{P} \in M$ of size $\leq \kappa$ (in M) then $\mathbb{P}$ has the κ^+-c.c. and hence M κ^+-globally covers

V by Theorem 2.1. V is κ-decomposable into M since, for any $a \in V$ with $a = \dot{a}^G$, we have $a = \bigcup\{\{m \in M : p \Vdash_{\mathbb{P}} "m \in \dot{a}"\} : p \in G\}$.

Suppose now that $M \kappa^+$-globally covers V and V is κ-decomposable into M. By Theorem 2.3, there is a κ^+-c.c. partial ordering $\mathbb{P}$ in M and a $\mathbb{P}$-generic filter G over M such that $V = M[G]$. Without loss of generality, we may assume that $\mathbb{P}$ consists of the positive elements of a complete Boolean algebra $\mathbb{B}$ (in M).

By κ-decomposability, G can be decomposed into κ sets $G_i \in M$, $i < \kappa$. Without loss of generality, we may assume that $\mathbb{1}_{\mathbb{P}}$ forces this fact. So letting $\dot{G}$ be the standard name of G and $\dot{G}_i$, $i < \kappa$ be names of G_i, $i < \kappa$ respectively, we may assume

(2.16) $\Vdash_{\mathbb{P}} "\dot{G} = \bigcup_{i<\kappa} \dot{G}_i"$.

Working in M, let $X_i \subseteq \mathbb{P}$ be a maximal pairwise incompatible set of conditions p which decide $\dot{G}_i$ to be $G_{i,p} \in M$ for each $i < \kappa$. By the κ^+-c.c. of $\mathbb{P}$, we have $|X_i| \leq \kappa$. Clearly, we have $p \leq_{\mathbb{P}} \prod^{\mathbb{B}} G_{i,p}$ for all $i < \kappa$ and $p \in X_i$. Let $\mathbb{P}' = \bigcup\{X_i : i < \kappa\}$. Then $|\mathbb{P}'| \leq \kappa$.

Claim 2.7.1 $\mathbb{P}'$ is dense in $\mathbb{P}$.

$\vdash$ Suppose $p \in \mathbb{P}$. Then there is $q \leq p$ such that q decides some $\dot{G}_i$ to be $G_{i,q}$ and $p \in G_{i,q}$. Let $r \in X_i$ be compatible with q. Then we have $r \leq_{\mathbb{P}} \prod^{\mathbb{B}} G_{i,r} = \prod^{\mathbb{B}} G_{i,q} \leq p$. $\dashv$ (Claim 2.7.1)

Thus V is a $\mathbb{P}'$-generic extension over M. $\square$ (Theorem 2.7)

3 A Formal Deductive System for $\mathcal{L}_\infty(\mu)$

In the proof of Lemma 2.4, we used a formal deductive system of $\mathcal{L}_\infty(\mu)$ without specifying exactly which system we are using. It is enough to consider a system of deduction which contains all logical axioms we used in the course of the proof together with modus ponens and some infinitary deduction rules like:

$$\frac{\varphi_i \to \psi, \quad i \in I}{\bigvee\{\varphi_i : i \in I\} \to \psi}$$

What we need for such a system is that its correctness and upward absoluteness hold while we do not make use of any version of completeness of the system.

Formal deduction systems for infinitary logics have been studied extensively in 1960s and 1970s, see e.g. [14, 15, 20]. Nevertheless, to be concrete, we shall introduce below such a deductive system S for $\mathcal{L}_\infty(\mu)$.

One peculiar task for us here is that we have to make our deduction system S such that S does not rely on AC so that we can apply it in an inner model M which does not necessarily satisfy AC to obtain Corollary 2.5.

Recall that we have introduced $\mathcal{L}_\infty(\mu)$ as the smallest class containing the sets $\langle \alpha, 0 \rangle$, $\alpha \in \mu$ as the codes of the prediactes "$\alpha \in \dot{A}$" for $\alpha \in \mu$ and closed with respect to $\langle \varphi, 1 \rangle$ for $\varphi \in \mathcal{L}_\infty(\mu)$ and $\langle \Phi, 2 \rangle$ for all sets $\Phi \subseteq \mathcal{L}_\infty(\mu)$ where $\langle \varphi, 1 \rangle$ and $\langle \Phi, 2 \rangle$ represent $\neg\varphi$ and $\bigvee\Phi$ respectively. Here, to be more precise about the role of the infinite conjunction we add the infinitary logical connective $\bigwedge$, and assume that $\bigwedge\Phi$ is coded by $\langle \Phi, 3 \rangle$ and thus $\mathcal{L}_\infty(\mu)$ is also closed with respect to $\langle \Phi, 3 \rangle$ for all sets $\Phi \subseteq \mathcal{L}_\infty(\mu)$.

The axioms of S consist of the following formulas:

(A1) $\varphi(\varphi_0, \varphi_1, \ldots, \varphi_{n-1})$

> for each tautology $\varphi(A_0, A_1, \ldots, A_{n-1})$ of (finitary) propositional logic and $\varphi_0, \varphi_1, \ldots, \varphi_{n-1} \in \mathcal{L}_\infty(\mu)$;

(A2) $\varphi \rightarrow \bigvee\Phi$ and $\bigwedge\Phi \rightarrow \varphi$

> for any set $\Phi \subset \mathcal{L}_\infty(\mu)$ and $\varphi \in \Phi$;

(A3) $\neg(\bigwedge\Phi) \leftrightarrow \bigvee\{\neg\varphi : \varphi \in \Phi\}$ and
$\neg(\bigvee\Phi) \leftrightarrow \bigwedge\{\neg\varphi : \varphi \in \Phi\}$

> for any set $\Phi \subseteq \mathcal{L}_\infty(\mu)$; and

(A4) $\varphi \wedge (\bigvee\Psi) \leftrightarrow \bigvee\{\varphi \wedge \psi : \psi \in \Psi\}$ and
$\varphi \vee (\bigwedge\Psi) \leftrightarrow \bigwedge\{\varphi \vee \psi : \psi \in \Psi\}$

> for any $\varphi \in \mathcal{L}_\infty(\mu)$ and any set $\Psi \subseteq \mathcal{L}_\infty(\mu)$.

Deduction Rules:

(Modus Ponens) $\qquad \dfrac{\{\varphi, \varphi \rightarrow \psi\}}{\psi}$

(R1) $\qquad \dfrac{\{\varphi \rightarrow \psi : \varphi \in \Phi\}}{\bigvee\Phi \rightarrow \psi}$ (R2) $\dfrac{\{\varphi \rightarrow \psi : \psi \in \Psi\}}{\varphi \rightarrow \bigwedge\Psi}$

A proof of $\varphi \in \mathcal{L}_\infty(\mu)$ from $\Gamma \subseteq \mathcal{L}_\infty(\mu)$ is a labeled tree $\langle T, f \rangle$ such that

(3.1) $T = \langle T, \leq \rangle$ is a tree growing upwards with its root r_0 and T with $(\leq)^{-1}$ is well-founded;

(3.2) $f : T \rightarrow \mathcal{L}_\infty(\mu)$;

(3.3) $f(r_0) = \varphi$;

(3.4) if $t \in T$ is a maximal element then either $f(t) \in \Gamma$ or t is one of the axioms of S;

(3.5) if $t \in T$ and $P \subseteq T$ is the set of all immediate successors of t, then

$$\frac{\{f(p) : p \in P\}}{f(t)}$$

is one of the deduction rules.

We have to stress here that, in (3.5), we do not assume that the function f is one-to-one since otherwise we have to choose a proof for each formula in the set in the premises of (R1) and (R2). Thus, for example, we can deduce $T \vdash \bigwedge \Phi$ in S from $T \vdash \varphi$ for all $\varphi \in \Phi$ without appealing to AC.

Now the proof of the following is an easy exercise:

Proposition 3.1

(1) For any $B \subseteq \mu$, $T \subseteq \mathcal{L}_\infty(\mu)$ and $\varphi \in \mathcal{L}_\infty(\mu)$, if $T \vdash \varphi$ and $B \models T$, then we have $B \models \varphi$.

(2) For transitive models M, N of ZF such that M is an inner model of N, if $M \models$ "$\langle T,f \rangle$ is a proof of φ in $\mathcal{L}_\infty(\mu)$", then $N \models$ "$\langle T,f \rangle$ is a proof of φ in $\mathcal{L}_\infty(\mu)$".

Proof

(1) By induction on cofinal subtrees of a fixed proof $\langle T,f \rangle$ of φ.

(2) Clear by definition. ☐ (Proposition 3.1)

An alternative setting to the argument by means of a deductive system is to make use of the following definition of $M \models$ "$\Gamma \vdash \varphi$" in the proof of Lemma 2.4:

$M \models$ "$\Gamma \vdash \varphi$" iff for any $B \subseteq \mu$ in some set-forcing extension $M[G]$ of M, $M[G] \models B \models \psi$ for all $\psi \in \Gamma$ always implies $M[G] \models B \models \varphi$.

Note that this is definable in M using the forcing relation definable on M. It remains to verify that this notion has the desired degree of absoluteness. Actually we can easily prove the full absoluteness, that is, if N is a transitive model containing M with the same ordinals as those of M then, for Γ, $\varphi \in M$ with $M \models \Gamma \subseteq \mathcal{L}_\infty(\mu)$ and $M \models \varphi \in \mathcal{L}_\infty(\mu)$, $\Gamma \vdash \varphi$ holds in M iff $\Gamma \vdash \varphi$ holds in N.

First suppose that $B \subseteq \mu$ is a set of ordinals in a set-generic extension $N[G]$ of N such that B witnesses the failure of $\Gamma \vdash \varphi$ in N. Let x be a real which is generic over N for the Lévy collapse of a sufficiently large ν to ω such that Γ and μ become countable in the generic extension $N[x]$. Then x is also Lévy generic over M and $M[x]$ is a submodel of $N[x]$. By Lévy Absoluteness, it follows that that there exists $B' \subseteq \mu$ in $M[x]$ which also witnesses the failure of $\Gamma \vdash \varphi$ in M.

Conversely, suppose that $\Gamma \vdash \varphi$ holds in N and let $B \subseteq \mu$ be a set of ordinals in a set-generic extension $M[G]$ of M such that B witnesses the failure of $\Gamma \vdash \varphi$ in M. Then B also belongs to an extension of M which is generic for the Lévy collapse of sufficently large ν to ω; choose a condition p in this forcing which forces the existence of such a B. Now if x is Lévy-generic over N and contains the condition p, we see that there is a counterexample to $\Gamma \vdash \varphi$ in N witnessed in $N[x]$, contrary to our assumption.

With both of the interpretations of $\vdash$ we can check that the arguments in Sect. 2 go through.

4   An Axiomatic Framework for the Set-Generic Multiverse

In this section, we consider some possible axiomatic treatments of the set-generic multiverse. Such axiomatic treatments are also discussed e.g. in [9, 19, 22]. We introduce a conservative extension MZFC of ZFC in which we can treat the multiverse of set-generic extensions of models of ZFC as a collection of countable transitive models. This system or some further extension of it (which can possibly also treat tame class forcings) may be used as a basis for direct formulation of statements concerning the multiverse.

The language $\mathcal{L}_{\mathrm{MZF}}$ of the axiom system MZFC consists of the ϵ-relation symbol 'ϵ', and a constant symbol 'v' which should represent the countable transitive "ground model".

The axiom system MZFC consists of

(4.1) all axioms of ZFC;
(4.2) "v is a countable transitive set";
(4.3) "$v \models \varphi$" for all axioms φ of ZFC;

By (4.1), MZFC proves the (unique) existence of the closure $\mathcal{M}$ of "$\{v\}$" under forcing extension and definable "inner model" of "ZF" (here 'ZF' is set in quotation marks since we can only argue in metamathematics that such "inner model" satisfies each instance of replacement). Note that $\mathcal{M} \subseteq \mathcal{H}_{\aleph_1}$. Here "inner model" is actually phrased in $\mathcal{L}_{\mathrm{ZF}}$ as "transitive almost universal subset closed under Gödel operations". If we had $v \models$ ZFC, we would have $w \models$ ZF for any inner model w of v in this sense by Theorem 13.9 in [13]. In MZFC, however, we have only $v \models \varphi$ for each axiom φ of ZFC (in the meta-mathematics). Nevertheless, for all such "inner model" w and hence for all $w \in \mathcal{M}$, we have $w \models \varphi$ for all axiom φ of ZF by the proof of Theorem 13.9 in [13] and the Forcing Theorem. Apparently, this is enough to consider $\mathcal{M}$ in this framework as the set-generic multiverse.

Similarly, we can also start from any extension of ZFC (e.g. with some additional large cardinal axiom) and make $\mathcal{M}$ closed under some more operations such as some well distinguished class of class forcing extensions.

The following theorem shows that we do not increase the consistency strength by moving from ZFC to MZFC.

Theorem 4.1 *MZFC is a conservative extension of ZFC: for any sentence ψ in $\mathcal{L}_{\mathrm{ZF}}$, we have ZFC $\vdash \psi \Leftrightarrow$ MZFC $\vdash \psi$. In particular, MZFC is equiconsistent with ZFC.*

Proof "$\Rightarrow$" is trivial.

For "$\Leftarrow$", suppose that MZFC $\vdash \psi$ for a formula ψ in $\mathcal{L}_{\mathrm{ZF}}$. Let $\mathcal{P}$ be a proof of ψ from MZFC and let T be the finite fragment of ZFC consisting of all axioms φ of ZFC such that $v \models \varphi$ appears in $\mathcal{P}$. Let $\Phi(x)$ be the formula in $\mathcal{L}_{\mathrm{ZF}}$ saying

"x is a countable transitive set and $x \models \bigwedge T$".

By the Deduction Theorem, we can recast $\mathcal{P}$ to a proof of ZFC $\vdash \forall x(\Phi(x) \to \psi)$. On the other hand we have ZFC $\vdash \exists x \Phi(x)$ (by the Reflection Principle, Downward Löwenheim-Skolem Theorem and Mostowski's Collapsing Theorem). Hence we obtain a proof of ψ from ZFC alone. $\square$ (Theorem 4.1)

It may be a little bit disappointing if each set-theoretic universe in the multiverse seen from the "meta-universe" is merely a countable set. Of course if M is an inner model of a model W of ZFC (i.e. M is a model which is a transitive class $\subseteq W$ and $M, W \models$ ZFC) there are always partial ordering $\mathbb{P}$ in M for which there is no $(M, \mathbb{P})$-generic set in W (e.g. any partial ordering collapsing a cardinal of W cannot have its generic set in W).

However, if we are content with a meta-universe which is not a model of full ZFC, we can work with the following setting where each of the "elements" of the set-generic multiverse is an inner model of a meta-universe: starting from a model V of ZFC with an inaccessible cardinal κ, we generically extend it to $W = V[G]$ by Lévy collapsing κ to ω_1. Letting $M = \mathcal{H}(\kappa)^V$, we have $M \models$ ZFC and M is an inner model of $W = \mathcal{H}(\kappa)^{V[G]} = \mathcal{H}(\omega_1)^{V[G]}$. $W \models$ ZFC—the Power Set Axiom and for any partial ordering $\mathbb{P}$ in M there is a $(M, \mathbb{P})$-generic set in W. Thus an NBG-type theory of W with a new unary predicate corresponding to M can be used as a framework of the theory for the set-generic multiverse (which is obtained by considering all the set-generic grounds of M, and then all the set generic extensions of them, etc.) as a "class" of classes in W. A setting similar to this idea was also discussed in [19].

5   Independent Buttons

The multiverse view sometimes highlights problems which would be never asked in the conventional context of forcing constructions. The existence of infinitely many independent buttons which arose in connection with the characterization of the modal logic of the set-generic multiverse (see [12]) is one such question.

A sentence φ in $\mathcal{L}_{ZF}$ is said to be a *button* (for set-genericity) if any set-generic extension $V[G]$ of the ground model V has a further set-generic extension $V[G][H]$ such that φ holds in all set-generic extensions of $V[G][H]$. Let us say that a button φ is pushed in a set-generic extension $V[G]$ if φ holds in all further set-generic extensions $V[G][H]$ of $V[G]$ (including $V[G]$ itself).

Formulas φ_n, $n \in \omega$ are *independent buttons*, if φ_n, $n \in \omega$ are unpushed buttons and for any set-generic extension $V[G]$ of the ground model V and any $X \subseteq \omega$ in $V[G]$,

(5.1) if $\{n \in \omega : V[G] \models \varphi_n$ is pushed$\} \subseteq X$ then there is a set-generic extension $V[G][H]$ such that $\{n \in \omega : V[G][H] \models \varphi_n$ is pushed$\} = X$.

In [12], it is claimed that formulas b_n, $n \in \omega$ form an infinite set of independent buttons over $V = L$ where b_n is a formula asserting: "$\omega_n{}^L$ is not a cardinal". This is used to prove that the principles of forcing expressible in the modal logic of the set-theoretic multiverse as a Kripke frame where modal operator $\Box$ is interpreted as:

(5.2) $M \models \Box\varphi \Leftrightarrow$ in all set-generic extensions $M[G]$ of M we have $M[G] \models \varphi$

coincides with the modal theory S4.2 (Main Theorem 6 in [12]).

Unfortunately, it seems that there is no guarantee that (5.1) holds in an arbitrary set-generic extension $V[G]$ for these b_n, $n \in \omega$.

In the following, we introduce an alternative set of infinitely many formulas which are actually independent buttons for any ground model of ZFC + "GCH below $\aleph_\omega$" + "$\aleph_n = \aleph_n^L$ for all $n \in \omega$" which can be used as b_n, $n \in \omega$ in [12].

We first note that, for Main Theorem 6 in [12] we actually need only the existence of an arbitrary finite number of independent buttons. In the case of $V = L$ the following formulas can be used for this: Let ψ_n be the statement that $\aleph_n^L$ is a cardinal and the L-least $\aleph_n^L$-Suslin tree T_n^L in L (i.e., the L-least normal tree of height $\aleph_n^L$ with no antichain of size $\aleph_n^L$ in L) is still $\aleph_n^L$-Suslin. If M is a set-generic (or arbitrary) extension of L in which the button $\neg\psi_n$ has not been pushed, then by forcing with T_n^L over M we push this button and do not affect any of the other unpushed buttons $\neg\psi_m$, $m \neq n$, as this forcing is $\aleph_n$-distributive and has size $\aleph_n$. Rittberg [18] also found independent buttons under $V = L$.

Now we turn to a construction of infinitely many independent buttons for which we even do not need the existence of Suslin trees. For $n \in \omega$, let φ_n be the statement:

(5.3) there is an injection from $\aleph_{n+2}^L$ to $\mathcal{P}(\aleph_n^L)$.

Note that φ_n is pushed in a set-generic extension $V[G]$ if and only if it holds in $V[G]$. Thus φ_n for each $n \in \omega$ is a button provided that φ_n does not hold in the ground model. We show that these φ_n, $n \in \omega$ are independent buttons (over any ground model where they are unpushed—e.g., when $V = L$).

Suppose that we are working in some model W of ZFC. In W, let $A = \{n \in \omega : \Box\varphi_n \text{ holds}\}$ and $B \subseteq \omega$ be arbitrary with $A \subseteq B$. It is enough to prove the following

Proposition 5.1 *We can force (over W) that φ_n holds for all $n \in B$ and $\neg\varphi_n$ for all $n \in \omega \setminus B$.*

Proof In W, let $\kappa_n = |\aleph_n^L|$ for $n \in \omega$. We use the notation of [16] on the partial orderings with partial functions and denote with $\mathrm{Fn}(\kappa, \lambda, \mu)$ the set of all partial functions from κ to λ with cardinality $< \mu$ ordered by reverse inclusion. By Δ-System Lemma, it is easy to see that $\mathrm{Fn}(\kappa, \lambda, \mu)$ has the $(\lambda^{<\mu})^+$-c.c. Let

(5.4) $\mathbb{P}_n = \begin{cases} \mathrm{Fn}(\kappa_{n+2}, 2, \kappa_n) & \text{if } n \in B \setminus A \\ \mathbb{1} & \text{otherwise.} \end{cases}$

Let $\mathbb{P} = \prod_{n \in \omega} \mathbb{P}_n$ be the full support product of $\mathbb{P}_n$, $n \in \omega$. Then we clearly have $\Vdash_\mathbb{P}$ "φ_n" for all $n \in B$. Thus to show that $\mathbb{P}$ creates a generic extension as desired, it is enough to show that $\Vdash_\mathbb{P}$ "$\neg\varphi_n$" for all $n \in \omega \setminus B$.

Suppose that

(5.5) $n \in \omega \setminus B$.

Then we have

(5.6) $\mathbb{P}_n = \mathbb{1}$.

Since φ_n does not hold in W, we have $\kappa_n < \kappa_{n+1} < \kappa_{n+2}$ and $2^{\kappa_n} = \kappa_{n+1}$ in W. By (5.6), $\mathbb{P}$ factors as $\mathbb{P} \sim \mathbb{P}(< n) \times \mathbb{P}(> n)$ where $\mathbb{P}(< n) = \prod_{k<n} \mathbb{P}_k$ and $\mathbb{P}(> n) = \prod_{k>n} \mathbb{P}_k$.

We show that both $\mathbb{P}(> n)$ and $\mathbb{P}(< n)$ over $\mathbb{P}(> n)$ do not add any injection from κ_{n+2} into $\mathcal{P}(\kappa_n)$.

$\mathbb{P}(> n)$ is κ_{n+1}-closed. Thus it does not add any new subsets of κ_n. So if it added an injection from κ_{n+2} into $\mathcal{P}(\kappa_n)$ then it would collapse the cardinal κ_{n+2}. Since $\mathbb{P}(> n)$ further factors as $\mathbb{P}(> n) \sim \mathbb{P}_{n+1} \times \prod_{k>n+1} \mathbb{P}_k$ and $\prod_{k>n+1} \mathbb{P}_k$ is κ_{n+2}-closed the only way $\mathbb{P}(> n)$ could collapse κ_{n+2} would be if $\mathbb{P}_{n+1}$ did so. But then, since $\mathbb{P}_{n+1}$ has the $(2^{<\kappa_{n+1}})^+$-c.c. with $(2^{<\kappa_{n+1}})^+ = (2^{\kappa_n})^+$, we would have $2^{\kappa_n} \geq \kappa_{n+2}$. This is a contradiction to the choice (5.5) of n. So $\mathbb{P}(> n)$ forces φ_n to fail.

In the rest of the proof, we work in $W^{\mathbb{P}(> n)}$ and show that $\mathbb{P}(< n)$ does not add any injection from κ_{n+2} into $\mathcal{P}(\kappa_n)$. Note that, by κ_{n+1}-closedness of $\mathbb{P}(> n)$, we have $\mathrm{Fn}(\kappa_{m+2}, 2, \kappa_m)^W = \mathrm{Fn}(\kappa_{m+2}, 2, \kappa_m)^{W^{\mathbb{P}(> n)}}$ for $m < n$.

We have the following two cases:

Case I. $n - 1 \in A \cup (\omega \setminus B)$. Then $\mathbb{P}(< n) \sim \mathbb{P}(< m)$ for some $m < n$ and $\mathbb{P}(< m)$ has the $(2^{\kappa_{m-1}})^+$-c.c. with $(2^{\kappa_{m-1}})^+ \leq \kappa_n$.

Case II. $n - 1 \in B \setminus A$. Then $2^{<\kappa_{n-1}} = \kappa_n$ and $\mathbb{P}(< n)$ has the κ_{n+1}-c.c.

In both cases the partial ordering $\mathbb{P}(< n)$ has κ_{n+1}-c.c. and hence the cardinals κ_{n+1} and κ_{n+2} are preserved. Since $\mathbb{P}(< n)$ has at most cardinality $2^{\kappa_{n-1}} \cdot \kappa_{n+1} = \kappa_{n+1}$, it adds at most $\kappa_{n+1}^{\kappa_n} = \kappa_{n+1}$ new subsets of κ_n and thus the size of $\mathcal{P}(\kappa_n)$ remains unchanged. This shows that $\Vdash_{\mathbb{P}} \text{``} \neg \varphi_n \text{''}$.

$\square$ (Proposition 5.1)

Acknowledgements The first author would like to thank the FWF (Austrian Science Fund) for its support through project number P 28420.

The second author is supported by Grant-in-Aid for Scientific Research (C) No. 21540150 and Grant-in-Aid for Exploratory Research No. 26610040 of the Ministry of Education, Culture, Sports, Science and Technology Japan (MEXT).

The third author is supported by Grant-in-Aid for Young Scientists (B) No. 23740076 of the Ministry of Education, Culture, Sports, Science and Technology Japan (MEXT).

References

1. T. Arrigoni, S.-D. Friedman, Foundational implications of the inner model hypothesis. Ann. Pure Appl. Logic **163**, 1360–1366 (2012)
2. T. Arrigoni, S.-D. Friedman, The hyperuniverse program. Bull. Symb. Log. **19**(1), 77–96 (2013)

3. L. Bukovský, Characterization of generic extensions of models of set theory. Fundam. Math. **83**, 35–46 (1973)
4. L. Bukovský, Generic Extensions of Models of ZFC, a lecture note of a talk at the Novi Sad Conference in Set Theory and General Topology, Novi Sad, August 18–21 (2014)
5. S.-D. Friedman, Strict genericity, in *Models, Algebras and Proofs*. Proceedings of the 1995 Latin American Logic Symposium (1999), pp. 129–139
6. S.-D. Friedman, *Fine Structure and Class Forcing*, de Gruyter Series in Logic and Its Applications, vol. 3 (Walter de Gruyter, Berlin, 2000)
7. S.-D. Friedman, Internal consistency and the inner model hypothesis. Bull. Symb. Log. **12**(4), 591–600 (2006)
8. G. Fuchs, J.D. Hamkins, J. Reitz, Set theoretic geology. Ann. Pure Appl. Logic **166**(4), 464–501 (2015)
9. V. Gitman, J. Hamkins, A natural model of the multiverse axioms. Nortre Dame J. Formal Logic **51**(4), 475–484 (2010)
10. S. Grigorieff, Intermediate submodels and generic extensions in set theory. Ann. Math. Second Series **101**(3), 447–490 (1975)
11. J.D. Hamkins, The set-theoretical multiverse. Rev. Symb. Log. **5**, 416–449 (2012)
12. J.D. Hamkins, B. Löwe, The modal logic of forcing. Trans. Am. Math. Soc. **360**(4), 1793–1817 (2008)
13. T. Jech, *Set Theory*, The Third Millennium Edition (Springer, Berlin, 2001/2006)
14. C. Karp, *Languages with Expressions of Infinite Length* (North-Holland, Amsterdam, 1964)
15. H.J. Keisler, *Model Theory for Infinitary Logic* (North-Holland, Amsterdam, 1974)
16. K. Kunen, *Set Theory, An Introduction to Independence Proofs* (North-Holland, Amsterdam, 1980)
17. R. Laver, Certain very large cardinals are not created in small forcing extensions. Ann. Pure Appl. Logic **149**, 1–6 (2007)
18. C.J. Rittberg, On the modal logic of forcing. Diploma Thesis (2010)
19. J.R. Steel, Gödel's program, in *Interpreting Gödel: Critical Essays*, ed. by J. Kennedy (Cambridge University Press, Cambridge, 2014), pp. 153–179
20. G. Takeuti, *Proof Theory*, 2nd edn. (North-Holland, Amsterdam, 1987)
21. T. Usuba, The downward directed grounds hypothesis and very large cardinals, preprint
22. J. Väänään, Multiverse set theory and absolutely undecidable propositions, in *Interpreting Gödel: Critical Essays*, ed. by J. Kennedy (Cambridge University Press, Cambridge, 2014), pp. 180–208
23. W.H. Woodin, Recent developments on Cantor's continuum hypothesis, in *Proceedings of the Continuum in Philosophy and Mathematics, 2004*, Carlsberg Academy, Copenhagen, November (2004)
24. W.H. Woodin, The realm of the infinite, in *Infinity: New Research Frontiers*, ed. by M. Heller, W.H. Woodin (Cambridge University Press, Cambridge, 2011)

On Strong Forms of Reflection in Set Theory

Sy-David Friedman and Radek Honzik

Abstract In this paper we review the most common forms of reflection and introduce a new form which we call *sharp-generated reflection*. We argue that sharp-generated reflection is the strongest form of reflection which can be regarded as a natural generalization of the Lévy reflection theorem. As an application we formulate the principle *sharp-maximality* with the corresponding hypothesis IMH$^\#$. IMH$^\#$ is an analogue of the IMH (Inner Model Hypothesis, introduced in Friedman (Bull Symb Log 12(4):591–600, 2006)) which is compatible with the existence of large cardinals.

1 Introduction

Vertical reflection for the universe V can be intuitively formulated as the following principle, denoted (Refl):

Any property which holds in V already holds in some initial segment of V.

(Refl)

(Refl) says that V cannot be described as the unique initial segment of the universe satisfying a given property. The strength of reflection depends on what we consider by *property*; by varying the notion of property we obtain a hierarchy of reflection principles. We say that a given V is *vertically maximal* if it satisfies a formalization of (Refl) which can be viewed, arguably, as being the strongest possible.[1]

Originally published in S.-D. Friedman, R. Honzik, On strong forms of reflection in set theory. Math. Log. Quart. **62**(1–2), 52–58 (2016).

[1]Indeed, we propose the notion of sharp-generation developed in Sect. 2.2 as a candidate for an ultimate form of (Refl).

S.-D. Friedman • R. Honzik (✉)
Kurt Gödel Research Center for Mathematical Logic, Universität Wien, Währinger Strasse 25, 1090 Vienna, Austria
e-mail: sdf@logic.univie.ac.at; radek.honzik@univie.ac.at

© Springer International Publishing AG 2018

C. Antos et al. (eds.), *The Hyperuniverse Project and Maximality*,
https://doi.org/10.1007/978-3-319-62935-3_6

The weakest form of reflection, with first-order notion of property, is Lévy's theorem which is provable in ZF.

Theorem 1.1 (Lévy) *Let* $\varphi(x_1, \ldots, x_n)$ *be a first-order formula with free variables shown. Then the following is a theorem of* ZF:

$$\forall \alpha \ \forall x_1, \ldots, x_n \in V_\alpha \ \exists \beta \geq \alpha \ \big(\varphi(x_1, \ldots, x_n) \leftrightarrow (V_\beta, \in) \models \varphi(x_1, \ldots, x_n)\big). \quad (1)$$

Since the language of ZF is first-order, there is no direct way of generalizing Lévy's theorem to higher-order formulas applied to V. Lévy resolved this problem by studying structures of the form $(V_\kappa, \in, R)$, with R ranging over subsets of V_κ. We say that $\varphi(R)$ true in $(V_\kappa, \in, R)$ reflects if there is some $\alpha < \kappa$ such that $(V_\alpha, \in, R \cap V_\alpha)$ satisfies $\varphi(R \cap V_\alpha)$. For an inaccessible κ, V_κ is thus identified with an approximation of the universe V and higher-order properties attributable to V_κ are expressed as first-order properties in V. It is known that by postulating a range of reflection principles for $(V_\kappa, \in, R)$, one can obtain large cardinals compatible with L (such as weakly compact cardinals).[2]

Reflection principles discussed in the previous paragraph allow φ to be higher-order, but the parameter R itself is always just second-order. Our motivation in this paper is to look for strengthenings of reflection with potential to yield vertical maximality, and which in particular should allow parameters of order higher than 2. For instance, for a third-order parameter $\mathscr{R} \subseteq \mathscr{P}(V_\kappa)$ one is tempted to formulate the following natural-looking principle:

> If $\varphi(\mathscr{R})$ is true in $(V_\kappa, \in, \mathscr{R})$, then for some $\alpha < \kappa$, $(V_\alpha, \in, \bar{\mathscr{R}})$ satisfies $\varphi(\bar{\mathscr{R}})$, where $\bar{\mathscr{R}} = \{R \cap V_\alpha \mid R \in \mathscr{R}\}$. $\quad$ (*)$_3$

However, an easy example shows that (*)$_3$ is inconsistent.[3] In order to retain some sort of reflection with higher-order parameters, we need to tread more carefully. First in Sect. 2.1, we reformulate (*)$_3$ (and its generalizations) using elementary embeddings internal to V (see Definition 2.1). Seeing that this reformulation has certain drawbacks (in particular it is not compatible with L), we will develop the idea of elementary embeddings in a different way, making the resulting notion compatible with L. This construction—based on indiscernibles and sharp-generation—is described in Sect. 2.2. An application of a sharp-generated reflection is given in Sect. 3.

---

[2] Instead of working with V_κ, one can work directly with V in theories with classes, such as GB. Let R range over classes. We say that $\varphi(R)$ true in V reflects if for some $\alpha < \kappa$, $(V_\alpha, V_{\alpha+1})$ satisfies $\varphi(R \cap V_\alpha)$.

[3] Consider the following example. For any infinite ordinal κ, let $\mathscr{R}$ be the collection of all $\alpha < \kappa$ (viewed as subsets of κ), and consider $\varphi(\mathscr{R})$ which says that every element of $\mathscr{R}$ is bounded in κ (φ is first-order with a third-order parameter $\mathscr{R}$). Clearly, $\varphi(\mathscr{R})$ is true in V_κ. However, $\varphi(\bar{\mathscr{R}})$ is false in V_α for every $\alpha < \kappa$. See [6] for more discussion of reflection with higher-order parameters.

2 Reflection with Elementary Embeddings

2.1 Embeddings Internal to V

To make the following discussion more standard, we will work with structures of the form $H(\kappa)^{+n}$, $0 < n < \omega$. Let $\mathscr{R}$ range over subsets of $H(\kappa^{+n})$; we write

$$(H(\kappa^{+n}), \in, \mathscr{R}) \models \varphi(\mathscr{R}) \tag{2}$$

instead of $(H(\kappa), \in, \mathscr{R}) \models \varphi(\mathscr{R})$ to express that $\varphi(\mathscr{R})$ holds in $H(\kappa)$ with appropriately interpreted higher-order quantifiers.[4] The notation in (2) has the advantage that it emphasizes that the properties of order $n + 1$ over $H(\kappa)$ actually reduce to first-order properties over $H(\kappa^{+n})$, with $\mathscr{R}$ being second-order over $H(\kappa^{+n})$.

The known concept of a *subcompact* cardinal can be used to make sense of reflection for higher-order parameters:

Definition 2.1 Let κ be an uncountable regular cardinal. We say that κ *satisfies reflection with parameters of order* $n + 2$, $0 < n < \omega$, if for every $\mathscr{R} \subseteq H(\kappa^{+n})$ there are a regular uncountable cardinal $\bar{\kappa} < \kappa$, $\bar{\mathscr{R}} \subseteq H(\bar{\kappa}^{+n})$, and an embedding $\pi : H(\bar{\kappa}^{+n}) \to H(\kappa^{+n})$ with critical point $\bar{\kappa}$, $\pi(\bar{\kappa}) = \kappa$, such that

$$\pi : (H(\bar{\kappa}^{+n}), \in, \bar{\mathscr{R}}) \to (H(\kappa^{+n}), \in, \mathscr{R}) \tag{3}$$

is elementary.

Note that demanding $(H(\bar{\kappa}^{+n}), \in, \bar{\mathscr{R}}) \prec (H(\kappa^{+n}), \in, \mathscr{R})$ is contradictory[5]; thus the requirement that π is not the identity is essential.

Remark 2.2 For n, $0 < n < \omega$, κ is κ^{+n}-subcompact iff κ satisfies reflection for parameters of order $n + 2$ according to Definition 2.1. Subcompact cardinals were defined by Jensen,[6] and apparently for different reasons than the study of reflection (Jensen isolated the concept of subcompact cardinals for his study of the failure of the square). α-subcompact cardinals can be defined for any cardinal $\alpha > \kappa$, not just the κ^{+n}'s for $n < \omega$, and are therefore suitable for expressing reflection with parameters of transfinite order. For more details about subcompact cardinals, see [2].

Definition 2.1 forces no "canonicity" on π; any embedding which satisfies the requirements will do. One might wonder whether more stringent requirements on π, such as demanding constructibility in some sense, might give the definition more structure. However, this cannot be done if by canonicity we mean constructibility

---

[4]For simplicity, we restrict our attention in this section to higher-order properties of finite order.

[5]Set $n = 1$ and choose $\mathscr{R}$ as in the example in Footnote 3. By elementarity, $\bar{\mathscr{R}}$ is equal to $H(\bar{\kappa}^+) \cap \mathscr{R}$, which leads to contradiction as in Footnote 3.

[6]Jensen defined κ to be subcompact if it is κ^+-subcompact according to our definition.

in L-like models: by Theorem 2.3, reflection for parameters of order three implies failure of square and for higher orders we get supercompact cardinals (of specific degrees):

Theorem 2.3 (GCH) *The following hold:*

(i) *For all n, $0 < n < \omega$: κ satisfies reflection with parameters of order $n + 4$ iff κ is κ^{n+2}-subcompact iff κ is κ^{+n}-supercompact.*

(ii) *κ satisfies reflection for parameters of order 4 iff κ is κ^{++}-subcompact iff κ is measurable.*

(iii) *If κ satisfies reflection for parameters of order 3 (which is the same as being κ^{+}-subcompact), then $\square_\kappa$ fails.*

Proof For proofs, see for instance [2].   □

There are other versions of strong forms of reflection implying transcendence over L; see for instance [7].

Definition 2.1 seems very natural, but—in our opinion—-the postulation of non-canonical elementary embeddings as elements of the universe V turned out to make the resulting principle too strong. Theorem 2.3 contradicts our original intuition regarding (Refl) and its formalization: while we would like to extend the usual form of reflection to higher-order parameters, we wish to retain compatibility with L (see Remark 2.8). A more suitable form of reflection compatible with L is described in next section.

2.2   Sharp-Generated Reflection

Let us start with V which we view as a transitive set which approximates the real universe. This viewpoint allows us to consider end-extensions $V \subseteq V^*$ of a larger ordinal length. Constructions of this type can be carried out in certain axiomatic theories more complicated than ZF or GB (for example Ackermann's, or theories developed by Reinhardt; see [5], Section 23, for more details). However we think that by treating V as a transitive set model (often countable), we obtain a much stronger (indeed the strongest possible) form of reflection.[7]

Let us extrapolate from the usual reflection and see where it takes us. It is natural to strengthen the reflection of individual first-order properties from V to some V_α to the simultaneous reflection of all first-order properties of V to some V_α, even with parameters from V_α. Thus V_α is an elementary submodel of V. Repeating this process suggests that in fact there should be an increasing, continuous sequence of ordinals $(\kappa_i \mid i < \infty)$ such that the models $(V_{\kappa_i} \mid i < \infty)$ form a continuous chain

---

[7]Recall that standard forms of reflection are also formulated with set approximations of the form $(V_\kappa, \in, R)$; however, we do not require V to be a rank-initial segment of the universe which makes it possible to consider countable V's.

$V_{\kappa_0} \prec V_{\kappa_1} \prec \cdots$ of elementary submodels of V whose union is all of V (where ∞ denotes the ordinal height of the universe V).

But the fact that for a closed unbounded class of κ's in V, V_κ can be "lengthened" to an elementary extension (namely V) of which it is a rank initial segment suggests via reflection that V itself should also have such a lengthening V^*. But this is clearly not the end of the story, because we can also infer that there should in fact be a continuous increasing sequence of such lengthenings $V = V_{\kappa_\infty} \prec V^*_{\kappa_\infty+1} \prec V^*_{\kappa_\infty+2} \prec \cdots$ of length the ordinals. For ease of notation, let us drop the $*$'s and write W_{κ_i} instead of $V^*_{\kappa_i}$ for $\infty < i$ and instead of V_{κ_i} for $i \le \infty$. Thus V equals W_∞.

But which tower $V = W_{\kappa_\infty} \prec W_{\kappa_\infty+1} \prec W_{\kappa_\infty+2} \prec \cdots$ of lengthenings of V should we consider? Can we make the choice of this tower "canonical"?

Consider the entire sequence $W_{\kappa_0} \prec W_{\kappa_1} \prec \cdots \prec V = W_{\kappa_\infty} \prec W_{\kappa_\infty+1} \prec W_{\kappa_\infty+2} \prec \cdots$. The intuition is that all of these models resemble each other in the sense that they share the same first-order properties. Indeed by virtue of the fact that they form an elementary chain, these models all satisfy the same first-order sentences. But again in the spirit of "resemblance", it should be the case that any two pairs $(W_{\kappa_{i_1}}, W_{\kappa_{i_0}})$, $(W_{\kappa_{j_1}}, W_{\kappa_{j_0}})$ (with $i_0 < i_1$ and $j_0 < j_1$) satisfy the same first-order sentences, even allowing parameters which belong to both $W_{\kappa_{i_0}}$ and $W_{\kappa_{j_0}}$. Generalising this to triples, quadruples and n-tuples in general we arrive at the following situation:

Our approximation V to the universe should occur in a continuous elementary chain $W_{\kappa_0} \prec W_{\kappa_1} \prec \cdots \prec V = W_{\kappa_\infty} \prec W_{\kappa_\infty+1} \prec W_{\kappa_\infty+2} \prec \cdots$ of length the ordinals, where the models W_{κ_i} form a *strongly-indiscernible chain* in the sense that for any n and any two increasing n-tuples $\vec{i} = i_0 < i_1 < \cdots < i_{n-1}$, $\vec{j} = j_0 < j_1 < \cdots < j_{n-1}$, the structures $W_{\vec{i}} = (W_{\kappa_{i_{n-1}}}, W_{\kappa_{i_{n-2}}}, \cdots, W_{\kappa_{i_0}})$ and $W_{\vec{j}}$ (defined analogously) satisfy the same first-order sentences, allowing parameters from $W_{\kappa_{i_0}} \cap W_{\kappa_{j_0}}$.

$$(*)$$

But this is again not the whole story, as we would want to impose higher-order indiscernibility on our chain of models. For example, consider the pair of models $W_{\kappa_0} = V_{\kappa_0}, W_{\kappa_1} = V_{\kappa_1}$. Surely we would want that these models satisfy the same second-order sentences; equivalently, we would want $H(\kappa_0^+)^V$ and $H(\kappa_1^+)^V$ to satisfy the same first-order sentences. But as with the pair $H(\kappa_0)^V$, $H(\kappa_1)^V$ we would want $H(\kappa_0^+)^V$, $H(\kappa_1^+)^V$ to satisfy the same first-order sentences *with parameters*. How can we formulate this? For example, consider κ_0, a parameter in $H(\kappa_0^+)^V$ that is second-order with respect to $H(\kappa_0)^V$; we cannot simply require $H(\kappa_0^+)^V \vDash \varphi(\kappa_0)$ iff $H(\kappa_1^+)^V \vDash \varphi(\kappa_0)$, as κ_0 is the largest cardinal in $H(\kappa_0^+)^V$ but not in $H(\kappa_1^+)^V$. Instead we need to replace the occurrence of κ_0 on the left side with a "corresponding" parameter on the right side, namely κ_1, resulting in the natural requirement $H(\kappa_0^+)^V \vDash \varphi(\kappa_0)$ iff $H(\kappa_1^+)^V \vDash \varphi(\kappa_1)$. More generally, we should be able to replace each parameter in $H(\kappa_0^+)^V$ by a "corresponding" element of $H(\kappa_1^+)^V$ and conversely, it should be the case that, to the maximum extent possible, all elements of $H(\kappa_1^+)^V$ are the result of such a replacement. This also should be possible for $H(\kappa_0^{++})^V, H(\kappa_0^{+++})^V, \ldots$ and with the pair κ_0, κ_1 replaced by any pair κ_i, κ_j with $i < j$.

It is natural to solve this parameter problem using embeddings, as in the last subsection. But the difference here is that there is no assumption that these embeddings are internal to V; they need only exist in the "real universe", outside of V. In this way we will arrive at a principle compatible with $V = L$ in which the choice of embeddings is indeed "canonical".

Thus we are led to the following.

Definition 2.4 Let W be a transitive set-size model of ZFC of ordinal height ∞. We say that W is *indiscernibly-generated* iff W satisfies the following:

(i) There is a continuous sequence $\kappa_0 < \kappa_1 < \ldots$ of the length ∞ such that $\kappa_\infty = \infty$ and there are commuting elementary embeddings $\pi_{ij} : W \to W$ where π_{ij} has critical point κ_i and sends κ_i to κ_j.

(ii) For any $i \leq j$, any element of W is first-order definable in W from elements of the range of π_{ij} together with κ_k's for k in the interval $[i, j)$.

The last clause in the above definition formulates the idea that to the maximum extent possible, elements of W are in the range of the embedding π_{ij} for each $i \leq j$; notice that the interval $[\kappa_i, \kappa_j)$ is disjoint from this range, but by allowing the κ_k's in this interval as parameters, we can first-order definably recover everything.

Indiscernible-generation as formulated in the above definition does indeed give us our advertised higher-order indiscernibility: For example, in the notation of the definition, if $\vec{i} = i_0 < i_1 < \ldots < i_{n-1}$ and $\vec{j} = j_0 < j_1 < \ldots < j_{n-1}$ with $i_0 \leq j_0$, and $x_k \in H(\kappa_{i_0}^+)^W$ for $k < n$ then the structure $W_{\vec{i}}^+ = (H(\kappa_{i_{n-1}}^+)^W, H(\kappa_{i_{n-2}}^+)^W, \cdots, H(\kappa_{i_0}^+)^W)$ satisfies a sentence with parameters $(\pi_{i_0, i_{n-1}}(x_{n-1}), \ldots, \pi_{i_0, i_0}(x_0))$ iff $W_{\vec{j}}^+$ satisfies the same sentence with corresponding parameters $(\pi_{i_0, j_{n-1}}(x_{n-1}), \ldots, \pi_{i_0, j_0}(x_0))$. There is a similar statement with W^+ replaced by higher-order structures $W^{+\alpha}$ for arbitrary α.

Indiscernible-generation has a clearer formulation in terms of #-*generation*, which we explain next.

Definition 2.5 A structure $N = (N, U)$ is called a *sharp with critical point κ*, or just a #, if the following hold:

(i) N is a model of ZFC$^-$ (ZFC minus powerset, with replacement replaced by the collection principle) in which κ is the largest cardinal and κ is strongly inaccessible.

(ii) (N, U) is amenable (i.e. $x \cap U \in N$ for any $x \in N$).

(iii) U is a normal measure on κ in (N, U).

(iv) N is iterable, i.e., all of the successive iterated ultrapowers starting with (N, U) are well-founded, yielding iterates (N_i, U_i) and Σ_1 elementary iteration maps $\pi_{ij} : N_i \to N_j$ where $(N, U) = (N_0, U_0)$.

We will use the convention that κ_i denotes the the largest cardinal of the i-th iterate N_i.

If N is a # and λ is a limit ordinal then $LP(N_\lambda)$ denotes the union of the $(V_{\kappa_i})^{N_i}$'s for $i < \lambda$. (*LP* stands for "lower part".) $LP(N_\infty)$ is a model of ZFC.

Definition 2.6 We say that a transitive model V of ZFC is *#-generated* iff for some sharp $N = (N, U)$ with iteration $N = N_0 \to N_1 \to \cdots$, V equals $LP(N_\infty)$ where ∞ denotes the ordinal height of V.

Fact 2.7 *The following are equivalent for transitive set-size models V of* ZFC:

  (i) V is indiscernibly-generated.
  (ii) V is #-generated.

Proof The last clause in the definition of indiscernible-generation ensures that the embeddings π_{ij} in that definition in fact arise from iterated ultrapowers of the embedding π_{01}, itself an ultrapower by the measure U_0 on κ_0 given by $X \in U_0$ iff $\pi_{01}(X)$ contains κ_0 as an element. Conversely, if (N, U) generates V, then the chain of embeddings given by iteration of (N, U) witnesses that V is indiscernibly-generated. □

In our opinion, #-generation fulfils our intuition for being vertical maximal, with powerful consequences for reflection. L is #-generated iff $0^\#$ exists, so this principle is compatible with $V = L$. If V is #-generated via (N, U) then there are embeddings witnessing indiscernible-generation for V which are canonically-definable through iteration of (N, U). Although the choice of # that generates V is not in general unique, it can be taken as a fixed parameter in the canonical definition of these embeddings. Moreover, #-generation evidently provides the maximum amount of vertical reflection: If V is generated by (N, U) as $LP(N_\infty)$ where ∞ is the ordinal height of V, and x is any parameter in a further iterate $V^* = N_{\infty^*}$ of (N, U), then any first-order property $\varphi(V, x)$ that holds in V^* reflects to $\varphi(V_{\kappa_i}, \bar{x})$ in N_j for all sufficiently large $i < j < \infty$, where $\pi_{j,\infty^*}(\bar{x}) = x$. This implies any known form of vertical reflection and summarizes the amount of reflection one has in L under the assumption that $0^\#$ exists, the maximum amount of reflection in L.

Thus #-generation tells us what lengthenings of V to look at, namely the initial segments of V^* where V^* is obtained by further iteration of a # that generates V. And it fully realises the idea that V should look exactly like closed unboundedly many of its rank initial segments as well as its "canonical" lengthenings of arbitrary ordinal height.

Therefore we believe that #-generated models are the strongest formalization of the principle of reflection (Refl)—we call this form of reflection *sharp-generated reflection*, and we shall call these models *vertically maximal*.

Remark 2.8 Notice that a sharp-generated model can satisfy $V = L$, and hence our reflection principle is compatible with L. The reason is that the non-trivial embeddings obtained from the sharp-iteration are external to the model in question. This contrasts with the use of nontrivial embeddings in Sect. 2.1. Compatibility with L agrees with our intuition that a natural formulation of vertical reflection (Refl) should be determined by the *height* of the universe, and not its *width* (and L has the same height as V).

3   An Application

We now apply sharp-generated reflection to formulate an analogue of the IMH principle in [3].

3.1   Vertically Maximal Models and IMH

The Hyperuniverse is the collection of all countable transitive models of ZFC. We view members of the Hyperuniverse as possible pictures of V which mirror all possible first-order properties of V. The Hyperuniverse Programme, which originated in [3], is concerned with the formulation of natural criteria for the selection of preferred members of the Hyperuniverse. First-order sentences holding in the preferred universes can be taken to be true in the "real V"; in other words, preferred universes may lead to adoption of new axioms. Models satisfying IMH, and IMH$^{\#}$ introduced below, are examples of such preferred universes.

Definition 3.1 We say that a #-generated model M is #-maximal if and only if the following hold. Whenever M is a definable inner model of M' and M' is #-generated, then every sentence φ, i.e. without parameters, which holds in a definable inner model of M' already holds in some definable inner model of M.

We say that a #-generated model M satisfies IMH$^{\#}$ if it is #-maximal.[8]

Note that IMH$^{\#}$ differs from IMH by demanding that both M and M', the outer model, are of a specific kind, i.e. should be #-generated (while the outer models considered in IMH are arbitrary). The motivation behind this requirement is that not all outer models count as "maximal"; if our main motivation is formulated in terms of maximality, consideration of non-maximal models as the outer models seems counterintuitive. Indeed, inclusion of such non-maximal models leads to incompatibility of maximal universes satisfying IMH with inaccessible cardinals (see [3]).

The following theorem is a sharp-generated analogue of the argument in [4].

Theorem 3.2 *Assume there is a Woodin cardinal with an inaccessible above. Then there is a model satisfying IMH$^{\#}$.*

Proof For each real R let $M^{\#}(R)$ be $L_{\alpha}[R]$ where α is least so that $L_{\alpha}[R]$ is #-generated. Note that $R^{\#}$ exists for each $R \subseteq \omega$ by our large cardinal assump-

---

[8]We thus give two names two a single concept; denotation IMH$^{\#}$ is used to emphasize the family resemblance to the earlier principle IMH.

tion. The Woodin cardinal with an inaccessible above implies enough projective determinacy to enable us to use Martin's theorem, see [5] Proposition 28.4, to find $R \subseteq \omega$ such that the theory of $(M^\#(S), \in)$ for $R \leq_T S$ stabilizes. By this we mean that for $R \leq_T S$, where $\leq_T$ denotes the Turing reducibility relation, the theories of $(M^\#(R), \in)$ and $(M^\#(S), \in)$ are the same.[9]

We claim that $M^\#(R)$ satisfies IMH$^\#$: Indeed, let M be a #-generated outer model of $M^\#(R)$ with a definable inner model satisfying some sentence φ. Let α be the ordinal height of $M^\#(R)$ (= the ordinal height of M). By Theorem 9.1 in [1], M has a #-generated outer model W of the form $L_\alpha[S]$ for some real S with $R \leq_T S$. Of course α is least so that $L_\alpha[S]$ is #-generated as it is least so that $L_\alpha[R]$ is #-generated.So W equals $M^\#(S)$. By the choice of R, $M^\#(R)$ also has a definable inner model satisfying φ. So $M^\#(R)$ is #-maximal.[10] $\square$

3.2 IMH$^\#$ is Compatible with Large Cardinals

Finally, we show that—unlike IMH—IMH$^\#$ is compatible with large cardinals.

Theorem 3.3 *Assume there is a Woodin cardinal with an inaccessible above. Then for some real R, any #-generated transitive model M containing R also models IMH$^\#$.*

Proof Let R be as in the proof of Theorem 3.2. Thus $M^\#(R) = L_\alpha[R]$ is a #-generated model of IMH$^\#$. Now suppose that $M^* = L_{\alpha^*}[R]$ is obtained by iterating $L_\alpha[R]$ past α; we claim that M^* is also a model of IMH$^\#$: Indeed, suppose that W is a #-generated outer model of M^* which has a definable inner model satisfying some sentence φ. Again by Jensen's Theorem 9.1 in [1], we can choose W to be of the form $L_{\alpha^*}[S]$ for some real $S \geq_T R$. But then $L_{\alpha^*}[S]$ is an iterate of $M^\#(S)$ (via the iteration given by $S^\#$) and therefore $M^\#(S)$ also has a definable inner model of φ. By the choice of R, $M^\#(R)$, and therefore by iteration also $L_{\alpha^*}[R]$, has a definable inner model of φ. This verifies the IMH$^\#$ for M^*.

Now any #-generated transitive model M containing R is an outer model of such a model of the form $L_{\alpha^*}[R]$ as above and therefore is also a model of IMH$^\#$. $\square$

---

[9]In more detail, given a sentence σ in the language with $\{\in\}$ consider the set of Turing degrees $X_\sigma = \{S \mid (M^\#(S), \in) \models \sigma\}$. X_σ has a projective definition (Δ_2^1). By Martin's theorem, X_σ or $X_{\neg\sigma}$ contains a cone of degrees. Denote Y_{σ^*} the unique set of the two X_σ and $X_{\neg\sigma}$ which contains the cone. Then $\bigcap_\sigma Y_{\sigma^*}$ contains a cone. Take R to be the base of this cone.

[10]Woodin noticed that this theorem and also Theorem 3.3 can be proved without recourse to Jensen's coding theorem: let R be a real such that every nonempty lightface Σ_3^1 set contains a member recursive in R. Then any M which is #-generated and contains R satisfies IMH$^\#$. However, Jensen's coding theorem does seem necessary for a modification of IMH$^\#$ which is formulated for ω_1-preserving #-generated extensions (this modification is not discussed in this paper).

Corollary 3.4 *Assume the existence of a Woodin cardinal with an inaccessible above and suppose that φ is a sentence that holds in some V_κ with κ measurable. Then there is a transitive model which satisfies both the $IMH^\#$ and the sentence φ.*

Proof Let R be as in Theorem 3.3 and let U be a normal measure on κ. The structure $N = (H(\kappa^+), U)$ is a #; iterate N through a large enough ordinal ∞ so that $M = LP(N_\infty)$, the lower part of the model generated by N, has ordinal height ∞. Then M is #-generated and contains the real R. It follows that M is a model of the $IMH^\#$. Moreover, as M is the union of an elementary chain $V_\kappa = V_\kappa^N \prec V_{\kappa_1}^{N_1} \prec \cdots$ where φ is true in V_κ, it follows that φ is also true in M. $\qquad\square$

Note that in Corollary 3.4, if we take φ to be any large cardinal property which holds in some V_κ with κ measurable, then we obtain models of the $IMH^\#$ which also satisfy this large cardinal property. This implies the compatibility of the $IMH^\#$ with arbitrarily strong large cardinal properties.

Acknowledgements The authors acknowledge the generous support of JTF grant Laboratory of the Infinite ID35216.

References

1. A. Beller, R. Jensen, P. Welch, *Coding the Universe*. London Math. Society Lecture Note Series, vol. 47 (Cambridge University Press, Cambridge, 1982)
2. A.D. Brooke-Taylor, S.-D. Friedman, Subcompact cardinals, squares, and stationary reflection. Isr. J. Math. **197**(1), 453–473 (2013)
3. S.-D. Friedman, Internal consistency and the Inner Model Hypothesis. Bull. Symb. Log. **12**(4), 591–600 (2006)
4. S.-D. Friedman, P. Welch, W. Hugh Woodin, On the consistency strength of the Inner Model Hypothesis. J. Symb. Log. **73**(2), 391–400 (2008)
5. A. Kanamori, *The Higher Infinite* (Springer, Berlin, 2003)
6. P. Koellner, On reflection principles. Ann. Pure Appl. Log. **157**(2–3), 206–219 (2009)
7. P. Welch, Global reflection principles, in *Logic, Methodology and Philosophy of Science - Proceedings of the 15th International Congress, 2015*, ed. by H. Leitgeb, I. Niiniluoto, P. Seppälä, E. Sober (College Publications, 2017), pp. 18–36

Definability of Satisfaction in Outer Models

Sy-David Friedman and Radek Honzik

Abstract Let M be a transitive model of ZFC. We say that a transitive model of ZFC, N, is an outer model of M if $M \subseteq N$ and ORD $\cap M =$ ORD $\cap N$. The outer model theory of M is the collection of all formulas with parameters from M which hold in all outer models of M (which exist in a universe in which M is countable; this is independent of the choice of such a universe). Satisfaction defined with respect to outer models can be seen as a useful strengthening of first-order logic. Starting from an inaccessible cardinal κ, we show that it is consistent to have a transitive model M of ZFC of size κ in which the outer model theory is lightface definable, and moreover M satisfies $V = HOD$. The proof combines the infinitary logic $L_{\infty,\omega}$, Barwise's results on admissible sets, and a new forcing iteration of length strictly less than κ^+ which manipulates the continuum function on certain regular cardinals below κ. In the Appendix, we review some unpublished results of Mack Stanley which are directly related to our topic.

1 Introduction

Let V be the universe of sets and let $M \in V$ be a transitive model of ZFC. We say that $N \supseteq M$, $N \in V$, is an *outer model* of M if it is a transitive model of ZFC and the ordinals in N are the same as the ordinals in M.[1] Examples of outer models range from set and class forcing extensions to outer models obtained by means of large cardinal concepts (such as $0^\sharp$). In this paper, we study the outer models from the

---

Originally published in S. Friedman, R. Honzik, Definability of satisfaction in outer models. J. Symb. Log. **81**(03), 1047–1068 (2016).

[1] We may also consider a stronger form of an outer model: with the notation as above, we say that N is a strong outer model of M if N satisfies ZFC with M as an additional predicate; alternatively, we can demand that M is a definable class in N. These stronger notions are not the main focus of this paper; we briefly comment on them in Sect. 4.4.

S.-D. Friedman (✉) • R. Honzik
Kurt Gödel Research Center for Mathematical Logic, Universität Wien, Währinger Strasse 25, 1090 Vienna, Austria
e-mail: sdf@logic.univie.ac.at; radek.honzik@ff.cuni.cz

© Springer International Publishing AG 2018
C. Antos et al. (eds.), *The Hyperuniverse Project and Maximality*,
https://doi.org/10.1007/978-3-319-62935-3_7

logical point of view and ask whether it is possible to define in M the satisfaction relation with respect to all outer models of M, thus strengthening the notion of first-order logic relative to M.

(Q1) Let V be the universe of sets. Suppose $M \in V$ is a transitive model of ZFC: is it possible to define in M the collection of all formulas with parameters in M which hold in some outer model of M in V? Can we also require that M is "nice" in the sense that it satisfies $V = HOD$?[2]

A related question is:

(Q2) Does the answer to (Q1) depend on the ambient universe V?

The existence of a model which gives a positive answer to (Q1) may seem improbable because the quantification over outer models is essentially higher-order over M (note that unlike in the case of forcing, there is no way to quantify over outer models by quantifying over elements in M). However, an analogy with first-order logic suggests that the definability level is more tractable than it at first appears:

Theorem 1.1 (First-Order Completeness) *Let V be the universe of sets. Suppose $M \in V$ is a transitive model of ZFC. Let φ be a first-order sentence with parameters in M. Then the following are equivalent:*

(i) *$ZFC + \varphi + \mathrm{AtDiag}(M)$ is consistent.*
(ii) *$M \models$ "$ZFC + \varphi + \mathrm{AtDiag}(M)$ is consistent".*
(iii) *There is $N \in V$ such that N contains M as a substructure, and $N \models ZFC + \varphi$,*

where $\mathrm{AtDiag}(M)$ is the collection of all atomic sentences and their negations with parameters in M. In particular, the set of formulas with parameters in M satisfied in a model extending M in the inclusion relation is definable in M.

By Theorem 1.1, we can refer to satisfaction in models containing M as a substructure by means of a syntactical property of having (or not having) in M a proof of a contradiction from a certain theory.

Is there some extension of first-order logic which provides an analogue of Theorem 1.1 for outer models? In principle there may be many, but one naturally looks for the weakest one because it may retain some of the desirable properties of first-order logic. It turns out that the infinitary logic $L_{\infty,\omega}$, which allows infinite conjunctions and disjunctions of any ordinal length, but only finitely many free variables in a formula, is the right framework.[3] Let us denote by $\mathrm{Hyp}(M)$ the least *admissible* set containing M as an element, where a transitive set N as an admissible set if it satisfies the axioms of KP, Kripke-Platek set theory. $\mathrm{Hyp}(M)$ is of the form $L_\alpha(M)$, where α is the least β such that $L_\beta(M)$ is a model of KP. Barwise developed

---

[2]If we drop the condition on $V = HOD$, the problem becomes easier; see Section "Forcing Omniscience" in Appendix. See also Remark 1.3 for more details.

[3]We identify the formulas in $L_{\infty,\omega}$ with sets under some reasonable coding; for instance if φ contains parameters from $H(\kappa)$ (the collection of sets whose transitive closure has size $< \kappa$) and has length less than κ, then we think of φ as an element of $H(\kappa)$. This convention makes it possible to refer to fragments of $L_{\infty,\omega}$; e.g. $L_{\infty,\omega} \cap M$ is the collection of all infinitary formulas which are elements of M.

the notion of proof (and therefore of syntactical consistency) for the infinitary logic $L_{\infty,\omega}$. An application of Barwise's Completeness theorem (see [1], Theorem 5.5) gives the following:

Theorem 1.2 (Barwise) *Let V be the universe of sets. Let $M \in V$ be a transitive model of ZFC, and let φ be an infinitary sentence in $L_{\infty,\omega} \cap M$ in the language of set theory. Then for a certain infinitary sentence φ^* in $L_{\infty,\omega} \cap \mathrm{Hyp}(M)$ in the language of set theory, the following are equivalent:*

(i) *$ZFC + \varphi^*$ is consistent.*
(ii) *$\mathrm{Hyp}(M) \models$ "$ZFC + \varphi^*$ is consistent".*
(iii) *In any universe W with the same ordinals as V which extends V and in which M is countable, there is an outer model N of M, $N \in W$, where φ holds.*

In particular, the set of formulas with parameters in M satisfied in an outer model M in an extension where M is countable is definable in $\mathrm{Hyp}(M)$.

The statement of Theorem 1.2 is less elegant than of Theorem 1.1 because it concerns a logic with expressive strength greater than first-order logic. Most importantly, we do not get a first-order definition by means of the notion of consistency as in Theorem 1.1(ii): in Theorem 1.2(ii), we leave M, and refer to $\mathrm{Hyp}(M)$, the smallest admissible set which contains M as an element. Thus the higher-order quantification over outer models is reduced to first-order quantification over $\mathrm{Hyp}(M)$, but not over M.

Furthermore, N—the model of the consistent theory $ZFC + \varphi^*$—may not exist in V, but only in some extension $W \supseteq V$ where M is countable. Theorem 1.2 therefore suggests an answer to (Q2): if we wish to answer (Q1) in the framework of $L_{\infty,\omega}$ and retain the straightforward correspondence between the consistency of a certain theory and existence of an outer model for that theory, M should be countable in V. However, the countability of M introduces technical issues in other respects, so we will not take this approach in the paper: instead, in Definition 2.1, we will define the notion of an outer model by referring to an extension $W \supseteq V$ where M is countable; or equivalently, by referring to consistency of a certain theory in $\mathrm{Hyp}(M)$. See Sect. 2.1 for more discussion of outer models, and some comments regarding the proof of Theorem 1.2.

We prove in this paper that one can construct by forcing over L a model M of size κ, κ inaccessible in L, in which the satisfaction for outer models is definable not only in $\mathrm{Hyp}(M)$ (which is ensured already by Theorem 1.2), but even in M, thus answering (Q1) positively for that M. Moreover M carries a definable wellorder (i.e. satisfies $V = HOD$). Our initial starting assumptions are minimal: we need just one inaccessible cardinal.

Remark 1.3 In his unpublished work [3], Mack Stanley proved that if M contains many Ramsey cardinals, then the answer to (Q1) for M is positive (see the Appendix for more details and the exact statement of the result). Later, he independently found a proof that only an inaccessible is enough to get a positive answer to the first part of

(Q1) (see section "Forcing Omniscience" in Appendix). However, his method does not seem to give our stronger result that M can satisfy $V = HOD$.[4]

The general outline of the paper is as follows:

In Sect. 2, we discuss the meaning of the notion of an outer model and its (apparent) dependence on the ambient universe. In Sect. 3, we give the proof of a weaker result which works for first-order formulas without parameters, or with a small number of parameters. In Sect. 4, we define the notion of a good iteration, prove the main theorem, and discuss some of its generalizations. In Sect. 5 we state some open questions. Finally, in the Appendix we briefly review the unpublished Stanley result.

2 Preliminaries

2.1 A Theorem of Barwise and the Notion of an Outer Model

Let V denote the universe of sets, and suppose $M \in V$ is a transitive model of ZFC. If we enlarge V to some V', for instance by forcing, then new outer models of M may appear in $V' \setminus V$. This process may be repeated indefinitely, with no natural stopping point.

However, perhaps the situation changes when we focus our attention on outer models which are relevant for *satisfaction* of formulas (with parameters). More precisely, there may be an extension V' of V (with the same ordinals) such that if there is no outer model of M in V' satisfying a given formula φ, then there will be no outer model satisfying φ in any further extension of V'. By a theorem of Barwise, see Theorem 1.2, this is in fact true; indeed V' can be taken to be a generic extension by the Levy collapsing forcing, which collapses $|M|$ to ω.[5] In particular, if M is already countable in V, then V itself can be taken for V'. Thus Barwise's theorem allows us to define the notion of an outer model in a robust way which does not depend on the ambient universe.

We will not prove Theorem 1.2 (see [1], in particular Chapter III and Theorems 5.5, 5.6, and 5.7 for details), but shall at least make some comments.

First, it may be illustrative to give some details regarding the sentence φ^* in the theorem because it shows how $L_{\infty,\omega}$ captures the notion of an outer model. φ^* is built up of constants $\bar{a}$ for every $a \in M$, and can look for instance as follows (we

---

[4]Notice that L can never define satisfaction in outer models otherwise it would be possible to define the satisfaction predicate in L. On the other hand with a proper class of Ramsey cardinals, M always defines satisfaction over its outer models. Thus the Dodd-Jensen K with large cardinals defines satisfaction in outer models. This presents a natural question whether M can satisfy $V = HOD$, and define satisfaction in outer models without relying on large cardinals.

[5]This can also be seen by Lévy absoluteness, as the existence of an outer model of M satisfying φ is a Σ_1 statement with parameter R for any real R coding M.

view *ZFC* as a single infinitary sentence, and include it in φ^* for clarity):

$$\varphi^* = ZFC \ \& \ \bigwedge_{x \in M} (\forall y \in \bar{x})(\bigvee_{a \in x} y = \bar{a}) \ \&$$

$$\& \ [(\forall x)(x \text{ is an ordinal} \rightarrow \bigvee_{\beta \in M \cap \text{ORD}} x = \bar{\beta})] \ \& \ \text{AtDiag}(M) \ \& \ \varphi, \qquad (2.1)$$

where AtDiag(*M*), the atomic diagram of *M*, is the conjunction of all atomic sentences and their negations which hold in *M* (when the constants are interpreted by the intended elements of *M*).

Second, the importance of countability in item (iii) of the theorem is caused by the use of model-theoretic inductive constructions which in general work only for countably many formulas (such as the Omitting Types Theorem); thus, for an uncountable *M*, the theory $ZFC + \varphi^*$ may be consistent, but we may not find a model for it in the current universe (this, of course, is a major difference between first-order logic and the infinitary logic $L_{\infty,\omega}$).

Third, the properties of Hyp(*M*) are important for the result: Barwise proved that with a suitable notion of proof, $\varphi \in M \cap \text{Hyp}(M)$ is provable in the ambient universe iff it is provable in Hyp(*M*), thus making the notion of proof independent of the ambient universe. Behind this is of course the observation that Hyp(*M*) is absolute between all transitive models which contain *M*.

In view of Theorem 1.2, we define:

Definition 2.1 Let *V* denote the ambient universe, and let $M \in V$ be a transitive model of size κ. We say that a first-order formula[6] φ with parameters from *M* is satisfied in an outer model of *M* if there is an outer model *N* of *M* in a generic extension of *V* where κ is countable, such that $N \models \varphi$.

By Theorem 1.2, the definition is independent of the choice of the generic extension.

Remark 2.2 In principle, there may be other ways to formalize the notion of satisfaction in outer models. For instance we could require that an outer model *N* of *M* must exist in the current universe *V*, even if *M* is uncountable. However, we would lose the connection with the logic $L_{\infty,\omega}$ and the corresponding completeness Theorem 1.2, making the problem less tractable. For instance, if $M = V_\kappa$, then *V* sees no non-trivial outer models of *M*, and the outer model theory of *M* cannot be definable in *M* in this case. It seems that for a meaningful analysis, the collection of outer models of *M* which we consider must be reasonably large. Definition 2.1 is a canonical way of ensuring this largeness condition. See Sect. 5 for open questions.

---

[6]In general, φ can be an infinitary formula as well; we consider first-order formulas here for concreteness.

2.2   The Outer Model Theory

Let V be the universe of sets and let $M \in V$ be a transitive model of ZFC of size κ, for some regular infinite κ.

Definition 2.3 We define the *outer model theory of M*, denoted OMT(M), as follows

$$\text{OMT}(M) = \{\varphi \mid \text{there is no outer model } N \text{ of } M,$$

$$N \in W, \text{ such that } N \models \neg\varphi\}, \qquad (2.2)$$

where φ is an infinitary formula in $L_{\infty,\omega} \cap M$ in the language of set theory with parameters in M and W is the model $V[G]$, where G is a generic filter for the Levy collapsing forcing which collapses κ to ω.

See Sect. 2.1, in particular Definition 2.1, for the legitimacy of W in this definition.

By Theorem 1.2, we can describe OMT(M) equivalently by means of the syntactical properties of $L_{\infty,\omega}$ and avoid talking about W:

$$\text{OMT}(M) = \{\varphi \mid ZFC + (\neg\varphi)^* \text{ is inconsistent in } \text{Hyp}(M)\}, \qquad (2.3)$$

where φ^* is the sentence described in Sect. 2.1.

Definition 2.4 Let M be as above. If OMT(M) is lightface definable in M, we say that M is *omniscient*.

The term *omniscience* is meant to indicate that M "knows about truth in all of its outer models". We view omniscience as a maximality property of M (it maximizes expressive power). Perhaps surprisingly, this maximality property is not a large cardinal property (as Stanley's result 1 would seem to indicate). By Theorem 4.18, an upper bound on its consistency strength is just one inaccessible cardinal (see Sect. 5 with open questions).

Often, it is more convenient to consider the following collection of sentences (with the notation of (2.2)):

$$\text{dOMT}(M) = \{\varphi \mid \text{there exists an outer model } N \in W \text{ of } M$$

$$\text{such that } N \models \varphi\}. \qquad (2.4)$$

We call dOMT(M) the *dual of the outer model theory of M*. Since $\varphi \in$ OMT(M) iff $\neg\varphi \notin$ dOMT(M), the outer model theory OMT(M) and its dual dOMT(M) are mutually inter-definable (but the former is a consistent theory, while the latter not). When we refer to the "outer model theory" below, for the purposes of definability, we can refer to either of these two collections.

2.3   Notational Conventions

Our notation is standard, following for instance [2]. In particular, if $P = \langle (P_i, \dot{Q}_i) \mid i < \mu \rangle$ is a forcing iteration of length μ and G is P-generic, let us write G_i, $i < \mu$, for G restricted to P_i. Further, if p is a condition in P_i, $i < \mu$, let us write $p^\frown 1$ for the condition in P which is the same as p at coordinates $j < i$, and at coordinates $j \in [i, \mu)$ is equal to the weakest condition in the respective forcing.

3   A Simplified Case

Let M be a transitive model of ZFC. Let us denote by $\mathrm{OMT}(M)^0$ (or $\mathrm{dOMT}(M)^0$) the intersection of $\mathrm{OMT}(M)$ (or $\mathrm{dOMT}(M)$) with the set of all first-order formulas with no parameters. As a warm-up, we construct a model in which the outer model theory for first-order formulas without parameters is lightface definable.

Theorem 3.1 *Assume κ is an inaccessible cardinal, $M = V_\kappa$. Then there exists a set-generic extension $V[G]$ of V by a forcing in V_κ such that the theory $\mathrm{OMT}(M[G])^0$ is lightface definable in $M[G]$.*

Proof Denote, using the notation in (2.4):

$$A_0 = \mathrm{dOMT}(M)^0$$

and identify A_0 with a subset of ω. We know that A_0 is interdefinable with $\mathrm{OMT}(M)$, and is therefore an element of $\mathrm{Hyp}(M)$; by the inaccessibility of κ, A_0 is an element of M because M contains $\mathscr{P}(\omega)$. However, A_0 may not be lightface definable in M (if it is, then the proof is finished).

Let Q_0 be an Easton-product forcing which codes A_0 by the pattern of GCH at the first ω many uncountable cardinals; more precisely $Q_0 = \prod_{n \in A_0} \mathrm{Add}(\aleph_{n+1}, \aleph_{n+3})$. Thus

$$1_{Q_0} \Vdash (\forall n < \omega)(n \in A_0 \leftrightarrow 2^{\aleph_{n+1}} = \aleph_{n+3}), \tag{3.5}$$

which makes A_0 lightface definable in $M[G_0]$, where G_0 is a Q_0-generic. Let A_1 denote $\mathrm{dOMT}(M[G_0])^0$.

Crucially,

$$A_1 \subseteq A_0 \tag{3.6}$$

because every outer model of $M[G_0]$ is by definition an outer model of M. If we have $A_0 = A_1$, the proof is finished, and $M[G_0]$ is the desired model.

Suppose we have strict inclusion in (3.6), then we will continue by defining Q_1 in $M[G_0]$ to code A_1 by a GCH pattern on the cardinals in the interval $[\aleph_{\omega+1}, \aleph_{\omega+\omega})$.

Continue in this fashion to define Q_α's and A_α's till the dual of the outer model theory stabilizes, i.e. until $A_\alpha = A_{\alpha+1}$ for some $\alpha < \omega_1$ (note that such an $\alpha < \omega_1$ must exist as otherwise we would shrink A_0 uncountably many times, which contradicts the countability of A_0).

Formally, define in the ambient universe V by induction a full support iteration

$$\mathbb{P}_{\omega_1} = \text{the inverse limit of } \langle (\mathbb{P}_\alpha, \dot{Q}_\alpha) \,|\, \alpha < \omega_1 \rangle$$

as follows:

(i) $\mathbb{P}_0 = \{\emptyset\}$.
(ii) If α is a limit ordinal, let $\mathbb{P}_\alpha$ be the inverse limit of $\langle (\mathbb{P}_\beta, \dot{Q}_\beta) \,|\, \beta < \alpha \rangle$.
(iii) If $\alpha = \beta + 1$, let $\dot{A}_\beta$ be a $\mathbb{P}_\beta$-name for $\text{dOMT}(M[\dot{G}_\beta])^0$, where $\dot{G}_\beta$ is a name for the $\mathbb{P}_\beta$-generic filter. Set $\mathbb{P}_\alpha = \mathbb{P}_\beta * \dot{Q}_\beta$, where $\dot{Q}_\beta$ is the name for $\prod_{n \in \dot{A}_\beta} \text{Add}(\aleph_{\omega\beta+n+1}, \aleph_{\omega\beta+n+3})$.

Notice that the definition in (iii) makes sense because for every $\beta < \omega_1$, the forcing $\mathbb{P}_\beta$ preserves the inaccessibility of κ, and hence the outer model theory of $M[G_\beta]$ is an element of $M[G_\beta]$ and can therefore be coded.

Let G_{ω_1} be $\mathbb{P}_{\omega_1}$-generic. As we noted above there is some $\alpha < \omega_1$ such that the dual of the outer model theory of $M[G_\alpha]$ equals the dual of the outer model theory of $M[G_{\alpha+1}]$ because it cannot shrink properly uncountably many times. For any such α, $\bar{M} = M[G_{\alpha+1}]$ is the desired model: $\text{dOMT}(\bar{M})^0$ can be read off from the continuum function on the last ω-segment of successor cardinals, where the GCH fails cofinally often.

Remark 3.2 Note that $\mathbb{P}_{\omega_1}$ may not be lightface definable in M, but by the inaccessibility of κ, $\mathbb{P}_{\omega_1}$ is an element of $M = V_\kappa$. $M[G]$ is therefore a legitimate generic extension of M, in particular $M[G]$ is a model of ZFC.

Remark 3.3 It is easy to see that the argument in the proof of Theorem 3.1 easily generalizes to situations where the relevant outer model theories are elements of the respective generic extensions. Thus the outer model theory of all first-order formulas with parameters from some $H(\mu)$, $\mu < \kappa$, can be coded by a variant of the construction in the proof of Theorem 3.1. We will not give the details because the relevant results are easy and moreover follow from the main Theorem 4.18. Finally notice that by further MacAloon coding, we can easily arrange that the resulting $M[G]$ carries a definable wellorder (i.e. satisfies $V = HOD$); see the main Theorem 4.18.

4   Main Result

As before, let κ be an inaccessible cardinal and $M = V_\kappa$. Suppose now we wish to define the outer model theory of M with formulas (in the language of set theory) which allow all parameters from M, or with infinitary formulas in $M \cap L_{\infty,\omega}$. Then the coding method from Theorem 3.1 is no longer applicable because the outer model theory of M is not an element of M.

Instead we will define an iteration of length $< \kappa^+$ which will contain as its initial segments witnesses (i.e. forcings), which will attempt to stabilize the membership or non-membership of a given formula to the outer model theory of the final generic extension. Since such witnesses need to be of length at least κ (because we need to decide the membership of κ many formulas), the whole iteration needs to be longer than κ as we are going to compose κ-many iterations of length at least κ. The length of the final iteration is not given in advance, but will be determined by an inductive definition. As in Theorem 3.1, we will code the theory—and also a definable wellorder of the universe (thus ensuring $V = HOD$ in the final model)— by the GCH pattern at some regular cardinals $< \kappa$. However, since the iteration is now longer than κ, we cannot choose these cardinals in increasing order.

We call such iterations *good iterations*.

4.1   Good Iterations

Assume $V = L$. Let κ be the least inaccessible cardinal and let X be the set of all singular (i.e. uncountable limit) cardinals below κ. Fix a partition $\langle X_i \mid i < \kappa \rangle$ of X into κ pieces, each of size κ, such that $X_i \cap i = \emptyset$ for every $i < \kappa$.

Definition 4.1 Let μ be an ordinal less than κ^+. We say that (P, f) is a *good iteration of length* μ if it is an iteration $P_\mu = \langle (P_i, \dot{Q}_i) \mid i < \mu \rangle$ with $< \kappa$ support of length μ, $f : \mu \to X$ is an injective function in L and the following hold:

(i) $\mathrm{rng}(f) \cap X_i$ is bounded in κ for every $i < \kappa$,
(ii) For every $i < \mu$, P_i forces that $\dot{Q}_i$ is either $\mathrm{Add}(f(i)^{++}, f(i)^{+4})$ or $\mathrm{Add}(f(i)^{+++}, f(i)^{+5})$.

Remark 4.2 The properties of the good iterations discussed below, and in particular Theorem 4.13, would still be true if we specified in Definition 4.1(ii) that $\dot{Q}_i$ is one of a family of forcings which are all $f(i)^{++}$-closed, non-collapsing, and of size $< f(i)^{+\omega}$. There is nothing special about $\mathrm{Add}(f(i)^{++}, f(i)^{+4})$ and $\mathrm{Add}(f(i)^{+++}, f(i)^{+5})$ except that we use these forcings in the main Definition 4.24.

We are going to show that good iterations preserve cofinalities. For the usual reverse Easton iterations $\langle (\mathbb{P}_i, \dot{Q}_i) \mid i < \mu \rangle$, this is done by dividing the iteration at stage i into a lower part $\mathbb{P}_i$ which has a small chain condition, and the tail which is sufficiently closed. However, this easy division assumes that the cardinals are used in increasing order by $\mathbb{P}$. For good iterations, this is not the case: typically, $\mathbb{P}_i$ has just the κ^+-cc, and the tail may not be closed more than the first singular cardinal below κ. To overcome this problem, we need to define suitable notions of lower and upper part of a condition, which would enable us to carry out a similar kind of analysis as for the usual reverse Easton iteration. The analogy is not straightforward, though: we need to work with a quotient forcing (corresponding to the upper part) which is just distributive (see Lemma 4.8). The lower and upper part of a condition will be defined by means of a *good name* for an element of the forcing $\dot{Q}_i$ at stage i, which we define next.

Definition 4.3 By induction on $i < \mu$, define the notion of a *good name at i*:

(i) $LP(i)$, the *lower part of P_i*, is the collection $\{j < i \mid f(j) < f(i)\}$. Similarly, let $UP(i)$, the *upper part of P_i*, be the complement of $LP(i)$: $UP(i) = i \setminus LP(i)$.

(ii) σ is a *good name at i* if σ is a P_i-name for an element of $\dot{Q}_i$, which satisfies:

    (a) σ is a nice name for a subset of $f(i)^{+5}$; i.e. σ is of the form $\bigcup_{\alpha < f(i)^{+5}}(\{\alpha\} \times A_\alpha)$, where A_α is an antichain in P_i. Moreover, σ is forced by P_i to be in $\dot{Q}_i$.[7]

    (b) The conditions $p \in A_\alpha$ satisfy that $p(j)$ for $j < i$ may be different from $1_{\dot{Q}_j}$ only at coordinates in $LP(i)$, and for all $j < i$, $p(j)$ is a good name at j (we regard $1_{\dot{Q}_j}$ as a good name).

The intuition behind the definition of a good name at i is to make sure that the interpretation of σ depends only on the generic at coordinates in $LP(i)$. Let us denote as Good(i) the collection of all good names at i.

Lemma 4.4 (GCH) *The number of good names at $i < \mu$ is less than $f(i)^{+\omega}$:*

$$|\text{Good}(i)| < f(i)^{+\omega}. \tag{4.7}$$

Proof The proof is by induction. Suppose it holds for $j < i$. Then the number of conditions p which satisfy (b) in Definition 4.3 is at most $f(i)^{f(i)}$ because for every j such that $f(j) < f(i)$, the number of good names at j is by induction less than $f(j)^{+\omega}$, which is less than $f(i)$. The number of sets of these conditions (and so of antichains) is therefore at most $f(i)^{++}$, and hence the number of names satisfying (a) in Definition 4.3 is then certainly less than $f(i)^{+\omega}$. $\dashv$

Definition 4.5 Let p and q be in P_i. For $\lambda \in X$ define

$$p \leq_\lambda q \leftrightarrow p \leq q \text{ and } p(j) = q(j) \text{ for all } j < i \text{ such that } f(j) < \lambda. \tag{4.8}$$

Lemma 4.6 *Let $i \leq \mu$ be fixed. Suppose the following holds:*

$$(\forall p \in P_i)(\forall \lambda \in \text{rng}(f \restriction i))(\exists q \leq_\lambda p)(q(f^{-1}(\lambda)) \in \text{Good}(f^{-1}(\lambda))). \tag{4.9}$$

Then

$$(\forall p \in P_i)(\exists q \leq p)((\forall j < i)q(j) \in \text{Good}(j)). \tag{4.10}$$

---

[7] We identify conditions in the Cohen forcings $\text{Add}(f(i)^{++}, f(i)^{+4})$ and $\text{Add}(f(i)^{+++}, f(i)^{+5})$ with subsets of $f(i)^{+5}$.

Proof Denote by supp(p) the support of p, which by our definition has size $< \kappa$. Let $\langle \lambda_\alpha \mid \alpha < v \rangle$, $v < \kappa$, be the increasing enumeration of the range of f on supp(p). Using (4.9), define a decreasing sequence of conditions $p = q_0 \geq_{\lambda_0} q_1 \geq_{\lambda_1} \cdots$ of length v with limit q^0 such that $q_{\alpha+1} \leq_{\lambda_\alpha} q_\alpha$ has a good name at $f^{-1}(\lambda_\alpha)$. Note that the limit stages (including the last one) are defined because the conditions at the coordinate $f^{-1}(\lambda_\alpha)$, $\alpha < v$, are extended at most λ_α-many times, and the forcing at $f^{-1}(\lambda_\alpha)$ is forced to be λ_α^+-closed. By construction, q^0 satisfies (4.10) on supp(p) (and all $j < i$ outside the support of q^0). Repeat the construction ω-many times, obtaining a decreasing sequence $q^0 \geq q^1 \geq q^2 \geq \cdots$ with limit q. By the construction, q satisfies (4.10) as required because the support of q is the union of the supports of the q^i, $i < \omega$. $\dashv$

In Lemma 4.10, we will prove (4.9) by induction on $i \leq \mu$. For the argument at stage i, we will need to have some information about P_j for $j < i$ for which (4.9) already holds. This is the purpose of Lemma 4.8.

Definition 4.7 Let $i < \mu$ be given. Set

$$P_i^* = \{p \in P_i \mid (\forall j \in LP(i))(p(j) \in \text{Good}(j)) \ \&$$
$$(\forall k \in UP(i))(p(k) = 1_{\dot{Q}_k})\}. \quad (4.11)$$

Lemma 4.8 *Let $i < \mu$ be fixed. Assume (4.9), and therefore also (4.10), hold for P_i. Then:*

(i) *There is a projection $\pi_i : P_i \to P_i^*$.*
(ii) *Let P_i^+ be the quotient P_i/P_i^*.*

$$1_{P_i^*} \Vdash P_i^+ \text{ is } f(i)^{+\omega}\text{-distributive.} \quad (4.12)$$

Proof

(i) By (4.10), we can assume that $p \in P_i$ consists of good names at all $j < i$. Define $\pi(p)$ as the condition q such that $q(j) = p(j)$ for $j \in LP(i)$, and $q(j) = 1_{\dot{Q}_j}$ otherwise. It is easy to see that π is a projection.
(ii) Let p_0 in P_i force that $\dot{h}$ is a name for a function from $\eta < f(i)^{+\omega}$ to the ordinals; assume by (4.10) that p_0 contains just good names at all $j < i$. We wish to find $\tilde{r} \leq p_0$ in P_i such that over $V^{P_i^*}$, $\tilde{r}$ can be used to define the interpretation of h in V^{P_i} (for generics containing $\tilde{r}$). First note that by Lemma 4.4, the size of P_i^* is at most $f(i)^{f(i)} = f(i)^+$ because every $p \in P_i^*$ is determined by the sequence of good names in $LP(i)$. Also note that the forcings at $j \in UP(i)$ are at least $f(i)^{+\omega+2}$-closed because by the injectivity of f, $f(j) > f(i)$ for all $j \in UP(i)$.

Let Y be the family of all sequences of good names $p(j)$, where p is a condition in P_i^* and j is an index in $LP(i)$, i.e.

$$Y = \prod_{j \in LP(i)} \{p(j) \mid p \in P_i^* \text{ and } p(j) \text{ is a good name}\}.$$

If $p \in P_i$ and $y \in Y$, we write

$$p * y \tag{4.13}$$

to denote the condition which is obtained by replacing $p(j)$ by $y(j)$ for all $j \in LP(i)$. Note that $p * y$ is a legitimate condition because a good name at j is forced by 1_{P_j} to be in $\dot{Q}_j$.

To determine $\dot{h}(0)$, construct simultaneously a $\leq_{f(i)}$-decreasing sequence of conditions $\langle r_\alpha^0 \mid \alpha < \nu_0 \rangle$, $\nu_0 < f(i)^{++}$, below p_0, and an $\subseteq$-increasing sequence of antichains $\langle A_\alpha^0 \mid \alpha < \nu_0 \rangle$ of conditions in P_i^* below $\pi(p_0)$. Greatest lower bounds are taken to define r_α^0 for α limit, and unions of A_β^0, $\beta < \alpha$, to define A_α^0, for α limit. To define $r_{\alpha+1}^0$, find $a \leq \pi(p_0)$ such that

$$\{a\} \cup A_\alpha^0 \text{ is an antichain}, \tag{4.14}$$

and let $A_{\alpha+1}^0 = A_\alpha^0 \cup \{a\}$. Let $y \in Y$ be the sequence of good names in a at coordinates in $LP(i)$. Find

$$s_\alpha \leq r_\alpha^0 * y \tag{4.15}$$

which decides the value of $\dot{h}(0)$. Define $r_{\alpha+1}^0$ by induction on $j < i$ as follows:

$$r_{\alpha+1}^0(j) = \begin{cases} p_0(j) & \text{if } j \in LP(i) \\ \sigma_j & \text{otherwise}, \end{cases} \tag{4.16}$$

where $(r_{\alpha+1}^0 * y) \mid j$ forces that σ_j is equal to $s_\alpha(j)$, and simultaneously, $r_{\alpha+1}^0 \mid j$ forces $\sigma_j \leq r_\alpha^0(j)$.[8] Thus in particular $r_{\alpha+1}^0 \leq_{f(i)} r_\alpha^0$.

Since P_i^* has size at most $f(i)^+$, there is some $\nu_0 < f(i)^{++}$ such that there is no a satisfying (4.14). By the closure of the conditions at $UP(i)$, the sequence $\langle r_\alpha^0 \mid \alpha < \nu_0 \rangle$ has a lower bound, which we denote as $\tilde{r}^{\dot{h}(0)}$.

Carry out the above construction for every $\gamma < \eta$ and construct sequences $\langle r_\alpha^\gamma \mid \alpha < \nu_\gamma \rangle$ and $\langle A_\alpha^\gamma \mid \alpha < \nu_\gamma \rangle$ to determine $\dot{h}(\gamma)$, and to obtain a decreasing sequence $\tilde{r}^{\dot{h}(0)} \geq_{f(i)} \tilde{r}^{\dot{h}(1)} \geq_{f(i)} \cdots$ of length η. By the closure of the coordinates at $UP(i)$, the sequence has a lower bound. Denote the lower bound as $\tilde{r}$.

Let G be a P_i-generic filter over V which contains $\tilde{r}$. Let g be the derived generic for P_i^* via the projection π. In $V[g]$ one can define $\dot{h}^G$ as follows: $\dot{h}^G(\gamma)$ is the unique ordinal ξ such that for a unique $a \in \bigcup \{A_\alpha^\gamma \mid \alpha < \nu_\gamma\}$, $a \in g$, $\tilde{r} * a$ forces $\dot{h}(\gamma) = \xi$.

In Theorem 4.13, it will be useful to have the above Lemma 4.8 formulated for $i = \mu$ and some parameter $\zeta < \kappa$ (a regular cardinal). Given a regular cardinal

---

[8]This is the usual manipulation with names which can interpret differently below incompatible conditions.

$\zeta < \kappa$ and a condition p in P, let us define $LP_\zeta(\mu)$, the lower part of P with respect to ζ, as follows:

$$LP_\zeta(\mu) = \{j < \mu \,|\, f(j) < \zeta\}, \tag{4.17}$$

and $UP_\zeta(\mu) = \mu \setminus LP_\zeta(\mu)$. The notion of the lower part with respect to ζ is used to define the analogue of P_i^*:

$$P_\zeta^* = \{p \in P \,|\, (\forall j \in LP_\zeta(\mu))(p(j) \in \text{Good}(j)) \,\&$$

$$(\forall k \in UP_\zeta(\mu))(p(k) = 1_{\dot{Q}_k})\}. \tag{4.18}$$

Lemma 4.8 can be directly generalized as follows:

Corollary 4.9 *Assume (4.9), and therefore also (4.10), hold for P. Suppose $\zeta < \kappa$ is a regular cardinal and*
(*) there is no unique $j < \mu$ such that $f(j) < \zeta < f(j)^{+\omega}$.
Then:

(i) *There is a projection $\pi_\zeta : P \to P_\zeta^*$.*
(ii) *Let P_ζ^+ be the quotient P/P_ζ^*.*

$$1_{P_\zeta^*} \Vdash P_\zeta^+ \text{ is } \zeta^+\text{-distributive}. \tag{4.19}$$

Proof Since (4.10) is formulated also for $i = \mu$, the proof goes exactly as in Lemma 4.8. Regarding the distributivity in (ii), note that P_ζ^* has size at most ζ as by the condition (*), ζ is not an element of an ω-block of cardinals $[f(j), f(j)^{+\omega})$ for any j. Note that if (*) is not the case, we need to factor more carefully; see the proof of Theorem 4.13.

Finally we can prove the key property (4.9).

Lemma 4.10 *For all $i \leq \mu$, (4.9) holds for P_i.*

Proof The proof is by induction.

Assume that i is a limit ordinal and for every $j < i$, (4.9) holds for P_j. Let $p \in P_i$ and $\lambda \in X$ be given. Choose $j < i$ such that $f(j) = \lambda$. By the induction assumption applied to P_{j+1}, there is a $q' \leq_\lambda p|(j+1)$ such that $q'(j)$ is a good name. Clearly if we stretch q' to a condition q in P_i by substituting $p(k)$ for $k \in [j+1, i)$, we get the required $q \leq_\lambda p$.

Assume now that (4.9) holds for P_i, and we wish to show it for P_{i+1}. Let $p \in P_{i+1}$ and λ be given. Assume that $f(i) = \lambda$ (otherwise the lemma follows trivially). By Lemma 4.8 applied to P_i, all conditions in $\dot{Q}_i$ are added by P_i^*. It follows that $p|i$ can be extended to some $q_0 \leq p|i$ which forces that $p(i)$ is extended by some P_i^*-name σ_0, which can be taken to be a good name. The problem is that we only have $q_0 \leq p|i$, and not the required $q_0 \leq_\lambda p|i$. However, as in the proof of Lemma 4.8(ii), by diagonalizing over a maximal antichain in P_i^*, and taking lower bounds on the coordinates in $UP(i)$, one can find $q \leq_\lambda p|i$ and a good P_i^*-name σ such that q forces that σ extends $p(i)$.

Definition 4.11 Let (P, f) be a good iteration of length $\mu < \kappa^+$. We call $p \in P$ a *good condition* if for every $i < \mu$, $p(i)$ is a good name.

Lemma 4.12 *Let* (P, f) *be a good iteration of length* $\mu < \kappa^+$. *Then the collection of good conditions forms a dense subset of* P *of size at most* κ.

Proof By Lemma 4.10, the property (4.9) in Lemma 4.6 holds, and therefore there is a dense subset of P which contains conditions composed only of good names. By Lemma 4.4, the number of such sequences is at most κ.

The properties of good iterations identified in the previous lemmas provide a straightforward way to show that good iterations preserve cofinalities.

Theorem 4.13 (GCH) *Let* μ *be an ordinal less than* κ^+. *A good iteration* (P, f) *of length* μ *preserves cofinalities.*

Proof By Lemma 4.12, P has the κ^+-cc, and so all cofinalities $\geq \kappa^+$ are preserved.

It remains to show that cofinalities $\leq \kappa$ are preserved.

Assume by contradiction that for some regular $\zeta < \kappa$ there is $p \in P$ such that

$$p \Vdash (\exists \lambda > \zeta \text{ regular in } V) \, \mathrm{cf}(\lambda) = \zeta.$$

We need to distinguish two cases.

Case A There is no unique $j < \mu$ such that $f(j) < \zeta < f(j)^{+\omega}$.

In this case, we reach contradiction by applying Corollary 4.9. P_ζ^* has size at most ζ so it cannot cofinalize λ; the quotient forcing P_ζ^+ is forced to be ζ^+-distributive, so cannot cofinalize λ either.

Case B There is a unique $j < \mu$ such that $f(j) < \zeta < f(j)^{+\omega}$.

Fix such j and write P as $P_0 * P_1 * P_2$, where P_0 is the forcing P_j, P_1 is the Cohen forcing (either at $f(j)^{++}$ or $f(j)^{+++}$), and P_2 is the tail of the iteration.

P_0 does not cofinalize λ by Lemma 4.8. P_1 is cofinality-preserving. For P_2, apply Corollary 4.9 over $V^{P_0 * P_1}$; note that Corollary 4.9 now applies because in $V^{P_0 * P_1}$, the tail iteration satisfies Case A.

Remark 4.14 If we defined a good iteration exactly as in Definition 4.3, but with full support, then (P, f) would still preserve all cofinalities. This is not useful for the present paper, but it might be of interest for future applications. The preservation of cofinalities $\leq \kappa$ is exactly the same as for the $< \kappa$ support. The preservation of κ^+ can be argued as follows:

We need to show that if $p \in P$ forces that $\dot{h}$ is a function from κ to κ^+, then for some condition $q \leq p$, q forces a bound on the range of $\dot{h}$ of size at most κ. The argument is a diagonal version of the proof of Lemma 4.8(ii), and uses an inductive construction of length κ as in Lemma 4.6. In particular, let $\langle \lambda_\alpha \mid \alpha < \kappa \rangle$ be the increasing enumeration of X. Define a decreasing sequence of conditions $p =$

$q_0 \geq_{\lambda_0} q_1 \geq_{\lambda_1} q_2 \geq_{\lambda_2} \cdots$ of length κ with limit q such that $q_{\alpha+1}$ decides the value of $\dot{h}(\alpha)$ to be in some family Y_α of size $< \kappa$; to obtain this $q_{\alpha+1} \leq_{\lambda_\alpha} q_\alpha$, when q_α has already been constructed, carry out the argument in Lemma 4.8(ii) applied to P_i with $i = \mu$, and with the $LP(i)$ defined as $LP(i) = LP(i)_\alpha = \{j < i \,|\, f(j) < \lambda_\alpha\}$ (the definition of $LP(i)_\alpha$ using λ_α at stage α explains the use of the word "diagonal" in the description of the method). Note that the length of the inductive construction at stage α is at most $\lambda_\alpha^{\lambda_\alpha} = \lambda_\alpha^+$ (the number of sequences of good names at $j \in LP(i)_\alpha$), and the forcings at $j, f(j) \geq \lambda_\alpha$, are at least λ_α^{++}-closed, so $q_{\alpha+1}$ can be correctly defined.

4.2 Compositions of Good Iterations

Before we state the theorem we make a remark and state a lemma concerning a composition of good iterations.

Remark 4.15 Suppose (P,f) is a good iteration in the ground model V, and $(\dot{Q}, g)$ is forced by 1_P to be a good iteration in V^P.[9] If $\text{rng}(f) \cap \text{rng}(g) = \emptyset$, then $P * \dot{Q}$ is a good iteration with respect to a function h defined as follows: $h = f \cup g^{\text{shift}}$ where g^{shift} is defined on $[\alpha, \beta)$ where $\alpha = \min(\kappa^+ \setminus \text{dom}(f))$, $\beta = \alpha + \text{dom}(g)$, and for $\xi \in \text{dom}(g)$, $g^{\text{shift}}(\alpha + \xi) = g(\xi)$. For simplicity of notation, if $\text{rng}(f) \cap \text{rng}(g) = \emptyset$, we will write

$$f \uplus g \text{ to denote } f \cup g^{\text{shift}}. \tag{4.20}$$

In particular, we write $(P * \dot{Q}, f \uplus g)$ to denote the resulting composition.

Lemma 4.16 Let $P = \langle (P_i, \dot{Q}_i) \,|\, i < \mu \rangle$, $P_0 = \{\emptyset\}$, be an iteration with $< \kappa$-support such that for every $i < \mu$, 1_{P_i} forces that $(\dot{Q}_i, f_i')$ is a good iteration. Assume further that for every $i \neq j < \mu$, $\text{rng}(f_i') \cap \text{rng}(f_j') = \emptyset$. By induction define a sequence $\langle f_i \,|\, i < \mu \rangle$: $f_0 = f_0'$, $f_{i+1} = f_i \uplus f_{i+1}'$, and $f_i = \bigcup_{j<i} f_j$ for i limit. Denote $f = \bigcup_{i<\mu} f_i$.

(i) If $\mu < \kappa$, then (P,f) is a good iteration.

(ii) If $\mu = \kappa$, and moreover for every $i < \kappa$, $\text{rng}(f_i) \cap X_j = \emptyset$ for all $j < i$, then (P,f) is a good iteration.

Proof Taking into account Remark 4.15, (P,f) satisfies the requirements for being a good iteration; the only property worth mentioning in (i) is that because $\mu < \kappa$, the range of f is bounded in every X_i. For (ii), the property of f being bounded in every X_i is ensured by demanding $\text{rng}(f_i) \cap X_j = \emptyset, j < i$. $\blacksquare$

---

[9]Recall that g is a function in L by the definition of good iteration; further note that we use the convention that checked names are written without a dot: hence $(\dot{Q}, g)$ is the same as $(\dot{Q}, \check{g})$.

Remark 4.17 Note that Lemma 4.16 is formulated only for $\mu \leq \kappa$ because we will not need it for larger values of μ. In particular μ will be the cofinality of the good iteration P.

4.3 Main Theorem

4.3.1 Statement and Motivation

Theorem 4.18 *Assume $V = L$. Let κ be the least inaccessible, and let $M = L_\kappa$. There is a good iteration $(\mathbb{P}, h)$ in V such that if G is $\mathbb{P}$-generic over V, then for some set $\tilde{G}$, which is defined from G, $M[\tilde{G}]$ is a model of ZFC in which $\mathrm{OMT}(M[\tilde{G}])$ is lightface definable. Moreover, $M[\tilde{G}]$ is a model of $V = HOD$.*

The set $\tilde{G}$ is defined from G as follows:

Definition 4.19 Assume V satisfies GCH. If (P, f) is a good iteration and G is P-generic over V, let us write $\tilde{G}$ for the following object: $\tilde{G}$ is the collection of all generic subsets added by the generic G to cardinals ξ, where ξ is a double successor or a triple successor of a singular cardinal in the range of f.

Notice in particular that if κ is inaccessible, $V_\kappa = M$, then $\tilde{G}$ is a subset of $H(\kappa)^{V[G]}$, and in $V[G]$, $M[\tilde{G}]$ is the smallest model of ZFC which contains $M \cup \tilde{G}$ and has height κ; in fact, $M[\tilde{G}] = H(\kappa)^{V[G]}$. Note also that the continuum function below κ is the same in $M[\tilde{G}]$ and $V[G]$.

We start by explaining the idea behind the proof of the main theorem to motivate rather technical Definitions 4.22 and 4.24. First, we define the notion of "killing a formula".

Definition 4.20 Let M be as above. We say that a condition p in a good iteration (P, f) *kills a formula* φ (with parameters which are P-names for sets belonging to $M[\tilde{G}]$) if for every P-generic G containing p, there is no outer model of $M[\tilde{G}]$ where φ holds, i.e. $\varphi \notin \mathrm{dOMT}(M[\tilde{G}])$.

Note that if $p \in P$ kills φ, then for every good iteration $\dot{Q}$ in V^P and every $\dot{q} \in \dot{Q}$, $(p, \dot{q})$ also kills φ (or equivalently, $(p, 1_{\dot{Q}})$ kills φ). In other words: Any extension of an iteration which kills φ, also kills φ (killing of φ is upwards persistent). This simple observation allows us to compose together good iterations in Definition 4.24 below.

Remark 4.21 Recall that by Theorem 1.2, the existence of an outer model of $M[\tilde{G}]$ satisfying φ is equivalent to the consistency of a certain infinitary sentence φ^*, so the property of killing φ in Definition 4.20 is expressible as a property of the forcing (P, f).

Let us denote $M = V_\kappa = L_\kappa$. The main idea of the proof of Theorem 4.18 is as follows: we want to decide the membership or non-membership of κ-many formulas with parameters in the outer model theory of the final model. We are going to define an iteration of length κ, dealing with the ith formula at stage $\mathbb{P}_i$. Suppose at stage

i, it is possible to kill φ_i by a good iteration $\dot{W}_i$, i.e. ensure that in $V^{\mathbb{P}_i * \dot{W}_i}$ there is no outer model of φ_i. If such $\dot{W}_i$ exists, set $\mathbb{P}_{i+1} = \mathbb{P}_i * \dot{W}_i * \dot{C}_i$, where $\dot{C}_i$ codes this fact by means of a good iteration. In the final model $M[\tilde{G}]$, we can decide the membership of φ_i in $\mathrm{OMT}(M[\tilde{G}])$ by asking whether at stage i we have coded the existence a witness $\dot{W}_i$ which kills φ_i, arguing as follows: If there is no outer model of $M[\tilde{G}]$ where φ_i holds, then indeed we have coded this fact at stage i by using some $\dot{W}_i$ (because the tail of $\mathbb{P}$—itself a good iteration—from stage i did kill φ_i so some such $\dot{W}_i$ must have existed). Conversely, if there is an outer model of $M[\tilde{G}]$ where φ_i holds, then we could not have found a witness $\dot{W}_i$ because if we did, then its inclusion in $\mathbb{P}$ would ensure that φ_i is killed. Note that there is no bound on the length of $\dot{W}_i$, except that it must be less than κ^+ (by the injectivity of the function f which makes $(\dot{W}_i, f)$ a good iteration).

Several points must be resolved to make this rough idea work. (A) Although technically $\mathbb{P}$ is an iteration of length κ, its length as a good iteration (using the composition Lemma 4.16) is some ordinal of cofinality κ below κ^+. This has two consequences: (i) $\mathbb{P}$ cannot choose the regular cardinals $< \kappa$ at which it forces in the increasing order, and (ii) $\mathbb{P}$ is not a subset of M. We solve (i) by considering an injective function f in the definition of a good iteration (P, f) which enumerates singular cardinals $< \kappa$ in a non-monotonic way; note also that because we need to code information as we progress, f must have some flexibility: therefore, f enumerates singular cardinals in whose "neighbourhood" we do the coding at regular cardinals. Regarding (ii), we solve the problem by considering $M[\tilde{G}]$ described in Definition 4.19.

(B) Whether $\mathbb{P}_i * \dot{W}_i$ does or does not kill φ_i may depend on a particular condition $p \in \mathbb{P}_i * \dot{W}_i$ which forces it (the forcings are not homogeneous); to deal with this problem, we will need to use bookkeeping functions e^i, $i < \kappa$, to enumerate all good conditions in the initial segments $\mathbb{P}_i$ and add witnesses $\dot{W}_i$ with respect to these conditions (we will not need to enumerate the conditions in $\dot{W}_i$ itself because by $< \kappa$ support, any condition which forces killing of φ_i in the final iteration $\mathbb{P}$ has its support bounded in some $\mathbb{P}_i$).

(C) Since the formulas φ_i contain parameters from the final extension $M[\tilde{G}]$, we will need to enumerate them as the iteration progresses by means of names (these are the (names for) enumerations $\dot{d}^i$ in the proof).

Finally (D) to achieve light-face definability, the well-ordering of the formulas given by the $\dot{d}^i$'s needs to be coded. We will also need to code which good conditions are elements of the generic filter (recall that our witnesses will be added with respect to all good conditions and we will need to look at the right conditions to decode the outer model theory correctly). This is the purpose of the forcing $\dot{C}_i$ in $\mathbb{P}_i * \dot{W}_i * \dot{C}_i = \mathbb{P}_i * \dot{Q}_i$, which does the coding by means of further good iterations. As a bonus, by coding the $\dot{d}^i$'s, the final model $M[\tilde{G}]$ will be a model of $V = HOD$.

4.3.2 Proof

In this section, we give the proof of Theorem 4.18.

Assume $V = L$, and let κ be the least inaccessible cardinal and X the set of all singular cardinals below κ; let $\langle \lambda_\alpha \mid \alpha < \kappa \rangle$ be the increasing enumeration of X. Fix an L_κ-definable partition

$$\langle X_i \mid i < \kappa \rangle \tag{4.21}$$

of X into κ pieces, each of size κ, such that $X_i \cap i = \emptyset$ for every $i < \kappa$.

As mentioned in (D) in Sect. 4.3.1, we need to code some information. Now, we describe one particular way of coding using good iterations. Let us say that a good iteration P has *pattern A* on $i \in X$ if it forces with $\mathrm{Add}(i^{++}, i^{+4})$; it has *pattern B* if it forces with $\mathrm{Add}(i^{+++}, i^{+5})$. Given $i < \kappa$ and $\alpha \in X_i$, we say that P has pattern $ABAB$ on α if P forces on four successive singular cardinals in X_i, starting with α, and on α, it has pattern A, on the successor of α in X_i, pattern B, etc. This notation extends to any finite combination of A's and B's. For any α we refer to pattern AA as *bit 0*, and pattern BB as *bit 1*.

Definition 4.22 Let V be a ground model (a generic extension of L by a good iteration). Let a be in $H(\kappa)^V$. We say that a good iteration (P, f) *codes a on top of X_i* if the range of f is an interval of cardinals in X_i starting with $\alpha \in X_i$ which is the least such that GCH holds in V for all $\beta \geq \alpha$ in X_i. In order to mark the beginning of the coding, P forces pattern $ABAB$ on α. The coding itself starts by first coding the transitive closure of $\{a\}$ by a subset a' of some ordinal.[10] The set a' is then coded by means of a good iteration, which codes a' by a sequence of bits 0 and 1, starting with the first singular cardinal in X_i after pattern $ABAB$.

Definition 4.22 extends to coding a finite number[11] of sets $\{a_0, \ldots, a_n\}$: just code successively the sets a_i, $i \in \{0, \ldots, n\}$, and separate the coding intervals by, e.g., $ABABAB$. Note that one can decode the coded information as follows: given X_i, find the topmost occurrence of $ABAB$ and the next occurrence of $ABABAB$ (if any). Using the bits 0 and 1 in this interval, decode a_0, etc.[12]

In the course of the inductive definition of the iteration in Definition 4.24, we will use some bookkeeping of certain triples of parameters. The bookkeeping function will be an L_κ-definable surjective function $b : \kappa \to \kappa^3$ such that if

$$b(i) = \langle j, k, m \rangle, \tag{4.22}$$

---

[10]For concreteness, we take a' to be a subset of some cardinal η, $|a'| = \eta$, which via a pairing function codes (η, R), $R \subseteq \eta^2$, such that (η, R) is isomorphic to the transitive closure of $\{a\}$ with the membership relation. By Mostowski collapse theorem, a' is enough to recover a.

[11]Finite is enough for us here.

[12]Any other reasonable notion of coding can be used here.

then the following conditions are satisfied:

- $j \leq i$,
- If $i = 0$, then $m = 0$,
- If i is a singular cardinal, then $m < i$,
- If $i > 0$ is not a singular cardinal and $i < \lambda_0$, then $m < \lambda_0$ (where λ_0 is the least singular cardinal),
- If i is not a singular cardinal and is greater than the first singular cardinal, then $m < \bar{i}$, where $\bar{i}$ is the greatest singular cardinal below i.

We write $b(i)(0) = j$, $b(i)(1) = k$, and $b(i)(2) = m$ to express the components of $b(i)$.

Remark 4.23 The many cases of the definition of b are motivated by the fact that the coding of parameters in Definition 4.24 takes place only at singular cardinals i (which ensures that $H(i)$ of the relevant model has size i as under our assumptions every singular cardinal will always be strong limit); moreover, 0 will be treated similarly as the singular cardinals to prime the construction. At stages i which are not singular cardinals, we will use a wellordering defined at the greatest previous singular stage (or 0 if there is no smaller singular cardinal).

The following definition specifies the main forcing $(\mathbb{P}, h)$ in Theorem 4.18.

Definition 4.24 $(\mathbb{P}, h) = \langle (\mathbb{P}_i, \dot{Q}_i) \mid i < \kappa \rangle$ is going to be an iteration with $< \kappa$ support, $h \in L$, defined by induction together with names $\dot{d}^i$, i a singular cardinal, and sequences e^i, $i < \kappa$. The names $\dot{d}^i$ will interpret as wellorderings of rank-initial segments of the final model; it will be the case that $1_{\mathbb{P}_i}$ forces that $\dot{d}^i$ end-extends all $\dot{d}^j$, $j < i$, and that the ordertype of the ordering $\dot{d}^i$ is i. The sequences e^i, $i < \kappa$, will enumerate all good conditions in $\mathbb{P}_i$, $i < \kappa$.

Stage 0 To prime the construction let $d^0 : \lambda_0 \to H(\lambda_0)^L$ be an enumeration of $H(\lambda_0)^L$ (where λ_0 is the least singular cardinal). Set $\mathbb{P}_0 = \{\emptyset\}$, $h_0 = \emptyset$, and $e^0 = \{\langle \emptyset, \emptyset \rangle\}$.

Successor Stage
Suppose the good iteration $(\mathbb{P}_i, h_i)$ is defined, we wish to define $(\mathbb{P}_{i+1}, h_{i+1})$. Choose e^i to be an enumeration of all good conditions in $\mathbb{P}_i$ (there are at most κ many of these—for simplicity, we assume that for $i > 0$, the domain of e^i is always κ).

If i is a singular cardinal, fix a name $\dot{d}^i$ such that

$$1_{\mathbb{P}_i} \Vdash \text{``}\dot{d}^i : i \to \dot{H}(i)^{M[\tilde{G}_i]}$$

$$\text{is an enumeration of the elements of } \dot{H}(i)^{M[\tilde{G}_i]}, \text{''} \qquad (4.23)$$

and

$$1_{\mathbb{P}_i} \Vdash \text{``}(\forall j < i) \, \dot{d}^i \text{ end-extends } \dot{d}^j.\text{''} \qquad (4.24)$$

We identify the range of $\dot{d}^i$ with formulas in $L_{\infty,\omega}^{M[\tilde{G}_i]} \cap \dot{H}(i)^{M[\tilde{G}_i]}$.

$\dot{Q}_i$ will be of the form $\dot{W}_i * \dot{C}_i$, where $\dot{W}_i$ and $\dot{C}_i$ are both good iterations.

Let i be arbitrary (i.e. not necessarily a singular cardinal). Let $b(i) = \langle j, k, m \rangle$. $\dot{W}_i$ will be defined with respect to a good condition p_k^j which is the kth good condition of $\mathbb{P}_j$ in the enumeration e^j and with respect to the mth formula $\dot{\varphi}$ in $\dot{H}(i^*)^{M[\tilde{G}_{i*}]}$ enumerated by $\dot{d}^{i^*}$:

$$1_{\mathbb{P}_i} \Vdash \dot{d}^{i^*}(m) = \dot{\varphi}, \tag{4.25}$$

where (recalling Definition (4.22)):

- i^* is i if i is a singular cardinal,
- i^* is $\bar{i}$, if i is not a singular cardinal and $i > \lambda_0$,
- i^* is 0 if $i < \lambda_0$. If $i^* = 0$, we take $H(i^*)^{M[\tilde{G}_{i*}]}$ to denote $H(\lambda_0)^L$.

Suppose first that there is a pair $(\dot{W}, f)$ forced by $1_{\mathbb{P}_i}$ to be a good iteration, which satisfies

$$\mathrm{rng}(h_i) \cap \mathrm{rng}(f) = \emptyset, \text{ and } \mathrm{rng}(f) \cap X_j = \emptyset \text{ for all } j < i, \tag{4.26}$$

and such that the condition $p_k^j \hat{\ } 1$ in $\mathbb{P}_i * \dot{W}$ kills $\dot{\varphi}$. Define

$$\dot{W}_i = (\dot{W}, f), \text{ and } h_i' = h \uplus f. \tag{4.27}$$

If no such $(\dot{W}, f)$ exists, set

$$\dot{W}_i = \{\emptyset\}, \text{ and } h_i' = h. \tag{4.28}$$

Finally, let $(\dot{C}_i, f')$ code the following up to three pieces of information on top of X_i (see Definition 4.22).[13]

(i) If $\dot{W}_i$ is nonempty, code the killing of $\dot{\varphi}$ by forcing pattern $ABAB$. If $\dot{W}_i$ is empty, code the non-killing of $\dot{\varphi}$ by forcing pattern $ABABABAB$.
(ii) Code the set of all $j < i$ such that the good condition $p_{b(j)(1)}^{b(j)(0)}$ is in $\dot{G}_{b(j)(0)}$.
(iii) If i is a singular cardinal, code the enumeration $\dot{d}^i$.

Let $h_{i+1} = h_i' \uplus f'$.

Limit Stage

Define: $h_i = \bigcup_{j<i} h_j$. If i is a limit of singular cardinals, let $\dot{d}^i$ be a name for $\bigcup_{j<i} \dot{d}^j$. This completes Definition 4.24.

Note that by induction, $(\mathbb{P}, h)$ is a composition of κ-many good iterations, and by Lemma 4.16, it is a good iteration. The length of the good iteration $(\mathbb{P}, h)$ is some $\mu < \kappa^+$ of cofinality κ.

---

[13]By the definition of coding in Definition 4.22, f' automatically satisfies the conditions in (4.26), so the composition $\mathbb{P}_i * \dot{W}_i * \dot{C}_i$ is defined correctly.

Let G be a $\mathbb{P}$-generic filter. As $i \cap X_i = \emptyset$ for every $i < \kappa$ (see (4.21)), and from stage i on $\mathbb{P}$ does not use cardinals in $X_j, j < i$, it follows that

$$H(i)^{M[\tilde{G}_i]} = H(i)^{M[\tilde{G}]}. \tag{4.29}$$

Also, because $X_i \cap \lambda_0$ is empty for every i, $H(\lambda_0)^L = H(\lambda_0)^{M[\tilde{G}]}$, which explains the definition of d^0 in Stage 0.

By design, the $\dot{C}_i$'s of the forcing code in a lightface way in $M[\tilde{G}]$ the following objects:

- The set H of all $i < \kappa$ such that the condition $p_{b(i)(1)}^{b(i)(0)}$ is in $G_{b(i)(0)}$.
- The wellordering

$$D = \bigcup_{i \in X} (\dot{d}^i)^{G_i} \tag{4.30}$$

of all elements of $M[\tilde{G}] = \bigcup_{i \in X} H(i)^{M[\tilde{G}_i]}$.

By nature of the wellordering D, $M[\tilde{G}]$ is a model of $V = HOD$.

It remains to verify that in $M[\tilde{G}]$, the outer model theory with parameters $OMT(M[\tilde{G}])$ is lightface definable.

Claim 4.25 *Let φ be in $L_{\infty,\omega} \cap M[\tilde{G}]$. The following are equivalent:*

(i) *There is no outer model of $M[\tilde{G}]$ where φ holds.*
(ii) *There exists $i < \kappa$ such that i is in H and $b(i)(2)$ is the index of φ in the initial segment of the wellordering D coded at X_{i^*}, where i^* is defined from i as in the items below (4.25), and $\dot{C}_i$ codes the killing of φ.*

Proof Assume (i) holds. Then there is a condition $p_0 \in G$ which forces it. Let $\tilde{i} < \kappa$ be such that the support of p is bounded in $\tilde{i}$ in the sense that for every $j > \tilde{i}, j < \kappa$, $p(j)$ is the weakest condition in $\dot{Q}_j$. Choose i such that $b(i) = \langle j, k, m \rangle$ satisfies $\tilde{i} < j$, p_k^j is the restriction of p_0 to $\mathbb{P}_j$ and m is the index of $\dot{\varphi}$ in the enumeration $\dot{d}^{i^*}$.[14] Then at stage i, it was possible to choose $(\dot{W}, f)$ such that $p_k^j \hat{\ } 1$ kills φ—namely the tail of the iteration $\mathbb{P}$ from $\mathbb{P}_i$ is an example of such $\dot{W}$. Accordingly, $\dot{C}_i$ codes the killing of φ.

Assume the negation of (i). Then there can be no such i as in (ii): if there is such i, then $\dot{Q}_i$ contains $\dot{W}_i$ which kills φ. But this is impossible if the negation of (i) holds.

This ends the proof of Theorem 4.18. $\quad\square$

---

[14] As the referee remarked, it may happen that φ (if considered as a concrete element of $H(i)^{M[\tilde{G}_i]}$ for some i, as we do) may not be in the range of the d^i's; however, a formula forced by the empty condition to be equivalent to φ will always be in the range of the d^i's.

4.4 Some Generalizations

By definability of the outer model theory in $M[\tilde{G}]$ of Theorem 4.18, $M[\tilde{G}]$ is a model where one can define the generalized notion of satisfaction with respect to outer models.

Definition 4.26 Let T be a theory extending ZFC, and φ a sentence, both T and φ in $L_{\infty,\omega} \cap M[\tilde{G}]$. We write $T \models_{OM} \varphi$ iff φ holds in every outer model of $M[\tilde{G}]$ satisfying T.

Theorem 4.27 *The relation $\models_{OM}$ is definable in $M[\tilde{G}]$ of Theorem 4.18.*

Proof Let T, φ be in $L_{\infty,\omega} \cap M[\tilde{G}]$. View ZFC as a single infinitary sentence and denote by ψ the infinitary sentence such that $T = ZFC \& \psi$. In $M[\tilde{G}]$, $T \models_{OM} \varphi$ iff there is no outer model of $\psi \& \neg\varphi$ iff $(\neg\psi \vee \varphi)$ is in $OMT(M[\tilde{G}])$.

Instead of working with the outer models of M, one might also study the *inner models of M.*

Definition 4.28 Let V be the universe of sets and let $M \in V$ be a transitive model of ZFC. We say that $N \in V$ is an inner model of M if N is a transitive model of ZFC with the same ordinals as M, and $N \subseteq M$.

Note that we do not require that N be definable in M.

As in the case of outer models, there may be more inner models of M as the universe V enlarges. However, there is a sentence $\varphi^* \in L_{\infty,\omega} \cap \mathrm{Hyp}(M)$ such that there is an inner model $N \subseteq M$ of φ in some extension of V where M is countable iff $ZFC + \varphi^*$ is consistent.

In analogy with Definition 2.3, let us define:

Definition 4.29 Let $M \in V$ be a transitive model of ZFC of size κ. We define the *inner model theory of M*, denoted $\mathrm{IMT}(M)$, as follows

$$\mathrm{IMT}(M) = \{\varphi \mid \text{there is no inner model } N \text{ of } M,$$

$$N \in W, \text{ such that } N \models \neg\varphi\}, \qquad (4.31)$$

where φ is an infinitary formula in $L_{\infty,\omega} \cap N$ with parameters in N and W is the model $V[G]$, where G is a generic filter for the Levy collapsing forcing which collapses κ to ω.

Theorem 4.18 can be easily modified to yield:

Theorem 4.30 *Assume $V = L$. Let κ be the least inaccessible, and let $M = L_\kappa$. There is a good iteration $(\mathbb{P}, h)$ in V such that if G is $\mathbb{P}$-generic over V, then for some set $\tilde{G}$, which is defined from G, $M[\tilde{G}]$ is a model of ZFC in which $\mathrm{IMT}(M[\tilde{G}])$ is lightface definable. Moreover, $M[\tilde{G}]$ is a model of $V = HOD$.*

Proof The proof proceeds exactly as the proof of Theorem 4.18, with the following exception. At stage i, if possible choose a good iteration $\dot{W}_i$ such that $p_k^{j} \hat{\ } 1$ in $\mathbb{P}_i * \dot{W}_i$ forces that there is an inner model for $\dot{\varphi}$ (more precisely forces that $ZFC + \varphi^*$ is consistent, where φ^* is as in the paragraph preceding Definition 4.29). The property of having an inner model is upwards persistent, as is the property of being killed in the case Theorem 4.18. This persistence makes it possible to define the inner model theory in the final model $M[\tilde{G}]$ as in Claim 4.25.⊣

By routine modifications, one can get a model $M[\tilde{G}]$ where both $\mathrm{OMT}(M[\tilde{G}])$ and $\mathrm{IMT}(M[\tilde{G}])$ are lightface definable. Or even more generally, one can define the *compatible model theory* of M, $\mathrm{CMT}(M)$, which contains all formulas which hold in all models compatible with M, where M and N of the same ordinal height are compatible if there is N^* of the same ordinal height such that $M, N \subseteq N^*$. Again, Theorem 4.18 can be generalized so that in $M[\tilde{G}]$, $\mathrm{CMT}(M[\tilde{G}])$ is lightface definable.

5   Open Questions

One property of omniscience which we have not discussed yet is the robustness of the notion in terms of its preservation. Suppose for instance that M is omniscient and we extend M by a set-forcing in the sense of M. Is $M[G]$ still omniscient? If the omniscience of M is witnessed by large cardinals, then $M[G]$ remains omniscient by Stanley's result (see Theorem 1). We do not know whether this holds in general:

Q1. Suppose M is an omniscient model. Is a set-generic extension of M still omniscient?

More specifically, we can ask whether one can modify our forcing construction to obtain an omniscient model which remains omniscient in forcing extensions of a certain type. We may reformulate it as asking for a model whose omniscience is indestructible for some non-empty collection of forcings.

Q2. Can one modify the present forcing to obtain an omniscient model indestructible for a certain non-empty collection of forcing notions?

Using Tarski's undefinability of truth, it is easy to see that L cannot be omniscient. However this does not extend to inner models for large cardinals by Theorem 1. An obvious question is therefore the following:

Q3. Suppose $V = K$ is omniscient, where K is the Dodd-Jensen core model. Does there exist a proper class of ω_1-Erdős cardinals in V?

Q4. The iteration $\mathbb{P}$ in Theorem 4.18 has some length $< \kappa^+$. Is it possible to show that $\mathbb{P}$ is actually a subset of $\mathrm{Hyp}(M)$? Or more generally, can one define an iteration $\mathbb{R}$ which achieves the results of Theorem 4.18 and is a subset of $\mathrm{Hyp}(M)$?

Note that with regard to Q4, our construction shows that $\mathbb{P}$ is contained in $\mathrm{Hyp}_2(M)$, the least Σ_2-admissible set containing M as an element (we just need an oracle for consistency, which is Π_1).

In Remark 2.2 we said that there may be other ways to define the collection of outer models to which we refer in defining $\mathrm{OMT}(M)$. We also noted that there are some obvious restrictions which should be considered to have the notion behave reasonably. The following question is relevant in this respect:

Q5. From similar assumptions as in Theorem 4.18, is there a good iteration, or at least a cardinal-preserving iteration, $\mathbb{P}$ and $M \subseteq M^*$ in $V[G]$, where G is a $\mathbb{P}$-generic filter, such that the outer model theory of M^* is definable, with the outer models restricted to be elements of $V[G]$?

Q6. What is the consistency strength of having M in which OMT(M) is lightface definable? By Theorem 4.18, the upper bound is ZFC plus "there is an inaccessible cardinal." Can this be improved to ZFC + "there is a standard model of ZFC"?

With regard to Q6, note that our construction actually gives a better upper bound than inaccessibility—for the proof of Theorem 4.18, it suffices that κ is inaccessible in $\mathrm{Hyp}_2(V_\kappa)$, the least Σ_2 admissible set containing V_κ as an element.

Appendix

Omniscience from Large Cardinals

In unpublished work [3], Mack Stanley proved that if M contains many Ramsey cardinals, then M is omniscient. The argument uses Barwise's Theorem 1.2 and the theory of iterated ultrapowers (for measurable cardinals) or sharps (for Ramsey cardinals) to "stretch" properties from rank-initial segments of M to the whole of M, thus making it possible to capture a higher-order property of M in a rank-initial segment of M.

With the permission of Stanley, we give here an outline of his argument that the existence of many measurable cardinals in M implies omniscience.

If M is a transitive set, let Ord(M) denote the ordinal $\mathrm{Ord} \cap M$. Also, let us denote by *M-logic* the fragment $L_{\infty,\omega} \cap \mathrm{Hyp}(M)$.

Theorem 1 (M. Stanley) *Suppose that M is a transitive set model of ZFC. Suppose that in M there is a proper class of measurable cardinals, and indeed this class is $\mathrm{Hyp}(M)$-stationary, i.e. Ord(M) is regular with respect to $\mathrm{Hyp}(M)$-definable functions and this class intersects every club in Ord(M) which is $\mathrm{Hyp}(M)$-definable. Then OMT(M) is M-definable.*

Proof Using *M*-logic we can translate the statement that a first-order sentence φ (with parameters from M) holds in some outer model of M to the consistency of a sentence φ^* in *M*-logic, a fact expressible over $\mathrm{Hyp}(M)$ by a Π_1 sentence. Using this we show that the set of φ which hold in some outer model of M is M-definable, and from this it follows that OMT(M) is also M-definable.

As Ord(M) is regular with respect to $\mathrm{Hyp}(M)$-definable functions we can form a club C in Ord(M) such that for κ in C there is a Σ_1-elementary embedding from $\mathrm{Hyp}((V_\kappa)^M)$ into $\mathrm{Hyp}(M)$ (with critical point κ, sending κ to Ord(M)). Indeed C can be chosen to be $\mathrm{Hyp}(M)$-definable.

For any κ in C let φ_κ^* be a sentence of $(V_\kappa)^M$-logic such that φ holds in an outer model of $(V_\kappa)^M$ iff φ_κ^* is consistent (a Π_1 property of $\mathrm{Hyp}((V_\kappa)^M)$). By elementarity, φ_κ^* is consistent iff φ^* is consistent.

Now suppose that φ holds in no outer model of M, i.e. φ^* is inconsistent. Then φ_κ^* is inconsistent for all κ in C and since the measurables form a $\mathrm{Hyp}(M)$-stationary class, there is a measurable κ such that φ_κ^* is inconsistent.

Conversely, suppose that φ_κ^* is inconsistent for some measurable κ. Now choose a normal measure U on κ and iterate $\mathrm{Hyp}((V_\kappa)^M)$ using U for $\mathrm{Ord}(M)$ steps to obtain a structure $\mathrm{Hyp}(M^*)$. By elementarity, the sentence φ^* which asserts that φ holds in an outer model of M^* is inconsistent. But M^* is an inner model of M, so also the sentence asserting that φ holds in an outer model of M is inconsistent.

Thus φ^* is consistent exactly if φ_κ^* is consistent for all measurable κ, and this is first-order expressible. $\dashv$

Forcing Omniscience

Mack Stanley independently discovered an easier construction of an omniscient model. However, his proof does not ensure $V = HOD$ in the final model (compare with Theorem 4.18). For the benefit of the reader, we state the result.

Theorem 2 (M. Stanley) *Work in L and let κ be inaccessible. There exists $\mathbb{P}(A) \subseteq L_\kappa$ such that if G is $\mathbb{P}(A)$-generic over L, then $L[G]$ is a cofinality preserving extension in which κ remains inaccessible, and in $L_\kappa[G]$ the set of all sentences of the language of set theory with parameters in $L_\kappa[G]$ that hold in all outer models of $L_\kappa[G]$ (calculated in a universe in which κ is countable) is definable without parameters in $L_\kappa[G]$.*

Proof Set $M = L_\kappa$, where κ is inaccessible in L.

Working in L, define $\mathbb{P}(\kappa)$ to be the Easton support product of the Cohen forcing $\mathrm{Add}(\aleph_{2\alpha+1}, \aleph_{2\alpha+3})$ for $\alpha < \kappa$.

For $A \subseteq \kappa$ in L, set

$$\mathbb{P}(A) = \{p \in \mathbb{P}(\kappa) \mid p(\alpha) = \emptyset \text{ for all } \alpha \in \kappa \setminus A\}.$$

Note that if $A \subseteq B \subseteq \kappa$, then $\mathbb{P}(A) \subseteq \mathbb{P}(B)$ and $M^{\mathbb{P}(A)} \subseteq M^{\mathbb{P}(B)}$. Furthermore, if G is $\mathbb{P}(B)$-generic over L, then $G \cap \mathbb{P}(A)$ is $\mathbb{P}(A)$-generic over L.

Working in L, define $A_\alpha \subseteq \kappa$ by recursion on α. Start by setting $A_0 = \emptyset$. Then declare that β belongs to $A_{\alpha+1}$ when either $\beta \in A_\alpha$ or β codes a pair (p, φ) where $p \in \mathbb{P}(A_\alpha)$ and φ with parameters from $M^{\mathbb{P}(A_\alpha)}$ is such that $p \Vdash_{\mathbb{P}(A_\alpha)} \varphi \in$ $\mathrm{OMT}(L[\dot{G}])$.

If α is a limit, set $A_\alpha = \bigcup_{\gamma < \alpha} A_\gamma$. Finally, set $A = A_\alpha$ where $A_\alpha = A_{\alpha+1}$. Note that A is definable over $L[G]$ for $\mathbb{P}(A)$-generic G.

Claim 3 *For $\mathbb{P}(A)$-generic G, $\varphi(x)$ is in $\mathrm{OMT}(L[G])$ iff in $L[G]$ there are a $\mathbb{P}(A)$-name σ, a generic $\bar{G}$ for $\mathbb{P}_A|\delta$ for some singular δ and a condition p in $\bar{G}$ such that (a code for) the pair (p, σ) is in the (definable) predicate A and $\sigma^{\bar{G}}$ equals x.*

Proof The direction left-to-right is easy, as we can just take $\sigma^G = x$, $\bar{G}$ to be $G \cap \mathbb{P}_A|\delta$ for some large enough δ and p in G to force $\varphi(x)$ into $\mathrm{OMT}(L[\dot{G}])$. Conversely, the right-hand-side implies that $\varphi(x)$ belongs to $\mathrm{OMT}(L[G^*])$ where G^* agrees with $\bar{G}$ below δ and with G above δ (G^* is generic as G above δ does not add subsets of δ), and therefore to $\mathrm{OMT}(L[G])$ as $L[G]$ contains $L[G^*]$. ⊣

The Claim shows that $\mathrm{OMT}(L[G])$ is definable in $L[G]$ for $\mathbb{P}(A)$-generic G, and therefore finishes the proof of Theorem 2. ⊣

Acknowledgements Both authors acknowledge the support of JTF grant Laboratory of the Infinite ID35216. Sy Friedman acknowledges the support of FWF grant P25671.

References

1. J. Barwise, *Admissible Sets and Structures* (Springer, Berlin, 1975)
2. K. Kunen, *Set Theory: An Introduction to Independence Proofs* (North-Holland, Amsterdam, 1980)
3. M.C. Stanley, Outer model satisfiability, preprint

The Search for New Axioms
in the Hyperuniverse Programme

Sy-David Friedman and Claudio Ternullo

Abstract The Hyperuniverse Programme, introduced in Arrigoni and Friedman (Bull Symb Log 19(1):77–96, 2013), fosters the search for new set-theoretic axioms. In this paper, we present the procedure envisaged by the programme to find new axioms and the conceptual framework behind it. The procedure comes in several steps. Intrinsically motivated axioms are those statements which are suggested by the standard concept of set, i.e. the 'maximal iterative concept', and the programme identifies higher-order statements motivated by the maximal iterative concept. The satisfaction of these statements ($\mathbb{H}$-axioms) in countable transitive models, the collection of which constitutes the 'hyperuniverse' ($\mathbb{H}$), has remarkable first-order consequences, some of which we review in Sect. 5.

1 New Set-Theoretic Axioms

Over the last years, there has been an intense debate within the set-theoretic community concerning the acceptance or non-acceptance of several set-theoretic statements such as V=L, large cardinals, axioms of determinacy (AD, PD, $AD^{L(\mathbb{R})}$) or forcing axioms (MA, PFA, etc.) and the discussion seems to be nowhere near being settled.

The received view concerning an axiom is that it should be 'self-evident', i.e., that it should be immediately, and with little effort, acknowledged as true. If such a

Originally published in S. Friedman, C. Ternullo, The search for new axioms in the Hyperuniverse Programme, in *Philosophy of Mathematics: Objectivity, Cognition and Proof*, ed. by F. Boccuni, A. Sereni. Boston Studies in the History and Philosophy of Science (Springer, 2016), pp. 165–188.

S.-D. Friedman (✉) • C. Ternullo
Kurt Gödel Research Center for Mathematical Logic, University of Vienna, Vienna, Austria
e-mail: sdf@logic.univie.ac.at; claudio.ternullo@univie.ac.at

© Springer International Publishing AG 2018
C. Antos et al. (eds.), *The Hyperuniverse Project and Maximality*,
https://doi.org/10.1007/978-3-319-62935-3_8

161

view is still to be held, then there is no hope to accept the aforementioned statements as new axioms.[1]

But even if one discards the 'self-evidence view' as inapplicable, there are still deep issues which have to be addressed by anyone supporting the acceptance of one or more of the statements mentioned above and, more generally, of any axiom candidate.

First of all, there is often a lack of intrinsic motivation for such statements, where, by 'intrinsic', as explained at length in the sections below, we mean 'required by the concept of set'. Secondly, the view that a new axiom should be accepted as true of the realm of sets has been seriously challenged by the independence phenomenon and the related existence of a set-theoretic multiverse: it is often relatively easy to produce a universe of sets which contradicts a given set-theoretic statement. Finally, all new axiom candidates are first-order and one main worry we want to bring out in this paper is precisely that first-order principles may be too weak to capture further properties of the cumulative set-theoretic hierarchy.

One clear preliminary upshot of the informal considerations above is the following: it is unlikely that any new first-order axiom candidate will be accepted on its own as an intrinsically motivated principle of set theory. Granted, it might still be accepted on purely extrinsic grounds, but it is not clear that this would be sufficient evidence for its acceptance.

In this paper, we are going to propose an alternative way to identify new intrinsically motivated set-theoretic axioms, which originates from the conceptual framework of the Hyperuniverse Programme, as detailed in [3], and which fosters a revisionary conception of what a 'new' axiom is. In our view, new axioms are higher-order set-theoretic principles, more specifically principles expressing the maximality of the universe of sets (V). The latter are strong mathematical propositions, some of which have been gradually isolated and examined in recent years in work by the first author and others, and, more recently, by the first author and Honzik.[2] We believe that there is a sense in which such propositions, as will be presented in Sect. 5, can legitimately claim to be motivated by the concept of set and, by virtue of this, be viewed as intrinsically motivated new axioms.

It should be mentioned that all of these statements have striking first-order set-theoretic consequences, which we will describe in more detail in the next sections and this fact, although not representing an intrinsic justification for their acceptance, indisputably adds to their mathematical attractiveness.

One further goal of the Hyperuniverse Programme is to find one single 'optimal' maximality principle, whose acceptance would, thus, lead to identifying one single collection of first-order consequences. Therefore, our foundational project fosters

---

[1]Of course, it is also as much debatable that the standard axioms of set theory, that is, ZFC, are all 'self-evident'. A very natural case in point is the Axiom of Choice, but one may have equally reasonable reservations on the Axioms of Infinity, Replacement or Foundation. A thorough discussion of some of these issues can be found in [25, 26, 36] and [31].

[2]See, in particular, [2, 3, 10] and [12].

the view that the procedure described here might also count as a procedure to find solutions to the open problems of set theory. However, the notion of 'solution', here, is inevitably as much revisionary as that of 'new' axiom.

The structure of the paper is as follows. In Sects. 2–3, we briefly discuss the features of 'intrinsic' evidence and set forth our conception of the set-theoretic universe as being a 'vertical' multiverse. In Sect. 4, we introduce the hyperuniverse as our auxiliary multiverse, wherein one can investigate the consequences of the maximality of V, through the use of V-logic. In Sect. 5, we enunciate maximality principles for V which, in our view, are motivated by the concept of set. In Sect. 6 we discuss the notion of 'new' axiom *qua* $\mathbb{H}$-axiom, then, in Sect. 7, we proceed to make some final considerations. Finally, in Appendix, as an appendix, we present some results which show that a heavily investigated collection of new set-theoretic axioms, absoluteness axioms, which has recently received a lot of extrinsic support, may fall short of the requirements described above.

2 Intrinsic Evidence for New Axioms

2.1 Brief Remarks on Ontology and Truth

In the next pages, we will be making frequent reference to issues of ontology and truth and it is maybe appropriate to briefly address these issues before examining the notion of intrinsic evidence.

On the grounds of what the programme aims to yield, i.e. new set-theoretic axioms, it is entirely natural to ask whether new axioms should be seen as 'true' statements of set theory, and in what sense. We will make it clear in the next section in what sense they should be viewed as 'true', but first we want to say something more general about 'truth'.

The programme's position is that axioms do not reflect truth in an independent realm of mathematical entities. It is rather the *concept of set* that plays a key role in our foundational project. As we shall see in the next subsection, and as is commonly acknowledged in set theory, the concept of set is instantiated by a specific mathematical structure, the *cumulative hierarchy*, but it does not automatically provide us with a fully determinate collection of properties of sets in this structure. Now, we believe that it is possible to derive properties of the concept of set which provide us with an indication of what further properties the set-theoretic hierarchy should have.

There is possibly a hint of realism in this position, insofar as we view the concept of set as being a 'stable' feature of our experience of sets and we subscribe to its stability in the sense that we do not question the ZFC axioms which are true of it.

However our view follows an overall epistemological concern, that of securing the truth of new axioms and of their first-order consequences through setting forth an alternative evidential framework for them which does not imply a pre-formed ontological picture. Therefore, ontology, in the most robust sense of the word, does not play a pre-eminent role in our project.

If there is a detectable ontological framework within our account, that is the *core* structure we identify in Sect. 3, i.e. the tower-like multiverse of V_κ's, where κ is a strongly inaccessible cardinal. In turn, properties of this multiverse will motivate the adoption of one further ontological construct, the hyperuniverse, which consists of all countable transitive models. Neither multiverse is given *a priori*.

The maximality principles we will be concerned with quantify over extensions of V. However, our language is that of first-order ZFC, therefore maximality principles do not formally involve talk of classes. So, in the end, we have sets, and nothing else.

This ontological view might be seen as entailing a conception of truth that lacks the requisite strength to see axioms as 'true'. But in fact, as we will see, the concept of set is adequate to make strong claims about set-theoretic maximality, for instance alternative conceptions of *vertical* maximality are ruled out as unwarranted on the grounds of the concept itself. It is true, however, that in order to have models where new axioms 'live' one has to shift to countable transitive models and, thus, to a different framework of truth. But this is not overall necessary. One can still appreciate the force of maximality principles within the whole V and, thus, stick to a vision of truth and ontology entirely befitting the concept of set. Further details on our positions will be given in the next few sections.

2.2   Two Sources of Evidence

Although the distinction is not entirely perspicuous, since [14], it has become fairly commonplace in the literature to refer to two main different forms of evidence for the acceptance of an axiom as 'intrinsic' (internal) and 'extrinsic' (external) evidence. Very roughly, the distinction can be glossed as follows. Intrinsic evidence for an axiom is that following from the concept of set, whereas extrinsic evidence relates to the fruitfulness and success of an axiom, possibly also outside set theory. In other terms, an axiom may be accepted either because it expresses a 'necessary' property of sets or because it is corroborated by good results (and interesting practice) or for both reasons.

The issue of whether this distinction has any plausibility is beyond the scope of this article and, for the sake of our arguments, we will not challenge it. However, it should be noticed that, in our opinion, in opposition to the point of view expressed by some authors, 'intrinsicness' does not imply the view we have just informally rejected, that axioms should be 'self-evident'.[3] In fact, an axiom may be true of the concept of set and not be immediately graspable as true. This is because not all true

---

[3]For instance, the equating of 'intrinsicness' with 'self-evidence' is clearly hinted at in the following passage of [25, p. 482]: 'The suggestion is that the axioms of ZFC follow directly from the concept of set, that they are somehow 'intrinsic' to it (obvious, self-evident [...]'.

properties of the concept of set are immediately graspable. Therefore, arguably, it is our task to gradually uncover such properties, by clarifying the content of our intuitions. Ideally, we should be able to determine the properties of the concept of set, and possibly of other properties of clear set-theoretic relevance, by following what Potter calls the 'intuitive' method:

> The intuitive method invites us instead to clarify our understanding of the concepts involved to such an extent as to determine (some of) the axioms they satisfy. The aim should be to reach sufficient clarity that we become confident in the truth of these axioms and hence, but only derivatively, in their consistency. If the intuitive method is successful, then, it holds out the prospect of giving us greater confidence in the truth of our theorems than the regressive method.[4]

But just what form of *intuition* does the intuitive method presuppose? Is intuition alone sufficient to justify the adoption of set-theoretic axioms? These are no doubt vexing questions on which we cannot fully dwell in this paper. However, some considerations are in order.

Intuition is, sometimes, construed in the Gödelian sense, as a faculty of perception which provides us with detailed information on mathematical objects. However, as we have seen, the programme does not commit itself to any form of object-realism. Therefore, our appeal to intuition and the intuitive method should be construed in the following way: as hinted at by Potter, we seem to have the ability to single out the relevant concepts and properties that are derivable from the concept of set.

As we shall see, the cumulative hierarchy instantiates the concept of set (as described below) and its maximality seems to follow naturally. Now, does that mean that we need to have access to platonistic entities in order to successfully carry out this task? We do not have a definite answer to this question, but, on the grounds of the considerations made in the previous section, it seems natural to tentatively rule out such possibility: realism should not extend so far as to postulate the existence of an independent, pre-formed ontology, but rather only postulate a stable concept of set, from which further properties of sets can be derived.

It should be noticed that we do not hold that our maximality principles, such as the IMH, become thus straightforwardly 'intrinsically justified'. What we believe to be intrinsically justified by the concept of set is rather the feature of the maximality of the cumulative hierarchy and, consequently, its *maximal extendibility*. Maximality principles take different forms, so we could, at most, say that such forms are *intrinsically motivated*, insofar as maximality, in general, is an intrinsically justified feature of the concept of set.

---

[4]The 'regressive', as opposed to 'intuitive', method mentioned by Potter holds that '... the object of a good axiomatization is to retain as many as possible of the naive set-theoretic arguments which we remember with nostalgia from our days in Cantor's paradise, but to stop just short of permitting those arguments which lead to paradox' [31, p. 36].

There is, of course, further work to be done to establish the stronger claim that some of our maximality principles are intrinsically justified, and we can only hope that further intrinsic evidence may, 1 day, substantially help bolster this claim.

2.3 The Maximum Iterative Concept

Although it is not clear what Cantor took 'sets' to be at the beginning of his set-theoretic investigations, over the years increasingly wide agreement has been reached that the concept implies an account of the iterative formation of all sets along stages indexed by the ordinals. By this account, each set belongs to a stage of what has been called the *cumulative hierarchy*, starting with the empty set, and then iterating the power-set at all successor stages and the union of all sets formed in previous stages at limit stages of the hierarchy.

In fact, the iterative conception is more correctly referred to as the 'maximal iterative conception':

[MIC] (1) All sets which can be formed at each stage are actually formed. (2) The formation of sets should continue as far as possible.

The vocabulary used in the formulation of the [MIC] has been variously interpreted as hiding *temporal*, *modal* and, in general, *metaphysical* forms of mutual dependency among sets, elements and stages, and this aspect is responsible for some sort of conceptual opacity in the [MIC].[5]

However, leaving aside such troubles for the time being, it seems clear that the basic rationale underlying the [MIC] is that the procedures to form sets ought not to be constrained by 'internal' limitations, that is, by mathematical principles hindering the maximisation and the continuation of such formation. This line of thought has been distinctly referred to in a fortunate article by Bernays as 'quasi-combinatorialism', the conceptual attitude which would allow one to treat and manipulate all mathematical objects, both finite and infinite, and combinations thereof, as fully determinate objects of thought.[6] The maximal character of the [MIC], therefore, can be motivated using a 'quasi-combinatorial' conceptual framework, as, by this, one does not put any constraint on the class of producible sets.

These are well-known facts. Now, we want to take a step further. The intuitive method invites us to focus our attention on one specific feature of the [MIC] that we are going to use extensively in the rest of the paper. Suppose one takes the cumulative hierarchy to be a determined object of thought, V, the universe of sets. Then the [MIC] may also imply one further principle of 'plenitude', which can be formulated in the following way:

---

[5]For an exhaustive overview of these issues see, again, [31], in particular, pp. 34–41, or [21].

[6]In Bernays' own words, 'quasi-combinatorialism' ultimately refers to '…an analogy of the infinite with the finite.' See [7], reprinted in [6, p. 259].

- [MaxExt] Given a universe of sets, all possible extensions of it which can be formed are actually formed.

The 'extensions' referred to in [MaxExt] are given by the creation of 'new' sets in the only two ways we know, either adding new stages to the hierarchy or 'producing' new subsets at successor stages; thus the principle seems to be perfectly justified in light of the [MIC]. However, the principle seems to shed light on one further dimension in the iteration, insofar as it assumes that the latter should get us beyond the universe itself. This automatically introduces the issue of whether we have grounds to believe that the universe is a determinate (actual) object of thought. As we shall see, there is, in fact, a way to interpret [MaxExt] in a way which remains faithful to its nature, but does not imply this kind of actualism.

In any case, our goal, for the time being, is to re-state the notion of 'intrinsicness' in the following way: intrinsic evidence for the acceptance of an axiom is that related to the [MIC], which, in particular, implies [MaxExt] as one of its features.

3 Conceptions of V: The 'Vertical' Multiverse

3.1 What is V? The Actualism/Potentialism Dichotomy

[MaxExt] seems to imply that the universe can be 'extended' and that there is no limitation on how much it can be extended. Extensions of the universe, as we said, are given by further stages in the cumulative hierarchy or new subsets. However, as anticipated, there is a difficulty in this point of view: the literal sense of the notion of 'extension', implies, at the very least, that what is extended is an object with 'boundaries', that is, a 'delimited' object. Now, it is not clear that the cumulative hierarchy is one such object. As a matter of fact, in the [MIC] there is nothing which commits us to seeing V as a delimited object. On the contrary, it would seem that V is best construed as an open-ended sequence of stages. On the other hand, the standard interpretation of the first-order quantifiers is that they range over the class of *all* sets, as though *all* sets were made available to us by unbounded quantification. So, what (if any) is the fact of the matter?

Debates over the nature of the infinite, whether it be *actual* or *potential* or both, date already to antiquity. The two viewpoints we have summarised above extend this kind of debate to the nature of one specific instance of the infinite, the universe of all sets. *Potentialists* believe that this object is neither actual nor actualisable, whereas *actualists* do.[7]

---

[7] An examination of the potentialist and actualist positions, with reference to the justification of *reflection principles*, is carried out in [23], which also draws upon [33] and [34]. A thoroughly actualist point of view on reflection is expounded in [18]. A potentialist conception is described in [24], which provides a modal account of the axioms of set theory already explored in [17] and [30]. For the early debate on such issues as the nature of the universe of sets, the role of the absolute

Now, [MaxExt] seems to fit best the potentialist viewpoint, insofar as it posits the existence of extensions of V. For, if V is an actualised domain, how could it possibly be extended?

Therefore, on the potentialist viewpoint as befitting [MaxExt], one cannot even speak of V, as there is no such actualised and determinate object as V, but one should rather refer to an endless sequence of initial segments V_α's whose union can always be extended. This means that [MaxExt] implies, at least, that the cumulative hierarchy is open-ended and that new stages in the formation of sets can always be formed. However, as we have seen, this is just one way of extending the universe. Extensions of V are not only extensions of its *height*, but also of its *width*. The width of the universe is given by the power-set operation and, thus, an extension of V in width means that there is also a way to expand the range of the power-set operation.

So, now we have a more complex picture. One could be either an actualist or a potentialist in either height or width, as summarised below:

HEIGHT ACTUALISM: the height of V is fixed, that is, no new ordinals can be added,
WIDTH ACTUALISM: the width of V is fixed, that is no new subsets can be added.
HEIGHT POTENTIALISM: the height of V is not fixed, new ordinals can always be added,
WIDTH POTENTIALISM: the width of V is not fixed, new subsets can always be added.

and combinations thereof.

As said, [MaxExt] seems to commit us to a full-blown form of potentialism, both in height and width. This makes full sense, especially from the point of view of 'quasi-combinatorialism': by this attitude, no internal limitation of the procedures to form new sets should be applicable and this also extends to such large-scale objects as V.

However, there is one difficulty with this view. While height potentialism seems to be robustly supported by our idea of 'adding' new ordinal-indexed stages to the cumulative hierarchy, so that we can always form a sequence of V_α's increasing with α, it is far more problematic to see how extensions of the width of the universe may come in 'stages'. In fact, such extensions as, for instance, the possible set-generic extensions of the universe are not organised in stages at all.

Therefore, whereas our intuitions about the [MIC] seem to suggest that the universe is a fully potential hierarchy of sets, in both height and width, it could be argued that it is simply not possible to make sense of extensions of the universe in width in a way which is in line with the iterative, stage-like character of the [MIC].

The Hyperuniverse Programme has recently fostered a conception which acknowledges the significance of this objection,[8] and that, therefore, follows a conception alternative to full-blown potentialism which historically was first

---

infinite and proper classes, all of which are relevant to the *actualism/potentialism* debate, also see the indispensable [15], as well as [19], and [37], which contains Gödel's late conceptions on V.
[8] See, in particular, [1] and [11].

brought forward by Zermelo. We now proceed to briefly review Zermelo's conception.

3.2 Zermelo's Account: A 'Vertical' Multiverse

As is known, in his seminal paper [39], Zermelo investigates 'natural models' of his axioms, that is, models indexed by *boundary numbers* (fixed ordinals). Zermelo also proves that natural models form a linear hierarchy by inclusion. An example of a natural model of ZFC is given by V_κ, where κ is a *strongly inaccessible* cardinal.

Now, as said, we construe Zermelo's position as a specific one in the actualism *vs* potentialism debate: the Zermelian account is *potentialist in height* and *actualist in width*.

Zermelo's actualism in width follows from the presence of second-order quantifiers in (some of) his axioms. In fact, Zermelo's 1930 axiomatisation is, essentially, second-order. It is this fact that allows him to establish the quasi-categoricity of set theory or, in more rigorous terms, that:

Theorem 1 *Given any two extensional and well-founded structures M_1 and M_2, such that $M_1 \models Z_2$ and $M_2 \models Z_2$ (where Z_2 denotes the axioms of second-order set theory), only three cases can occur: M_1 is isomorphic to M_2, M_1 is isomorphic to a proper initial segment of M_2, or M_2 is isomorphic to a proper initial segment of M_1.*

A trivial consequence of quasi-categoricity is the absoluteness of the power-set operation, which automatically leads one to width actualism. However, our emphasis, here, is on the 'quasi-' bit of his result, since models may still differ in height and, thus, be extendible in a way which clearly suggests height potentialism. In particular, Zermelo construed the sequence of V_α's as stopping points in an *endless* process of potentialisation of an only temporarily actualised universe.

Zermelo vividly recapitulates his approach in the following manner:

> To the unbounded series of Cantor's ordinals there corresponds a similarly unbounded double-series of essentially different set-theoretic models, in each of which the whole classical theory is expressed. The two polar opposite tendencies of the thinking spirit, the idea of creative advance and that of collection and completion [*Abschluss*], ideas which also lie behind the Kantian 'antinomies', find their symbolic representation and their symbolic reconciliation in the transfinite number series based on the concept of well-ordering. This series reaches no true completion in its unrestricted advance, but possesses only relative stopping-points, just those 'boundary numbers' [*Grenzzahlen*] which separate the higher model types from the lower. Thus the set-theoretic 'antinomies', when correctly understood, do not lead to a cramping and mutilation of mathematical science, but rather to an, as yet, unsurveyable unfolding and enriching of that science. ([39], in Ewald [9, p. 1233])

Zermelo's sequence of natural models can also be viewed as a tower-like multiverse, a 'vertical' multiverse, a collection of universes linearly ordered by inclusion.

Unfortunately, at the practical level, the 'vertical' multiverse fits only half of [MaxExt]: extensions in height are now incorporated within this picture, whereas extensions in width are banned. However, as we said, this seems to be more in line with some worries concerning the impossibility, from a mathematical point of view, to account for extensions in width in an orderly fashion.

Therefore, if we want to keep full potentialism and Zermelo's account, we have to find a way to address also extensions in width within this account. This task we accomplish in the second half of the next section, by introducing V-logic.

4   The Hyperuniverse ($\mathbb{H}$): V-Logic

In the previous sections we have established two facts: (1) intrinsic evidence relates to the [MIC], in particular, to one of its features, that is [MaxExt]; (2) as we have seen, [MaxExt] seems to be more in line with a full-blown potentialist picture of the universe. However, there is no way to address extensions of the width of the universe in a way which suits the iterative character of the [MIC], therefore we ought to settle on an account of V wherein the width of the universe is fixed. Such an account is very fittingly provided by the Zermelian 'vertical' multiverse.

Before turning to the programme's maximality principles in the next section, we first have to carry out two tasks: we have to show that there is indeed a way to formulate principles addressing extensions of the universe not only in height but also in width within a Zermelian conceptual framework and, secondly, we have to identify universes where first-order consequences of such principles hold. We start with the latter goal: the hyperuniverse provides an ontological environment where one can investigate consequences of our maximality principles.

4.1   The Hyperuniverse

Let us leave aside, for a moment, the concept of set, the [MIC] which constitutes its full expression, the ensuing picture of the realm of sets as the cumulative hierarchy and let us turn our attention to the techniques used by set-theorists to establish results concerning set-theoretic truth.

As is known, there is only one way to establish the independence of set-theoretic statements from the axioms, i.e. through finding two models wherein that statement and its negation are, respectively, true. If the axioms are consistent, then they cannot prove or disprove such a statement.

There is a wide variety of models that set-theorists investigate: e.g., the constructible universe L, core models K, HOD, $M[G]$ (where G is a generic filter on a forcing poset $\mathbb{P} \in M$) and so forth. The main techniques employed consist in the construction of an *inner model* and of a *forcing extension* of a ground model M. Almost invariably, the ground model used is a *countable transitive model*.

So, the problem is the following: how do all these models relate to the concept of set, which seemed to give rise to a unique picture of the realm of sets, that is,

V? Moreover, does each of these models constitute a separate and, to some extent, alternative ontological construct?

The situation we are to face up to here is direly ambivalent. On the one hand, one could legitimately claim that all model-theoretic constructions are in V, 'reflecting' the universe each in its own particular way. On the other hand, one could say that, if V is a fully determinate construct, something which seems plausible in light of our adoption of the [MIC] and of its associated Bernaysian 'quasi-combinatorialism', then all of these models represent different and, sometimes, mutually incompatible versions of set-theoretic truth, which cannot possibly be amalgamated into one single framework.

Now, call the view that there is a single universe of sets *monism*, whereas let *pluralism* be the view that there are many universes, and that V has no ontological priority. Our approach is alternative to both and may be legitimately called 'dualistic'. Within the programme, we are, in a sense, forced to postulate both the existence of one 'extendible' universe and, at the same time, that of a plural framework containing many universes, where properties of the universe allow the detection of further set-theoretic truth. Now, the models we want to confine our attention to are countable transitive models and our plural framework is defined as follows:

Definition 1 (Hyperuniverse) Let $\mathbb{H}^{ZFC}$ be the collection of all countable transitive models of ZFC. We call $\mathbb{H}^{ZFC}$ the hyperuniverse.[9]

But just why should one confine one's attention only to countable transitive models? Our choice is not related to the concept of set and rather originates from concerns arising from practice: we want to infer new truth (first-order statements) from intrinsically motivated new axioms (maximality principles) and, in order to do this, countable transitive models are not only suitable, but also necessary (more details on this are given below in our discussion of V-logic). Further reasons for adopting the hyperuniverse as a multiverse construct are more precisely substantiated in what follows:

(1) First of all, it should be noticed that $\mathbb{H}$ is *closed under forcing and inner models*, which, as we saw, are the main techniques in the current practice. In other terms, if we start with countable transitive models, the use of forcing and inner models does not require more than and leave us with countable transitive models.

(2) The satisfaction of maximality principles in countable transitive models is also already suggested by the Löwenheim-Skolem theorem: given a statement ϕ, if ϕ is true in V, then ϕ is true in *some* element of the hyperuniverse. However, the notion of 'satisfaction', here, has to be mathematically secured more robustly (see Sect. 4.2 below).

(3) In $\mathbb{H}$, as a consequence of its very definition, there is no ill-founded model, and this fact is perfectly in line with our motivating evidential framework, that is, the [MIC].

---

[9]Henceforth, we shall only use $\mathbb{H}$ to refer to it.

Therefore, the adoption of the hyperuniverse is entirely subservient to achieving the result we wish to attain, that of finding new set-theoretic truth, but, as we have seen, is also well justified in light of different concurrent considerations and, in particular, of the fact that countable transitive models constitute the main tool used by set-theorists to investigate set-theoretic truth, a tool whereby the iterative and well-founded character of the cumulative hierarchy expressed by the [MIC] can be very aptly reproduced in a small-scale context.

4.2   V-Logic

We now proceed to describe how one can make sense of width maximality using V-logic. Such width maximality principles include the IMH, SIMH, IMH# and SIMH#, all of which will be defined in the next section.

As we said, the Löwenheim-Skolem theorem allows one to argue that any first-order property of V reflects to a countable transitive model. However, on a closer look, one needs to deal with the problem that not all relevant properties of V are first-order over V. In particular, the property of V 'having an outer model (a 'thickening') with some first-order property' is a higher-order property. We show now that, with a little care, all reasonable properties of V formulated with reference to outer models are actually first-order over a slight extension ('lengthening') of V.

We first have to introduce some basic notions regarding the infinitary logic $L_{\kappa,\omega}$, where κ is a regular cardinal.[10] For our purposes, the language is composed of κ-many variables, up to κ-many constants, symbols $\{=,\in\}$, and auxiliary symbols. Formulas in $L_{\kappa,\omega}$ are defined by induction: (i) All first-order formulas are in $L_{\kappa,\omega}$; (ii) Whenever $\{\varphi\}_{i<\mu}$, $\mu < \kappa$ is a system of formulas in $L_{\kappa,\omega}$ such that there are only finitely many free variables in these formulas taken together, then the infinite conjunction $\bigwedge_{i<\mu} \varphi_i$ and the infinite disjunction $\bigvee_{i<\mu} \varphi_i$ are formulas in $L_{\kappa,\omega}$; (iii) if φ is in $L_{\kappa,\omega}$, then its negation and its universal closure are in $L_{\kappa,\omega}$. Barwise developed the notion of proof for $L_{\kappa,\omega}$, and showed that this syntax is complete, when $\kappa = \omega_1$, with respect to the semantics (see discussion below and Theorem 2).

Let us now consider a special case of $L_{\kappa,\omega}$, the so-called V-logic. Suppose V is a transitive set of size κ. Consider the logic $L_{\kappa^+,\omega}$, augmented by κ-many constants $\{\bar{a}_i\}_{i<\kappa}$ for all the elements a_i in V. In this logic, one can write a single infinitary sentence which ensures that if M is a model of this sentence (which is set up to ensure some desirable property of M), then M is an outer model of V (satisfying that desirable property). Now, the crucial point is the following: if V is countable, and this sentence is consistent in the sense of Barwise, then such an M really exists

---

[10]Full mathematical details are in [5]. We wish to stress that the infinitary logic discussed in this section appears only at the level of theory as a tool for discussing outer models. The ambient axioms of ZFC are still formulated in the usual first-order language.

in the ambient universe.[11] However, if V is uncountable, the model itself may not exist in the ambient universe, but, in that case, we still have the option of staying with the syntactical notion of a consistent sentence.[12]

We have to introduce one further ingredient, that of an *admissible set*. M is an admissible set if it models some very weak fragment of ZFC, called Kripke-Platek set theory, KP. What is important for us here is that for any set N, there is a smallest admissible set M which contains N as an element—M is of the form $L_\alpha(N)$ for the least α such that M satisfies KP. We denote this M as $\mathrm{Hyp}(N)$.

And we have the following crucial result:

Theorem 2 (Barwise) *Let V be a transitive set model of ZFC. Let $T \in V$ be a first-order theory extending ZFC. Then there is an infinitary sentence $\varphi_{T,V}$ in V-logic such that following are equivalent:*

(1) *$\varphi_{T,V}$ is consistent.*
(2) *$\mathrm{Hyp}(V) \models$ "$\varphi_{T,V}$ is consistent."*
(3) *If V is countable, then there is an outer model M of V which satisfies T.*

By Theorem 2, if we wish to talk about outer models of V ('thickenings', that is, extensions of the width of V), we can do it in $\mathrm{Hyp}(V)$—a slight lengthening of V—by means of theories, without really thickening our V, that is, without postulating that such extensions are real. However, if we wish to have models of the resulting consistent theories, then, using the Löwenheim-Skolem theorem, we can shift to countable transitive models. And this is precisely where the hyperuniverse comes into play.

Now, we also want to make sure that members of the hyperuniverse really witness statements expressing the width maximality of V. One such statement is the Inner Model Hypothesis or IMH (for whose full examination see next section).

V satisfies the IMH if for every first-order sentence ψ, if ψ is satisfied in some outer model W of V, then there is a definable inner model $V' \subseteq V$ satisfying ψ. The formulation of IMH requires the reference to all outer models of V, but with the use of infinitary logic, we can formulate IMH syntactically in $\mathrm{Hyp}(V)$ as follows: V satisfies IMH if for every $T = ZFC + \psi$, if $\varphi_{T,V}$ from Theorem 2 above is consistent in $\mathrm{Hyp}(V)$, then there is an inner model of V which satisfies T. Finally, with an application of the Löwenheim-Skolem theorem to $\mathrm{Hyp}(V)$, this becomes a statement about elements of the hyperuniverse.

---

[11] Again, for more details we refer the reader to [5].

[12] This means that the hyperuniverse, although fully justifiable in view of the use of V-logic, can be disposed of, if one only wants to keep the Zermelian multiverse (and its immediate connection with the [MIC] and [MaxExt]).

5   Maximality Principles for *V*

We have now arrived at the crux of our paper. Within the programme, we cast our new axioms as maximality principles about *V* and, after having established, using the notion of satisfaction in *V*-logic, that (1) these principles can be formulated in a Zermelian framework and (2) they are satisfied by members of $\mathbb{H}$, we can also see what first-order consequences they have through the study of countable transitive models, i.e. elements of the hyperuniverse.

First, there is one point which should be emphasised again: as the reader will see in a moment, the maximality principles that have been formulated within the programme all address extensions of *V* and, therefore, in our view, they specify ways such extensions, as postulated by [MaxExt], should be conceived of. Thus all such principles can be seen as specifications of [MaxExt]. As our evidential framework for the search for new axioms was given by the [MIC] and these principles follow from this evidential framework quite naturally, we believe that we have in this way found a source for new axioms based on the maximal interactive conception.

Predictably, some principles refer to extensions in height and others to extensions in width. Accordingly, we may say that the former address the vertical maximality and the latter the horizontal maximality of the universe.

Vertical maximality has been recently formulated by the first author and Honzik in terms of a strong form of reflection called #-*generation*. We do not discuss the details here, but refer the reader to their paper [12].

Let us instead examine horizontal maximality. In the programme, this property is expressed by the IMH.

Definition 2 (IMH) If for every first-order sentence ψ, if ψ is satisfied in some outer model *W* of *V*, then there is a definable inner model $V' \subseteq V$ satisfying ψ.

Just to make things as clear as possible, 'outer models', in the definition above, are precisely the formal equivalent of extensions of the universe in width. Moreover, in our view, IMH prescribes the maximality of the universe (by using the language of 'extensions'), insofar as it prescribes its maximality with respect to *inner models*. Universes satisfying the IMH exist in $\mathbb{H}$:

Theorem 3 *Assuming the consistency of large cardinals, there are members of the hyperuniverse which satisfy the IMH.*

The proof is in [13], where it is shown that the consistency of slightly more than the existence of a Woodin cardinal is sufficient. One might question the use of Woodin cardinals here, which may not be intrinsically justified. But note that it is not the *existence* of Woodin cardinals that is needed to obtain the existence of members of $\mathbb{H}$ satisfying the IMH. It is only the consistency of Woodin cardinals that is used as an *auxiliary mathematical tool* in order to construct universes satisfying IMH and we believe that this fact does not commit us to asserting the existence of such cardinals, as 'consistency' is far less than 'existence'. It should be noted, incidentally, that in all members of $\mathbb{H}$ satisfying IMH there are *no* large cardinals at all. Therefore, if one believes that IMH is a correct higher-order principle about V, then one obtains that there are no large cardinals in V.

But the IMH does not take vertical maximality into account. Let IMH# denote the IMH for vertically-maximal, i.e. for #-generated, universes. In other words, M satisfies the IMH# if M is #-generated and whenever a first-order sentence holds in a #-generated outer model of M, it also holds in a definable inner model of M.

Theorem 4 *There are members of the hyperuniverse which satisfy IMH#.*
For a proof see [12]. The attraction of IMH# is that it captures aspects of both vertical and horizontal maximality simultaneously.

We also mention some strengthenings of the principles given above. An *absolute parameter* is a set p which is uniformly definable over all outer models of V which 'respect p', i.e. which preserve cardinals up to and including the cardinality of the transitive closure of p. The SIMH (Strong IMH) is the IMH for sentences with absolute parameters relative to outer models which respect them: if a sentence with absolute parameters holds in an outer model which respects those parameters then it holds in a definable inner model.

A related principle is the CPIMH (Cardinal Preserving IMH). A *cardinal-absolute parameter* is a set p which is uniformly definable over all cardinal-preserving extensions of V. Then CPIMH asserts that if a sentence with cardinal-preserving parameters holds in a cardinal-preserving outer model of V it also holds in a definable inner model of V.

Restricting to #-generated universes yields corresponding principles SIMH# and CPIMH#.

We do not know whether there are elements of $\mathbb{H}$ satisfying SIMH, CPIMH or their #-versions, but it is reasonable to conjecture that they do.[13] We have:

Theorem 5 (see [10])

(a) In all universes satisfying IMH, PD is false, and there are no large cardinals.
(b) All universes which satisfy SIMH, CPIMH or their #-versions also satisfy $\neg$CH.

Thus maximality principles emanating from the Hyperuniverse Programme do indeed have striking first-order consequences.

6 New Axioms as $\mathbb{H}$-Axioms

6.1 *The Nature of $\mathbb{H}$-Axioms*

As we said at the beginning, we do not want to advocate any specific first-order new axiom in this paper, but rather present an alternative conceptual framework whereby higher-order statements are indeed new axioms, which also happen to

---

[13]In particular, there are universes which obey them restricted to the parameter ω_1.

have important first-order consequences. The framework we have presented, in particular the mathematical results detailed in the previous section, lend support to the following conclusion: members of $\mathbb{H}$ satisfying maximality principles have remarkable properties, e.g. in all countable transitive models satisfying IMH, PD is false and in all of them satisfying CPIMH#, CH is false.

Now, let us focus our attention for a moment on ¬PD and ¬CH. These first-order set-theoretic statements are consequences of new axioms that:

(1) hold in 'local' areas of $\mathbb{H}$
(2) are expressed in terms of intrinsically motivated maximality principles as, respectively, IMH and CPIMH#.

By virtue of this, we label IMH and CPIMH# $\mathbb{H}$-axioms, insofar as they hold in specific portions of $\mathbb{H}$ and are intrinsically motivated on the grounds of the [MIC] and [MaxExt].

Again, it is important to emphasise on what grounds our claim can be made: using V-logic, we can characterise the relationship between maximality principles and their consequences as mirroring that between higher-order properties of V and first-order truths in members of $\mathbb{H}$. In particular, in the Hyperuniverse Programme higher-order properties of V are, in a sense, turned into $\mathbb{H}$-axioms, properties of members of $\mathbb{H}$ expressible through (first-order) quantification over $\mathbb{H}$.

Furthermore, we also claim that ¬PD would be, in accordance with our conceptual presuppositions, an intrinsically motivated new set-theoretic truth insofar as IMH is an intrinsically motivated maximality principle.

Of course there are members of $\mathbb{H}$ which do not satisfy the IMH. Consequently, ¬PD is a statement holding only in a portion of $\mathbb{H}$, something which accounts for our idea that $\mathbb{H}$-axioms are 'local' axioms. This is inevitable if one wishes to be conceptually faithful to the multiverse phenomenon.

However, there is a global corrective to this 'pluralistic' view. The programme strives for the identification of an 'optimal' maximality principle ($\mathbb{H}$-axiom). Now, suppose that P were such a principle; we would then exclude any member of $\mathbb{H}$ which would not satisfy P and therefore P could be taken to be the 'new' $\mathbb{H}$-axiom we are searching for, derivable from the maximal iterative conception and with intrinsically justified first-order consequences.

It could be objected that viewing axioms as 'consequences' of more general principles implies that one accepts these 'consequences' without understanding their 'content', in particular whether they are 'intuitively true' and this would distance our methodology from a genuine search for 'meaningful' additions to ZFC. However, the methodology envisaged here precisely aims to provide an alternative notion of 'intuitively true' as based on the acceptance of the intuitive truth of maximality principles concerning V. Therefore, in our view, the 'meaningfulness' of the consequences of a maximality principle is guaranteed by the meaningfulness of the principle itself.

6.2   Alternative Approaches

Finally, we go back, again, to the issue we started with at the beginning of this paper: what new axioms should be. First of all, we will try to dispel one main worry about the methodology described, namely, that it could imply that all new axiom candidates other than $\mathbb{H}$-axioms should automatically fail to be viewed as plausible new axioms and, what is worse, as lacking any evidence in favour of their acceptance.

This would be a gross misrepresentation of our perspective. In the previous subsection, when we regarded ¬PD or ¬CH as consequences of new axioms, our aim was not to make a general argument in favour of the rejection of PD or CH. At the same time, nowhere in this paper have we suggested that the 'current' new axioms should *all* be rejected: the proof of this is that, again, PD, CH or their negations have already been subjected to extensive mathematical investigations as new axioms, and, in this respect, our programme has nothing new to add.

What we have tried to establish here is that, *if* our evidential framework is preferable to others, then there are reasons to think that PD might be rejected precisely on its grounds.

Leaving aside our framework for a moment, it is maybe appropriate to make a brief digression on the status of PD. Over the years, PD has been celebrated as a new axiom for which there is a significant body of evidence.[14] In particular, two aspects are almost invariably highlighted: (1) PD is successful, because it makes the theory of sets of reals up to and including the projective sets behave well (under PD, all projective sets of reals are Lebesgue measurable and every uncountable projective set of reals has a perfect subset, which means that CH cannot be projectively refuted); (2) PD remarkably connects two apparently distant areas of set theory, descriptive set theory and the theory of large cardinals, as it was proved that the existence of Woodin cardinals and PD have the same consistency strength.

However, arguments in favour of PD are mostly extrinsic and are based on the fact that it follows from large cardinals or from set-generic absoluteness principles, but justifications are lacking for both the existence of large cardinals and for a form of absoluteness which imposes an artificial restriction to set-forcings.[15] Also, advocates of PD often claim that truth is taken to be based solely on current set-theoretic practice, ignoring what is relevant for mathematics outside of set theory or for the maximal iterative conception. So arguing that PD can be inferred from current set-theoretic practice may be insufficient for claiming its truth.[16]

---

[14]For the full case for axioms of definable determinacy, such as PD, see, e.g., [22, 29, 38].

[15]On this, see Appendix of the present paper.

[16]To be fair, advocacy of PD along an alternative, intrinsic-evidence-based line of thought, has also been made. See, for instance, [16]: 'But aside from extrinsic evidence, there are other reasons to regard *PD* as the *correct* axiom for the projective sets. With the progress made in the theory of canonical models for large cardinals it has become clear that *PD* is implied by and is in fact equivalent to a vast number of *prima facie* unrelated combinatorial principles including large-

Now, returning to our main topic, why should all other proposed definitions of what a 'new axiom' should be like be replaced by ours? Because other approaches may be fraught with insurmountable difficulties. For instance, consider the following three alternatives:

A New Axiom Should be a First-Order Statement True of the Concept of Set As we have seen, true of the concept of set means true of the [MIC], but there might be quite a few set-theoretic statements for which such criterion cannot apply. For instance, is the Axiom of Choice true of the [MIC]? How about the Axiom of Determinacy? Even if such criterion is applicable, there might be cases where one intrinsically motivated first-order axiom may contradict another enjoying the same status.

A New Axiom Should be a First-Order Statement, Not Intrinsically, But Rather Extrinsically Justified Many new axioms such as forcing axioms or PD and, in general, definable determinacy axioms, have a lot of strong extrinsic support. However, this fact may not be sufficient and, in fact, too limited. For instance, in the Appendix, we present arguments showing that extrinsically supported absoluteness axioms may be inadequate.

A New Axiom is an Axiom Which Is 'Practically' Confirmed, that Is, Verified Empirically in Specific Areas of Set Theory' This is a refinement of the statement above. However, the definition is still problematic, as the notion of an axiom's being 'practically confirmed' is obscure and would require clarification.[17]

We do not know whether the above procedure to identify and justify $\mathbb{H}$-axioms and the notion of $\mathbb{H}$-axiom itself will become standard. It does seem to us that our proposal responds better to the conceptual difficulties of the aforementioned alternative approaches. In particular, after the substantial demise of 'Gödel's programme', the search for new intrinsically motivated new axioms is at a loss within all other current research programmes. The reasons have been amply considered above, especially in our introductory remarks: the notion of set-theoretic truth falls short of a unique characterisation, if it is to reflect a unique realm of objects, in particular as a consequence of the existence of the multiverse, and it does not seem that this situation can be easily repaired, unless one adopts higher-order principles motivated by the concept of set.

---

cardinal axioms. Still this may not establish their intrinsic necessity because the relevant large-cardinal axioms at present do not enjoy the same kind of intrinsic plausibility as for example Mahlo cardinals. However, the intrinsic necessity of an axiom need not be immediate and could depend on the discovery of additional facts' (p. 274). Of course, at present it is not clear what 'intrinsic' facts would add to the defensibility of PD and whether they will ultimately be discovered.

[17]For further details on these different approaches, see, respectively: (1) on the strength and value of extrinsic justifications, [22, 27–29]; (2) on second-order logic and set theory, [32], and [20]; (3) on the quasi-empirical view, again, [22], or [16].

7 Concluding Summary

In this paper, we have shown how the search for new axioms is carried out within the Hyperuniverse Programme. The methodology devised is motivated by the existence of three concurrent phenomena: (1) the set-theoretic multiverse; (2) the availability of higher-order principles describing forms of maximality of V in line with the [MIC], that is, $\mathbb{H}$-axioms; (3) a demonstrable link between such maximality principles and countable transitive models.

Maximality principles, that specify different notions of the maximality of V, also have, through the use of V-logic, robust consequences in countable transitive models. Obviously, different maximality principles may have different first-order consequences. So, the main shortcoming of this conception is that it is not sufficient to fix set-theoretic indeterminacy *uniquely*. However, we believe that the further development of the programme may establish the existence of an 'optimal' maximality criterion, which, in turn, may lead to the acceptance of one single, intrinsically justified collection of first-order statements to be added to ZFC.

The project is open to further generalisations and developments. New maximality principles will come out, helping us to identify further universes where certain set-theoretic statements do or do not hold. A more careful description of V, of different types of universes in $\mathbb{H}$, and axioms therein, may, therefore, be on its way.

Appendix: Absoluteness Axioms

In recent years, considerable attention has been paid by set-theorists to what we may call the *absoluteness programme*. The main goal of this programme is to foster suitable mathematical strategies and principles (absoluteness axioms) to 'induce' the absoluteness of certain set-theoretic statements across an appropriately selected collection of models (or set-theoretic multiverse).[18]

Although absoluteness axioms have received a lot of extrinsic support in recent years,[19] here we want to present evidence that no new first-order absoluteness axiom has good prospects to be viewed as a plausible axiom candidate extending ZFC. As far as the 'extrinsic' value of these axioms is concerned, the reasons for this claim are *structural*, that is, refer to internal features of the absoluteness phenomenon and do not depend upon the nature and the content of the axiom under consideration.

We now explain why this is so.

Recall the Lévy hierarchy of logical formulas: one starts with Δ_0-sentences, those with only *bounded* quantifiers. Σ_1- and Π_1- sentences contain, respectively, one block of existential or one block of universal quantifiers followed by bounded

---

[18]Given a formula ϕ and transitive models M and N, we say that ϕ is absolute between M and N iff $\phi^M(x_1, x_2, \cdots, x_n) \leftrightarrow \phi^N(x_1, x_2, \cdots, x_n)$.

[19]For an introductory overview of some of these see [4].

quantifiers and, in general, $\Sigma_{k+1} = \exists x_1 x_2 \ldots x_n \Pi_k$ and $\Pi_{k+1} = \forall x_1 x_2 \ldots x_n \Sigma_k$. Also, recall that $H(\kappa)$ denotes the union of all transitive sets of size less than κ.[20] The Σ_n-theory of $H(\kappa)$ is the set of Σ_n-sentences true in $H(\kappa)$.

Definition 3 We say that $M \sqsubseteq N$ if $M \subseteq N$ are transitive models of ZFC with the same ordinals.

Now, there exist trivial forms of absoluteness. For instance, as is known, if $M \sqsubseteq N$, where M and N are models of ZFC, the theory of $H(\omega)$ is the same in M and N. Going one level higher in the hierarchy of $H(\kappa)$, one finds the following seminal result due to Lévy and Shoenfield:

Theorem 6 *If $M \sqsubseteq N$ are models of ZFC, then the Σ_1-theory of $H(\omega_1)$ is the same in M and N.*

Now, what about the Σ_2-theory of $H(\omega_1)$? Climbing up the scale of complexity of set-theoretic sentences, absoluteness comes to a halt:

Theorem 7 *There are models $M \sqsubseteq N$ of ZFC such that the Σ_2-theory of $H(\omega_1)$ is not the same in M and N.*

Proof The statement "there is a nonconstructible real" is a Σ_2 property of $H(\omega_1)$. Take N to satisfy this and M to be L^N. □

This negative result may be circumvented via a two-step strategy: the first step consists in restricting the $\sqsubseteq$-relation in a suitable way. Consider the following definition:

Definition 4 $M \sqsubseteq^{set-generic} N$ iff N is a set-generic extension of M.

Theorem 8 (Bukovský[21]) $M \sqsubseteq^{set-generic} N$ *iff $M \sqsubseteq N$ and for some cardinal κ of M every function in N on a set in M into M is contained in a multi-valued function in M with fewer than κ values for each argument.*

One further refinement of this definition leads to the following notion:

Definition 5 $M \sqsubseteq^{stationary-preserving-set-generic} N$ iff N is a set-generic extension of M and any subset of ω_1^M which is stationary in M is also stationary in N.

In other terms, by restricting the $\sqsubseteq$-relation to, respectively, $\sqsubseteq^{set-generic}$ or, on the other hand, $\sqsubseteq^{stationary-preserving-set-generic}$ one only takes into account *generic extensions* of models obtained through *set-forcing* or *stationary-preserving set-forcing*.

The second step in the strategy consists in considering certain extensions of ZFC, say, ZFC + Ax., and then replacing the multiverse $\mathbb{M}^{ZFC}$ by the multiverse $\mathbb{M}^{ZFC+Ax.}$ associated to the stronger system ZFC + Ax.

Using this two-step strategy, Woodin and Viale have obtained results which are, no doubt, of mathematical significance,[22] but, with respect to our foundational

---

[20]That is, the union of all sets whose *transitive closure* has cardinality less then κ.

[21]See [8].

[22]See, in particular, [38] and [35]. Among other things, Woodin proved the following:

project, their results present some crucial shortcomings: (1) the axioms they consider, such as the existence of class-many Woodin cardinals, are not justified *intrinsically*; (2) the restriction of the $\sqsubseteq$-relation to set-generic extensions is unwarranted in view of our definition of $\mathbb{H}$. Furthermore, even leaving these issues aside, it is not clear how far the programme they have been carrying out can be extended, and with what results. We will come to this in a moment.

Alternatively, one could employ only the second step of the above strategy, by supporting the acceptance of axioms such as $V = L$. Gödel's work yields:

Theorem 11 *If $M \sqsubseteq N$ are models of ZFC + V = L, then M = N.*

However, promising though this strategy may seem, it reveals the same shortcoming as before, insofar as it is hinged upon the acceptance of a mathematical principle, $V = L$, which does not possess a sufficient degree of intrinsic motivation in view of our notion of 'intrinsicness' expounded in Sect. 5.

As said, in fact, there is strong evidence that the second step in the above two-step strategy, that of extending ZFC to a stronger first-order theory to obtain greater absoluteness, is doomed to failure. Consider the following:

Theorem 12 *Suppose T is a first-order theory, compatible with the following two statements:*

(1) *the class $\{\alpha : \alpha$ measurable $\}$ is stationary;*
(2) *the class $\{\alpha : V_\alpha \prec_{\Sigma_\omega} V\}$ is unbounded.*

Then, $\Sigma_2(H(\omega_1))$-absoluteness fails for models of T: there are models $M \sqsubseteq N$ of T such that the Σ_2-theory of $H(\omega_1)^M \neq \Sigma_2$-theory of $H(\omega_1)^N$. If T consists of only a finite set of axioms then (2) above is not needed.

Sketch of Proof The hypotheses imply that there is a model V of ZFC with a largest measurable κ such that T holds in V_κ. Now iterate the measure on κ through the ordinals, resulting in a model N. In N, there is a model V_0 like V but only satisfying KP, with an iterable top measure. Again iterate the top measure through the ordinals to form an inner model M. Then $M \sqsubseteq N$ are both models of T but by choosing V_0 minimally we can arrange that in M there is no iterable model P of KP with a top measurable κ_0 such that T holds in the V_{κ_0} of P. This $\Pi_2(H_{\omega_1})$ sentence fails in N and this gives the asserted failure of absoluteness.   □

The theorem asserts that any first-order theory which is compatible with the existence of *measurable cardinals* (in fact, a *stationary class* of measurable cardinals) fails to ensure $\Sigma_2(H(\omega_1))$ absoluteness for its models. This is very strong

---

Theorem 9 *If $M \sqsubseteq^{set-generic} N$ are models of ZFC+ large cardinals + CH, then the Σ_1-theory of $H(\omega_2)$ (with parameter ω_1) is the same in M and N.*
Viale has recently proved:

Theorem 10 *If $M \quad \sqsubseteq^{stationary-preserving-set-generic} \quad N$ are models of ZFC + large cardinals+MM^{+++}, then the theory of $H(\omega_2)$ is the same in M and N.*

evidence against the use of first-order axioms for obtaining convincing absoluteness principles for $\Sigma_2(H(\omega_1))$ statements.

To summarise, there is a network of results which seem to show that, through the adoption of absoluteness axioms, one can find new set-theoretic truth, by extending the absoluteness of set-theoretic statements to levels of increasing first-order complexity. However, first of all, none of the axioms adopted or used in the programme seems to be intrinsically motivated. Secondly, there is also some evidence that such an extension collides with the existence of measurable cardinals. As a consequence, one appears to be forced to artificial restrictions of the multiverse to only certain models or of the notion of absoluteness itself.

Consequently, we come to the following conclusion: no first-order *absoluteness axiom* has good prospects of being accepted as a new axiom on the grounds of both intrinsic or extrinsic justifications.

In our opinion, if one wants to spell out a plausible conception of 'truth in the multiverse', one has to proceed in the alternative way we propose, through the use of higher-order principles.

References

1. C. Antos, S.-D. Friedman, R. Honzik, C. Ternullo, Multiverse conceptions in set theory. Synthese **192**(8), 2463–2488 (2015)
2. T. Arrigoni, S. Friedman, Foundational implications of the inner model hypothesis. Ann. Pure Appl. Logic **163**, 1360–66 (2012)
3. T. Arrigoni, S. Friedman, The hyperuniverse program. Bull. Symb. Log. **19**(1), 77–96 (2013)
4. J. Bagaria, Bounded forcing axioms and generic absoluteness. Arch. Math. Log. **39**, 393–401 (2000)
5. J. Barwise, *Admissible Sets and Structures* (Springer, Berlin, 1975)
6. P. Benacerraf, H. Putnam (eds.), *Philosophy of Mathematics. Selected Readings* (Cambridge University Press, Cambridge, 1983)
7. P. Bernays, Sur le platonisme dans les mathématiques. L'Enseignement Mathématique **34**, 52–69 (1935)
8. L. Bukovsky, Characterization of generic extensions of models of set theory. Fundam. Math. **83**, 35–46 (1973)
9. W. Ewald (ed.), *From Kant to Hilbert: A Source Book in the Foundations of Mathematics*, vol. II (Oxford University Press, Oxford, 1996)
10. S. Friedman, Internal consistency and the inner model hypothesis. Bull. Symb. Log. **12**(4), 591–600 (2006)
11. S. Friedman, Evidence for set-theoretic truth and the hyperuniverse programme. IfCoLog J. Log. Appl. **3**(4), 517–555 (2016)
12. S. Friedman, R. Honzik, On strong forms of reflection in set theory. Math. Log. Q. **62**(1–2), 52–58 (2016)
13. S. Friedman, P. Welch, W.H. Woodin, On the consistency strength of the inner model hypothesis. J. Symb. Log. **73**(2), 391–400 (2008)
14. K. Gödel, What is cantor's continuum problem? Am. Math. Mon. **54**, 515–525 (1947)
15. M. Hallett, *Cantorian Set Theory and Limitation of Size* (Clarendon, Oxford, 1984)
16. K. Hauser, Is the continuum problem inherently vague? Philos. Math. **10**, 257–285 (2002)
17. G. Hellman, *Mathematics without Numbers. Towards a Modal-Structural Interpretation* (Clarendon, Oxford, 1989)

18. L. Horsten, P. Welch, Reflecting on absolute infinity. J. Philos. **113**(2), 89–111 (2016)
19. I. Jané, The role of the absolute infinite in cantor's conception of set. Erkenntnis **42**, 375–402 (1995)
20. I. Jané, Higher-order logic reconsidered, in *Oxford Handbook of Philosophy of Mathematics*, ed. by S. Shapiro (Oxford University Press, Oxford, 2005), pp. 747–774
21. I. Jané, The iterative conception of sets from a Cantorian perspective, in *Logic, Methodology and Philosophy of Science: Proceedings of the Twelfth International Congress*, ed. by P. Hájek, D. Westertål, L. Valdés-Villanueva (King's College Publications, London, 2005), pp. 373–393
22. P. Koellner, On the question of absolute undecidability. Philos. Math. **14**(2), 153–188 (2006)
23. P. Koellner, On reflection principles. Ann. Pure Appl. Log. **157**(2–3), 206–19 (2009)
24. Ø. Linnebo, The potential hierarchy of sets. Rev. Symb. Log. **6**(2), 205–228 (2013)
25. P. Maddy, Believing the axioms, I. Bull. Symb. Log. **53**(2), 481–511 (1988)
26. P. Maddy, Believing the axioms, II. Bull. Symb. Log. **53**(3), 736–764 (1988)
27. P. Maddy, Set-theoretic naturalism. Bull. Symb. Log. **61**(2), 490–514 (1996)
28. P. Maddy, *Naturalism in Mathematics* (Oxford University Press, Oxford, 1997)
29. D. Martin, Mathematical evidence, in *Truth in Mathematics*, ed. by H.G. Dales, G. Oliveri (Clarendon Press, Oxford, 1998), pp. 215–231
30. C. Parsons, What is the iterative conception of sets? in *Philosophy of Mathematics. Selected Readings*, ed. by P. Benacerraf, H. Putnam (Cambridge University Press, Cambridge, 1983), pp. 503–529
31. M. Potter, *Set Theory and its Philosophy* (Oxford University Press, Oxford, 2004)
32. S. Shapiro, *Foundations without Foundationalism. A Case for Second-Order Logic* (Oxford University Press, Oxford, 1991)
33. W.W. Tait, Foundations of set theory, in *Truth in Mathematics*, ed. by H.G. Dales, G. Oliveri (Oxford University Press, Oxford, 1998), pp. 273–290
34. W.W. Tait, Zermelo on the concept of set and reflection principles, in *Philosophy of Mathematics Today*, ed. by M. Schirn (Clarendon/Oxford Press, Oxford, 1998), pp. 469–483
35. M. Viale, Category forcings, MM^{+++} and generic absoluteness for the theory of strong forcing axioms. J. Am. Math. Soc. **29**(3), 675–728 (2016)
36. H. Wang, *From Mathematics to Philosophy* (Routledge and Kegan Paul, London, 1974)
37. H. Wang, *A Logical Journey* (MIT Press, Cambridge, 1996)
38. W.H. Woodin, The continuum hypothesis. Not. Am. Math. Soc. Part 1: **48**(6), 567–76; Part 2: **48**(7), 681–90 (2001)
39. E. Zermelo, Über Grenzzahlen und Mengenbereiche: neue Untersuchungen über die Grundlagen der Mengenlehre. Fundam. Math. **16**, 29–47 (1930)

Explaining Maximality Through the Hyperuniverse Programme

Sy-David Friedman and Claudio Ternullo

Abstract The (maximal) iterative concept of set is standardly taken to justify ZFC and some of its extensions. In this paper, we show that the maximal iterative concept also lies behind a class of further *maximality principles* expressing the maximality of the universe of sets V in height and width. These principles have been heavily investigated by the first author and his collaborators within the Hyperuniverse Programme. The programme is based on two essential tools: the *hyperuniverse*, consisting of all countable transitive models of ZFC, and *V-logic*, both of which are also fully discussed in the paper.

1 The Maximal Iterative Concept of Set

1.1 Generalities

In this paper, we will be pre-eminently dealing with maximality principles for the universe of sets, that is, principles which prescribe that the universe is *maximal*. Of course, it is far from obvious what 'maximal' means or implies here, and the next subsections aim to fully clarify what we mean by that.

Maximality principles may be seen as expressing a fundamental feature of the iterative concept of set. It is not too hard to see why, yet it is worth examining this in more detail.

The iterative concept of set consists in the idea that sets are generated in *stages*, starting with ur-elements or, possibly, with the empty set and, then, forming the power-set of the previous levels at stages indexed by successor-ordinals and the union of all previous levels at stages indexed by limit-ordinals. The resulting picture

Originally published in S. Friedman, C. Ternullo, Explaining Maximality Through the Hyperuniverse Programme, preprint.

The first author wishes to thank the FWF for its support through project P 28420.

S.-D. Friedman (✉) • C. Ternullo
Kurt Gödel Research Center for Mathematical Logic, University of Vienna, 25 Währinger Strasse, 1090 Wien, Austria
e-mail: sdf@logic.univie.ac.at; claudio.ternullo@univie.ac.at

185

is simply what is standardly acknowledged to be the universe of sets, the union of the V_α for all ordinals α, consisting of all sets formed through all stages.

The history of the progressive development of the axiomatisation of set theory and, in particular, of the emergence of ZFC has shown that all the most widely accepted axioms of set theory are true of the iterative concept and may, in fact, be motivated by it. This fact has gradually evolved into the more robust view that the concept of set is *essentially equivalent* to the iterative concept of set.

There are several issues with the iterative concept and its full justifiability.[1] However, in light of our goals, here we only wish to focus our attention on two prominent features of it: its connection to *maximality* and its closeness to a *platonistic* interpretation of mathematics.

Sometimes it is said that a guiding principle in the 'genetic' approach to sets is that one should form *as many of them as possible*. The principle seems equivalent to the idea that all logical and conceptual constraints on the formation of sets should be removed, and this leads to viewing the iterative concept as a maximum (or maximal) iterative concept. Here is how Wang comments on the principle with respect to the power-set axiom:

> The concept of all subsets is often thought to be opaque because we envisage all possibilities independently of whether we can specify each in words; for example, just as there are 2^{10} subsets of a set with 10 members, we think of 2^a subsets of a set with a members when a is an infinite cardinal number. In particular, we do not concern ourselves over how a set is defined, e.g. whether by an impredicative definition. This is the sense in which the individual steps of iteration are 'maximum'. (in [5], p. 532)

Two main features of the maximal approach are neatly highlighted in the passage above: the fact that (1) the 'infinite should be treated in a way analogous to the finite', a principle which allows us to extend certain set-theoretic operations holding in the finite to the transfinite, and (2) the fact that *impredicative* definitions are seen as entirely legitimate.

As is known, Bernays, in his [6], had construed the aforementioned principles as expressing Platonism in mathematics (set theory). In Bernays' view, central to mathematical (set-theoretic) Platonism would be a *quasi-combinatorial* conception, that is the view that mathematical (set-theoretic) operations, entities and concepts holding in the finite can (and should) be extended to the infinite, even in the absence of any available methods of 'construction'.[2]

---

[1]For further details, see the classical Boolos, [7], Parsons, [22] and Wang, [25]. Potter, [23] provides a more recent, but not less accurate, overview of the topic.

[2]Cf. the following two crucial passages of [6]: 'But analysis is not content with this modest variety of platonism [that of arithmetical platonism, *our note*]; it reflects it to a stronger degree with respect to the following notions: set of numbers, sequence of numbers, and function. It abstracts from the possibility of giving definitions of sets, sequences, and functions. These notions are used in a 'quasi-combinatorial' sense, by which I mean: in the sense of an analogy of the infinite to the finite' and later in the text: 'In Cantor's theories, platonistic conceptions extend far beyond those of the theory of real numbers. This is done by iterating the use of the quasi-combinatorial concept

For instance, on this view, the existence of the power-set of the integers does not follow from exploring all known methods for constructing subsets of the integers, but rather from an ideal intuitive grasp of the *fully given* collection of subsets of integers.

Now, if the maximal iterative concept really is the expression of a *platonistic* attitude in mathematics, in accordance with this fact, we might, as a consequence, want to (or be forced to) hold that V is *fully given* in an ideal sense, that no possible extension of it is conceivable (a position known as *actualism*) or that all statements of set theory have a unique truth-value, all of which statements do have notable bearings on the availability of the maximality principles we will be introducing. Therefore, all such issues will have to be carefully taken into account throughout the paper.[3]

1.2 Expanding on the Maximal Concept

Returning to the the maximal concept of set, its manifestations in set theory are manifold. Bernays, in the previous quote, was mentioning 'methods of collection', such as those permitted, for instance, by the Infinity, the Power-Set or the Replacement Axiom.

One of the most distinctive ways to construe the maximality inherent in the concept of set is the idea that the universe itself, V, be maximal. Again, Wang expounds this further characterisation of the meaning of 'maximal' in the following way:

> In a general way, hypotheses which purport to enrich the content of power sets (say that of integers) or to introduce more ordinals conform to the intuitive model. We believe that the collection of all ordinals is very 'long' and each power set (of an infinite set) is very 'thick'. Hence, any axioms to such effects are in accordance with our intuitive concept. [5, p. 553]

To rephrase Wang's quote, one could say: the iterative concept of set leads one to realise that there is a rich hierarchy of sets, whose formation is given by the (maximal) procedures associated ('methods of collecting'). Now, it is reasonable to ask whether such methods of collecting (e.g., the Power-Set Axiom) may themselves be maximised in some way. In simpler words, one could say that, according to the maximal iterative concept, the hierarchy of sets should be *as wide as possible* and extend *as far as possible*. However, it is not prima facie clear what 'as long as possible' and 'as far as possible' mean. It is therefore the task of the study of maximality principles to disclose (or clarify) the meaning of 'maximality'.

---

of a function and adding methods of collection. This is the well-known method of set theory' (both are reproduced in [5], p. 259–60).

[3] See, in particular, Sects. 2.2 and 7.

1.3 New Intrinsically Justified Axioms

The main rationale for exploring maximality principles is to extend ZFC, through, ideally, declaring these principles new set-theoretic *axioms*. One straightforward criterion to evaluate whether a new axiom is acceptable is to checking whether it decides set-theoretic statements which are not decided by ZFC. But there's something else which should guide us in finding new axioms, that is their conceptual 'aptness', measured against the maximal iterative concept.

Since Gödel, [13], it is customary, in the literature, to define these two forms of evidence for new axioms as, respectively, *extrinsic* and *intrinsic*. Intrinsic evidence relates to 'conformity to the intuitive model', as Wang would say, whereas extrinsic evidence to the success of an axiom. Maximality principles, as we formulate them, clearly obey the maximal iterative concept, and this will fully justify our view that the maximality principles described in the next sections are *intrinsically motivated*. Moreover, these principles should ultimately be viewed as new intrinsically justified *axioms*, but, as detailed in the last section, this will require the satisfaction of further epistemic desiderata.[4] It should be noticed that the maximality principles we introduce have also clearly proved to be able to reduce set-theoretic incompleteness, although we will not deal with this in the present paper.

To summarise, a well-established conception of the axioms of set theory holds that ZFC conforms to a maximal iterative concept, and that its extensions should follow suit.[5] Maximality principles are an expression of this attitude and, thus, can be viewed as being intrinsically motivated.

2 A Zermelian Approach to V

2.1 Height and Width Maximality

In this section, we introduce further key concepts concerning the relationship between maximality principles and the maximal iterative concept.

---

[4]The difference between an axiom's being *intrinsically motivated (plausible)* and *intrinsically justified* consists in the level of definitiveness conveyed by the justificatory process. Thus, an intrinsically motivated axiom (or principle) need not be a definitively accepted axiom (or principle) of set theory. Koellner, [20], p. 207–8, explains the difference as follows: '..the notion of intrinsic justification is intended to be more secure than mere 'intrinsic plausibility' [...], in that whereas the latter merely adds credence, the former is intended to be definitive (modulo the tenability of the conception).'

[5]Incidentally, such a view is already expressed (although very tersely) by Gödel when he discusses the prospects of deciding CH through a new axiom: '..from an axiom in some sense opposite to this one [$V = L$], the negation of Cantor's conjecture could perhaps be derived. I am thinking of an axiom which (similar to Hilbert's completeness axiom in geometry) would state some maximum property of the system of all sets, whereas axiom A [i.e. $V = L$] states a minimum property. Note that only a maximum property would seem to harmonize with the concept of set. . . ' [14, p. 262–63, footnote 23].

Set theorists have progressively formulated several maximality principles.[6] As has been clarified above, all of these prescribe, in some way, that the universe of sets is a very rich structure, in particular, the richest possible allowed by the set-theoretic axioms.

Now, as hinted at above in Wang's quote, the maximality of V has two dimensions, *maximality in height* and *maximality in width*, which can be characterised as follows:

Height Maximality The cumulative hierarchy should be as *tall* as possible.

Width Maximality The cumulative hierarchy should be as *wide* as possible.

Again, making sense of the statements above consists in understanding carefully what it means for V to be as tall as and as wide as possible. Let's start with height maximality.

There is a principle and, in fact, a class of principles, which has attracted set-theorists' attention in the last few decades, which seems to express height maximality very aptly, and this is the Reflection Principle. Very generally, reflection can be described as asserting that the universe cannot be uniquely characterised by any given collection of first-order properties. As is often said, the universe is indescribable (or ineffable).[7] Now, through reflection one can generate new ordinals α's and corresponding new levels V_α's of the hierarchy. Therefore, another way to construe reflection as a maximality principle is viewing it as inducing (maximal) 'lengthenings' of V, and that is precisely our construal of the principle.

Mathematically, in ZFC, reflection is a theorem which asserts that, if V has a first-order property ϕ, then for some ordinal α, there is a V_α which satisfies ϕ^α (that is, the relativisation of ϕ to V_α). A stronger version of reflection, in particular, implies that, given any arbitrarily high level α, there is always a $\beta > \alpha$ such that $V_\beta \models \phi^\beta$.

Strengthenings of first-order reflection, in particular, *second-order reflection*, including second-order parameters, are able to prove that there exist such large cardinals as *inaccessibles* and *Mahlos*, in fact, that there exist proper classes of them. Therefore, second-order reflection is strong enough to provide new ordinals and, consequently, if one construes height maximality in terms of producing 'lengthenings' of V, then one could say that second-order reflection induces a significant lengthening of V.

---

[6]Incurvati, in [17], makes an overview of different forms of maximality in set theory, and also provides a mathematically detailed account of some of the most important maximality principles in use. Among other things, the paper also includes a philosophical examination of the IMH, which is widely discussed in the present paper.

[7]It is widely known that the emergence of the principle is connected to Cantor's idea of the *absolute infinity* of V (for which see Cantor's renowned 1899 letter to Dedekind, in [8], p. 931–5). Gödel was one of the major advocates of reflection, to the point that he seems to have surmised that the axioms of set theory should be essentially reducible to one single *reflection axiom* (see Gödel, [13], Wang, [26] and Ternullo, [24] for this).

As far as width maximality is concerned, things are somewhat less intelligible. The width of the universe is given by the extent of the power-set operation. Now, it is unclear how one could vary the extent of such an operation. As for the height of the universe, one could try to maximise width by also adopting some form of *reflection*. Width reflection has been informally introduced by Koellner in [20],[8] and essentially arises from a construal of the *core model programme* in terms of reaching an approximation of V using L-like models $L[0^{\#}], L[0^{\#\#}], \ldots, L[0^{\infty}], \ldots$. Each of these, by width reflection, is taken to fail to approximate V by some specified property P. For instance, failure by L to approximate V leads to proving the existence of $0^{\#}$.[9] So this may overall be construed as a way to produce thickenings of L which fail to fully approximate V from *within* in the same way as V_{α} must fail to approximate V from *below*. However, it is controversial, to say the least, that this form of width reflection, heavily based on L, is in accordance with the maximum iterative concept.

2.2 Actualism and Potentialism: Zermelo's Conception

The aforementioned maximality principles for V, if all fully in line with the maximal iterative concept, are conducive to several issues concerning the correct conceptualisation of V, which need be taken into account.

For instance, it has been argued that if height maximality is essentially expressed by reflection principles construed as prescribing the *indescribability* of V, then one is more naturally inclined to see V as *fully given* in the sense of being inextensible. This is the *actualist* position, which had already been mentioned in Sect. 1.1. But if one adopts actualism, then higher-order quantification is less likely to be made sense of.[10]

A *potentialist* conception, on the other hand, construes V as non-fixed in height and width and, thus, can make sense of higher-order quantification and, in general, of lengthenings and thickenings more easily. What would be lost for the potentialist, though, is the availability of the full givenness of V, which appears to lie at the root of reflection.

One way to resolve this would be to declare that the Platonism inherent in the maximal iterative concept would automatically imply that one is supposed to be able to intuit V as a *completed* object. By this interpretation, full-fledged actualism

---

[8]The idea is also very carefully examined by Incurvati in the mentioned [17] and further explored in [10]. See also p. 17 of the present article.

[9]A brief description of the core model programme is in Jensen, [18].

[10]Again, see [20] for this.

would be the only option available and, furthermore, maximality principles referring to widenings and lengthenings of V would inevitably be viewed as meaningless.[11]

Interestingly enough, we will show that the maximality principles formulated within the Hyperuniverse Programme are compatible with both potentialism and actualism. In particular, even a radical form of potentialism can accommodate maximality. And actualism, accompanied by some class theory of the same streng-th as MK (but, in fact, less robust than that), also fully befits the programme.[12]

All this might lend support to the view that the issues of whether the maximal iterative concept is essentially platonistic in character, and of whether a platonist could only be an actualist about V are less relevant than it might seem at first glance.

In any case, the underlying conception in which the Hyperuniverse Programme is most fruitfully cast was the Zermelian conception, as described in [27]. In that work, Zermelo, after formulating the axioms of set theory (often labelled Z_2), proves that, for those axioms (some of which have a second-order characterisation), the power-set of V is fixed. More specifically, he proves that:

Theorem 1 *Any two models of the axioms of set theory Z_2 are either isomorphic, or one is isomorphic to a proper initial segment of the other.*

This settles things as far as the width of the universe is concerned. Concerning height, Zermelo introduces the concept of a *normal domain*. A normal domain is the least rank initial segment of the hierarchy which satisfies the (second-order) axioms of set theory. The least normal domain which satisfies (second-order) ZFC is, as is known, V_κ, where κ is the least inaccessible cardinal. But then one can iterate this, by considering (second-order) ZFC+'there is one inaccessible'. The least normal domain which satisfies (second-order) ZFC+'there is one inaccessible' is V_λ, where $\lambda > \kappa$ is the least inaccessible after κ. Thus, one obtains a vertical multiverse consisting of V_α's, where α is some large cardinal.

The Zermelian picture of the universe has some clear attractions, some of which can be described as follows:

(1) Height potentialism fully befits the form of reflection introduced in the next section. It is very comfortable to define *lengthenings* of the hierarchy required by this principle within the Zermelian picture.
(2) While height actualism seems counter-intuitive to some extent, width actualism would seem to be more easily justifiable insofar as there is no apparent way to address thickenings of V in a way which resembles the ordinal-indexed progress of stages in height.
(3) One can make full sense of *higher-order* quantification more easily within the Zermelian multiverse, insofar as the universe is non-categorical in height. Fully

---

[11]But things are a lot more subtle. Zermelo, for instance, who would seem to have been a platonist was the major proponent of a partly *potentialist* (if width actualist) conception, for which see the next few pages. For Zermelo's ideas on philosophy, set theory and the justifiability of the axioms, see, in particular, [21] and [19].

[12]A full account of this is provided by Antos, Barton and Friedman in [1]. For further details, also see footnote 21.

actualist (absolutist) versions of the universe struggle to provide an equally acceptable account of this.

These reasons may be insufficient for a full case in favour of the adoption of Zermelo's picture as the only correct picture of V, but are clearly sufficient to accommodate the maximality principles formulated within the Hyperuniverse Programme. However, although the Zermelian conception may be viewed as the correct conceptualisation of V, we will point the reader, when necessary, to alternative options.

3 Height Maximality: Reflection

In a sense, the principles we will be discussing improve and expand on those already mentioned in Sect. 2.1. For instance, our height maximality principle is a form of the reflection principle and, in our view, the strongest possible. In order to see this we have to recall some notions already briefly introduced in Sect. 2.1.

As said, height maximality in terms of reflection of the universe V can be intuitively formulated as follows:

(*Reflection*) Any property which holds in V already holds in some rank initial segment V_α of V.

In other words, V cannot be described as the unique initial segment of the universe satisfying a given property. The strength of such reflection depends on what we take the word 'property' to mean.[13] If this just means 'first-order property with set parameters' then we obtain Lévy reflection, a form of reflection provable in ZFC.

A priori, there is no need to limit ourselves to first-order properties of V. But to express second-order properties of V we need to move beyond ZFC to Gödel-Bernays class theory GB. The latter has variables ranging over sets and also variables ranging over the larger collection of classes (collections of sets: note that every set is also a class). The $\in$-relation applies between sets and classes and we impose the Comprehension Scheme for formulas with only set-quantifiers (but with both set and class variables). Thus in GB we can quantify over classes but cannot apply Comprehension to formulas containing such quantifiers. We also include *Global Choice* as an axiom, which says that there is a class function F such that $F(x)$ is an element of x for every nonempty set x.

---

[13]Properties are often formulated using higher-order quantification. Let M be a class. We say that a variable x is 1-st order (or of order 1) if it ranges over elements of M. In general, we say that a variable R is $n + 1$-st order (or of order $n + 1$), $0 < n < \omega$, if it ranges over $\mathcal{P}^n(M)$, where $\mathcal{P}^n(M)$ denotes the result of applying the powerset operation n times to M. A formula φ is Π_m^n if it starts with a block of universal quantifiers of variables of order $n + 1$, followed by existential quantification of variables of order $n + 1$, and these blocks alternate at most $m - 1$ times; the rest of the formula can contain variables of order at most $n + 1$, and quantifications over variables of order at most n. Σ_m^n is obtained by switching the words universal and existential.

GB is conservative over ZFC. However it can be strengthened by adding second-order reflection axioms to it, such as:

- Π_m^1 Reflection If $\varphi(R)$ is a Π_m^1 formula with a class variable R, then *reflection for* $\varphi(R)$ is the implication

$$\varphi(R) \rightarrow (V_\alpha, V_{\alpha+1}) \vDash \varphi(R \cap V_\alpha)$$

where on the right-hand-side the set variables range over V_α and the class variables over $V_{\alpha+1}$.

Even Π_1^1 Reflection for sentences (without the class variable R) is rather strong, as it implies the existence of an inaccessible cardinal. That is because the regularity of an ordinal α is equivalent to the truth of a Π_1^1 sentence in $(V_\alpha, V_{\alpha+1})$. By adding parameters we get stronger large cardinals such as Mahlo cardinals and weakly compact cardinals.

But just as ZFC is inadequate for second-order reflection, GB is inadequate for third-order reflection.[14]

Of course there is no reason to stop at third-order reflection, and in light of the Zermelian conception, it is meaningful to discuss 'α-th order' reflection for ordinals α in *lengthenings* of V, i.e. in models V^* which have V as a rank initial segment.

This naturally leads to the following form of higher-order reflection:

- Extended Reflection Axiom (ERA) V satisfies the ERA if V has a lengthening V^*, a model of ZFC, such that if φ is first-order and $\varphi(A)$ holds in V^* where A is a subclass of V, then $\varphi(A \cap V_\alpha)$ holds in V_β for some pair of ordinals $\alpha < \beta$ in V.

This allows us to reflect properties (with second-order parameters) that are α-th order, for all ordinals α appearing in the least ZFC model lengthening V. This embodies all of the classical froms of strong reflection and more.

---

[14]As an aside, it is worth noting that if formulated with third-order parameters, third-order reflection is in fact inconsistent! For instance, for a third-order parameter $\mathcal{R}$, i.e. a collection of classes, one is tempted by the following natural-looking principle:

- Third-order reflection If $\varphi(\mathcal{R})$ is true in $(V, \mathcal{R})$ then for some α, $\varphi(\bar{\mathcal{R}})$ is true in $(V_\alpha, \bar{\mathcal{R}})$, where $\bar{\mathcal{R}} = \{R \cap V_\alpha \mid R \in \mathcal{R}\}$.

But such a principle will fail if $\mathcal{R}$ consists of all bounded subsets of the ordinals (viewed as a collection of classes) and $\varphi(\mathcal{R})$ simply says that each element of $\mathcal{R}$ is bounded in the ordinals. Therefore when discussing third-order reflection it is customary to only allow *second-order*, and not third-order parameters. An alternative is to consider *embedding reflection* (see for example the discussion in Section 2.1 of [11], and in [16]) where $\bar{\mathcal{R}}$ results from applying the inverse of an elementary embedding to $\mathcal{R}$. This very strong form of reflection yields supercompact cardinals, however does not appear to be derivable from the maximal iterative conception, as are the forms of reflection consistent with $V = L$.

However, clearly the ERA can easily be strengthened further, by requiring the lengthening V^* of V to satisfy more than ZFC, such as ZFC + 'there is ZFC-lengthening of ZFC'. Indeed, it appears that there is no optimal form of reflection which can be described in terms of lengthenings of V, as we can always strengthen such a reflection principle further by requiring a lengthening V^* of V in which the principle holds with reference only to lengthenings of V appearing in V^*.

How are we then to achieve an optimal reflection principle? This problem is fully addressed mathematically in Section 2.2 of [11], where the principle of #-*generation* is introduced. This asserts the existence of a special kind of set called a # *(sharp)* that 'generates' V through iteration. An optimal form of reflection results as this iteration also produces a closed unbounded class of *indiscernibles* for V, adequate for witnessing any conceivable form of reflection. It is crucial that a # generating V cannot be an element of V, otherwise such optimality would not be possible.

We cannot provide the full details of #-generation here, but at least some notions will be briefly discussed.

First, imagine that V can be seen as being the last step in an elementary chain of universes $(V_{\kappa_i} \mid i < \infty)$ and we set $V = V_{\kappa_\infty}$. We can continue the construction of this chain 'beyond' V itself, producing an upwards elementary chain of universes $V = V_{\kappa_\infty} \prec V_{\kappa_\infty+1} \prec V_{\kappa_\infty+2} \prec \cdots$.

By elementarity, all of these universes will satisfy the same first-order sentences, but we want more. We want that any two pairs of universes 'resemble' each other, i.e. satisfy the same first-order sentences, and this can be extended to any pair of n-tuples of universes $W_{\vec{i}}$, where $\vec{i} = i_0 < i_1 < \cdots < i_{n-1}$ and $W_{\vec{j}}$, where $\vec{j} = j_0 < j_1 < \cdots < j_{n-1}$ (to simplify our notation, we use the symbol W_i for $V_{\kappa_i}^*$). But we want to impose an even higher level of resemblance, whereby all n-tuples of models satisfy the same second-order sentences and so on. In the end, the whole process can be seen as the construction of a series of embeddings $\pi_{ij} : V \to V$, leading to an *indiscernibly-generated* V. In more rigorous terms:

Definition 2 ([11], p. 6) V is indiscernibly-generated iff: (1) There is a continuous sequence $\kappa_0 < \kappa_1 < \cdots$ of length ∞ such that $\kappa_\infty = \infty$ and there are commuting elementary embeddings $\pi_{ij} : V \to V$, where π_{ij} has critical point κ_i and sends κ_i to κ_j. (2) For any $i \leq j$, any element of V is first-order definable in V from elements of the range of π_{ij} together with κ_k's for k in the interval $[i, j)$.

Indiscernible-generation has an equivalent but more useful formulation in terms of #-*generation* (for its definition see [11], p. 6). So we will use the term #-*generation* for this strong form of reflection.

Now, one can show that #-generation implies all forms of reflection which are compatible with $V = L$ (again see [11]).

As a consequence of this, we believe that #-generation expresses the strongest possible amount of vertical reflection and therefore can legitimately claim to be the optimal principle expressing the vertical maximality of V.

4 Width Maximality: V-Logic, IMH

4.1 The Strategy

From the Zermelian perspective, which incorporates height potentialism and width actualism, expressing principles of *width maximality* principles presents a real challenge. Whereas in the case of height maximality we made liberal use of *lengthenings of V*, no analogous notion of *thickening (or outer model) of V* is available.

Now, since [9], the programme has expressed width maximality in terms of the following principle:

- (The Inner Model Hypothesis, IMH) If a first-order sentence holds in an inner model of some outer model of V then it also holds in some inner model of V.

As is clear, the IMH is conceptually problematic for the Zermelian, as it explicitly refers to 'outer models' which are not available in the Zermelian picture. However, if the IMH were referring not to the whole V, but just to some countable transitive model (which we will mostly indicate as 'little-V') of ZFC, then the IMH would make perfect sense even within a Zermelian perspective.[15]

However recent developments, discussed in [2, 10] and [4], provide a solution to this problem. The introduction of *V-logic* enables one to express first-order properties of arbitrary outer models (almost) *internally* within V, in the same way as first-order properties of set-forcing extensions of V can be *internalised* using the forcing relation. The word 'almost' occurs because this new 'truth in outer models' relation will not in general be first-order definable over V, but rather over a small *lengthening* (not *thickening*) of V called $Hyp(V)$ (the least 'admissible set' containing V as an element). As lengthenings are available to the Zermelian, this enables her to express principles such as the IMH without loss of content.

Therefore, we shall scrutinise two approaches to width maximality: the first, through the use of V-logic, will allow one to make sense of IMH as if it were referring to the whole V, and the second will construe the IMH as referring to some countable model 'little-V'. The latter approach is particularly convenient, as it entirely befits our goal to reduce the study of the consequences of maximality principles to their consequences in *countable transitive models*.

Let us review the first approach. As we said, the case of IMH is analogous to that of Martin's Axiom (MA), a principle of set-forcing.[16] Several formulations of MA are available, in particular, $MA_{\aleph_1}$ asserts:

- (Outer Model $MA_{\aleph_1}$). Whenever $V[G]$ is a generic extension of V by a partial order $\mathbb{P}$ with the countable chain condition in V, and $\varphi(x)$ is a $\Sigma_1(\mathcal{P}(\omega_1))$ formula

---

[15]Note that IMH is also known to consistently hold for some choice of little-V. See [12].

[16]For further on this analogy, see [4].

(i.e. a Σ_1 formula with a subset of ω_1 as parameter), if in $V[G]$ there is a y such that $\varphi(y)$ holds, then there is also such a y in V.

Note the quantification in this definition over the (generic) outer models $V[G]$ of V. How can the width actualist possibly make sense of this? The answer is of course via the definable forcing relation:

- (Internal $MA_{\aleph_1}$). Whenever $\mathbb{P}$ is a partial order with the countable chain condition in V, and $\varphi(x)$ is a $\Sigma_1(\mathcal{P}(\omega_1))$ formula, if there is a forcing condition p in $\mathbb{P}$ forcing the existence of a y such that $\varphi(y)$ holds, then there is also such a y in V.

These two formulations of $MA_{\aleph_1}$ are equivalent when V is replaced by a countable transitive model little-V of ZFC. When little-V is not countable (and possibly equal to V), we use the latter internal formulation to express $MA_{\aleph_1}$.

Thus we convert a principle that makes reference to outer models of V to one which is internal, expressible within V.

4.2 V-Logic and IMH

The point of V-logic is that it provides a tool to enable us to do the analogous thing not for just generic outer models, but for outer models in general. *V-logic* has a symbol for $\in$, a predicate symbol $\bar{V}$ to denote V and a constant symbol $\bar{x}$ to denote x for each set x. The proof relation $\vdash_V$ of V-logic begins with axioms that assert that $\bar{x}$ belongs to $\bar{V}$ for each set x, together with the usual axioms of first-order logic and all quantifier-free sentences true in V. The rules of inference are modus ponens together with the infinitary rules:

- From $\varphi(\bar{y})$ for all y in x, infer $\forall y \in \bar{x}\varphi(y)$.
- From $\varphi(\bar{x})$ for all x in V, infer $\forall x \in \bar{V}\varphi(x)$.

Proofs are then well-founded trees which can be shown to belong to $Hyp(V)$, the least admissible set containing V as an element. Assuming height potentialism, (which is provided by the Zermelian conception), $Hyp(V)$ makes full sense.

As said, now we proceed in a way fully analogous to what we did above using the forcing relation. Reconsider the IMH:

- (The Inner Model Hypothesis, IMH) If a first-order sentence holds in an inner model of some outer model of V then it also holds in some inner model of V.

We then formulate an internal version of this as follows:

- (The *Internal* Inner Model Hypothesis, IMH) If the theory in V-logic T_φ asserting that the first-order sentence φ holds in an inner model of some outer model of $\bar{V}$ is consistent in V-logic, then there is an inner model of V in which φ holds.

The 'internal' IMH is expressible as a first-order property of $Hyp(V)$, using the fact that the consistency of T_φ in V-logic is equivalent to saying that there is no V-logic proof in $Hyp(V)$ of a contradiction using the axioms of T_φ. And as in the case of $MA_{\aleph_1}$, the two formulations of the IMH, the one using outer models and the internal one, are equivalent when V is replaced by a countable transitive model little-V of ZFC.

Thus V-logic opens the door to expressing a wide range of width maximality principles, even in the Zermelian, width actualist context. With rare exceptions, these principles are formalisable internally in $Hyp(M)$ for arbitrary transitive ZFC models M, and not just for countable ones. In fact, in almost all cases, the study of width maximality principles for V can be reduced to its study for *countable transitive models* of ZFC. We discuss this in the next section.

5 Reduction to $\mathbb{H}$

5.1 *Reduction of IMH*

Our introduction of V-logic was intended to deal with the problem that for an uncountable transitive model of ZFC (such as V itself) there may be no (proper) outer models available and therefore we are required to discuss width maximality in terms of the consistency of V-logic theories.

As promised, we shall now deal with the second approach, where V is taken to be a countable transitive model little-V. Moreover, in this section we show that we can *reduce* our study of width maximality, and to some extent of height maximality, to a study of countable transitive models. As the collection of countable transitive models carries the name *hyperuniverse*, we are led to what is known as the *Hyperuniverse Programme*.

First we illustrate the reduction to the hyperuniverse with the specific example of the IMH. Suppose that we formulate the IMH as above, using V-logic, and want to know what first-order consequences it has.

Fact 3 *Suppose that a first-order sentence φ holds in all countable models of the IMH. Then it holds in all models of the IMH.*

This is for the following reason: Suppose that φ fails in some model M of the IMH, where M may be uncountable. Now notice that the IMH is first-order expressible in $Hyp(M)$, the least admissible lengthening of M. But then apply the downward Löwenheim-Skolem theorem to obtain a countable little-v which satisfies the IMH, as verified in its associated little-$Hyp(V)$, yet fails to satisfy φ. But this is a contradiction, as by hypothesis φ must hold in all *countable* models of the IMH.

So *without loss of generality*, when looking at first-order consequences of width maximality criteria as formulated in V-logic, we can restrict ourselves to countable little-V's. The advantage of this is that, then, we can dispense with the little-V-logic as by the Completeness Theorem for little-V-logic, consistent theories in little-V-logic do have models, thanks to the countability of little-V. Thus for a countable little-V, the IMH simply says:

- (*IMH for little-V's*). Suppose that a first-order sentence holds in an inner model of an outer model of little-V. Then it holds in an inner model of little-V.

But, if V is taken to be 'little-V', then V can really be 'thickened', which means that the Zermelian picture collapses to a *radical potentialist* picture, wherein both height and width of V are not fixed.

As we have seen, the Zermelian and the radical potentialist versions of the IMH coincide on countable models.

5.2   Reduction of #-Generated V: #-Generation Revisited

As far as the case of #-generation is concerned, its reduction to the hyperuniverse is not so obvious, and we shall see that the choice of working either within a Zermelian perspective or a radical potentialist perspective makes a big difference.

First, consider the following encouraging analogue for #-generation of our earlier reduction claim for the IMH, which we state here without proof.

Fact 4 *Suppose that a first-order sentence φ holds in all countable models which are #-generated. Then it holds in all models which are #-generated.*

Now the difficulty is this: how do we express #-generation from a width actualist perspective? Recall that to produce a generating # for V we have to produce a set of rank less than $Ord(V)$ which does not belong to V, in violation of width actualism.

At this point we need to say a bit more about #'s and models generated by them. A pre-# is a structure (N, U) where U measures the subsets in N of the largest cardinal κ of N, meeting certain first-order conditions; it is a # if in addition it is *iterable*, i.e. for any ordinal α if we take iterated ultrapowerse of (N, U) for α steps then it remains wellfounded. V is #-generated if it results as the union of the lower parts of the α-iterates of some # as α ranges over $Ord(V)$.

But notice that to express the iterability of a generating # for V we are forced to consider theories T_α formulated in $L_\alpha(V)$-logic for *arbitrary* (Gödel-) lengthenings $L_\alpha(V)$ of V: T_α asserts that V is generated by a *pre-#* which is α-*iterable*, i.e. iterable for α-steps. Thus we have no fixed theory that captures #-generation, only a tower of theories T_α (as α ranges over ordinals past the height of V) which capture closer and closer approximations to #-generation.

Therefore, in order to overcome these difficulties, we need to introduce another form of #-generated V, that is, weakly #-generated V.

Definition 5 V is *weakly #-generated* if for each ordinal α past the height of V, the theory T_α which expresses the existence of an α-iterable pre-# which generates V is consistent.

Weak #-generation is meaningful for a width actualist who is also a height potentialist (that is, a Zermelian), as it is expressed entirely in terms of theories internal to lengthenings of V.

A countable little-V is weakly #-generated if it is α-generated for each countable ordinal α (where the witnesssing pre-# may depend on α). Little-V is #-generated iff it is α-generated when $\alpha = \omega_1$ iff it is α-generated for all ordinals α.

Now we have the following reduction to countable little-V's:

Fact 6 *Suppose that a first-order sentence φ holds in all countable little-V which are weakly #-generated, and this is provable in ZFC. Then φ holds in all models which are weakly #-generated.*

To summarise: as radical potentialists we can comfortably work with full #-generation as our principle of height maximality. But as width actualists, we instead work with weak #-generation, expressed in terms of theories inside Gödel lengthenings $L_\alpha(V)$ of V. Weak #-generation is sufficient to maximise the height of the universe. And properly formulated, the reduction to the hyperuniverse also applies to weak #-generation: to infer that a first-order statement follows from weak #-generation it suffices to show that in ZFC one can prove that it holds in all weakly #-generated countable models.[17]

In what follows we will primarily work with #-generation, as at present the mathematics of weak #-generation is poorly understood. Indeed, as we shall see in the next section, a synthesis of #-generation with the IMH is consistent, but this remains an open problem for weak #-generation.

6 $\mathbb{H}$-Axioms: Synthesis of #-Generation with IMH-Variants

In light of the reduction to the hyperuniverse ($\mathbb{H}$), we see that maximality features of V such as #-generation and the IMH can be expressed as axioms about countable models, i.e. as properties of members of $\mathbb{H}$ expressed through quantification over $\mathbb{H}$. We refer to these as $\mathbb{H}$-*axioms*.

An important step in the development of the Hyperuniverse Programme is the *synthesis* of the $\mathbb{H}$-axiom of #-generation, expressing vertical maximality, with $\mathbb{H}$-axioms which express horizontal maximality. The first example of such a synthesis is the *IMH#*, which asserts the IMH for vertically-maximal universes:

Definition 7 (IMH#) M satisfies the IMH# if M is #-generated and whenever a first-order sentence holds in a #-generated outer model of M, it also holds in a definable inner model of M.

---

[17]Weak #-generation is indeed strictly weaker than #-generation for countable models: Suppose that $0^\#$ exists and choose α to be least so that α is the α-th Silver indiscernible (α is countable). Now let g be generic over L for Lévy collapsing α to ω. Then by Lévy absoluteness, L_α is weakly #-generated in $L[g]$, but it cannot be #-generated in $L[g]$ as $0^\#$ does not belong to a generic extension of L.

IMH# captures both vertical maximality and aspects of horizontal maximality simultaneously. But the development of $\mathbb{H}$-axioms does not stop here. One may introduce further logical contraints, and derive further principles incorporating them.

An *absolute parameter* is a set p which is uniformly definable over all outer models of V which 'respect' them in the sense that they preserve cardinals up to and including the cardinality of the transitive closure of p. The *SIMH (Strong IMH)* is the IMH for sentences with absolute parameters relative to outer models which respect them:

Definition 8 (SIMH) If a sentence with absolute parameters holds in an outer model which respects those parameters then it holds in a definable inner model.

A related principle is the *CPIMH (Cardinal Preserving IMH)*. A *cardinal-absolute parameter* is a set p which is uniformly definable over all cardinal-preserving extensions of V. Then CPIMH asserts the following:

Definition 9 (CPIMH) If a sentence with cardinal-preserving parameters holds in a cardinal-preserving outer model of V it also holds in a definable inner model of V.

Restricting SIMH and CPIMH to #-generated universes yields corresponding principles SIMH# and CPIMH#.[18]

More recent work (see [10]) develops further $\mathbb{H}$-axioms, such as forms of *Cardinal Maximality* (for example: κ^+ of HOD is less than κ^+ for every infinite cardinal κ), *Width Reflection* (for each ordinal α there is an amenable elementary embedding of an inner model into V with critical point greater than α) and its associated analogue of #-generation for width called *Width Indiscernibility* and *Omniscience* (the first-order definability of satisfaction across outer models, see [11]).[19]

7 The Dynamic Search for Truth

We now proceed to review some of the issues we had briefly mentioned at the beginning, relating to the correct interpretation of maximality, and to whether and in what sense the maximal iterative concept should be construed as expressing a *platonistic* conception of mathematics.

It is important to recall once more the way we construe the maximality of V. We said that V can *literally* be maximised, through maximising the ordinals α indexing the V_α and the subsets in $V_{\alpha+1}$, for all ordinals α. In turn, this was conceptualised as corresponding to 'lengthening' and 'thickening' the universe. Whenever this was shown not to be possible within the Zermelian picture, we found a way to *internalise* the maximisation through the use of a powerful logic, V-logic.

---

[18]See Fig. 1 at the end of the paper.

[19]For the consequences of all of these in members of $\mathbb{H}$, see, in particular, [2, 3, 10].

Now, we have seen that there is an altogether different approach to the maximality of V, that is *full actualism*, whereby such *absolutely infinite* objects as V are viewed as already *maximal*, in a way which cannot be transcended. Full actualism befits *universism*, insofar as it also encourages the idea that there is a *fully determinate* universe of sets.

Universism, although not implausible, is, at least, epistemologically dubious, like the radical form of Platonism which underlies it. The main trouble with this conception is that the associated semantic determinacy (that is, the idea that, for all ϕ, ϕ is *uniquely* decided by a suitable collection of axioms) leaves a considerable portion of set-theoretic practice, dealing with different 'universes', entirely unaccountable. Furthermore, universism, unless it is endowed with a suitably strong class theory, is inadequate to express the myriad of valuable forms of width maximality that are otherwise available.

It is even more doubtful that universism stems from a correct interpretation of the maximal iterative concept, as proclaimed by its supporters.[20] But even if it were, we have seen that there are ways to incorporate 'thickenings' of the universe even within a width actualist picture and extend this to 'lengthenings' using MK.[21] Therefore, the *platonist absolutist* who believes in the existence of a preferred structure determinately encompassing all truths about sets would not have to abandon her position, even in the case maximality principles should be viewed as more correctly implying the idea of 'thickenings' in height and width (as in the Zermelian or fully potentialist picture).

Moreover, it is not clear what one gains epistemically from holding that universism is the only way to make sense of maximality. It is interesting to briefly take into account the discussion of this issue provided by Hauser. Hauser has, in our view, convincingly, shown that finding objective solutions to such undecidable statements as CH does not depend upon having a pre-formed picture of V, that is, from believing in the full determinacy of V itself. Rather, objective solutions of set-theoretic problems will most likely be the outcome of procedures conforming to particular *evidential* standards of proof. In the author's own words:

> [This position] can be characterized in a nutshell as *objectivity over objects* and involves a twofold inversion of priorities. The first one shifts the attention from ontology to epistemology, i.e., questions about the existence and nature of mathematical items are discussed exclusively in the context of mathematical truth. [...] In the second inversion, evidence is treated as the primary epistemological concept. This reflects the widespread agreement among philosophers (and mathematicians) about what counts as evidence for the truth of a proposition—regardless of their conflicting ideas about the nature of truth. [15, 265–66]

---

[20]For a defence of this position see [16].

[21]We come back to the issue briefly discussed in footnote 12. The overall strategy is to formulate the IMH in V-logic, as shown above in Sect. 4. Recall that V-logic proofs are carried out in $Hyp(V)$, the least admissible structure containing V as an element. Now, it can be proved that, in a sub-theory of MK, it is also possible to build a class coding $Hyp(V)$, and therefore fully make use of V-logic to handle width maximality.

The author also casts a hypothesis concerning the way the general acceptability of new axioms will be construed in view of such epistemic inversion. In his view, the latter

..may be characterized as a gradual convergence towards a *reflective equilibrium* of high-level convictions and their lower level and 'practical' consequences along the lines of the holistic views on theory formation [. . .]. (*ibid.*, p. 275)

Now, when we evoked 'optimality' with reference to the search for new axioms, we intended to refer precisely to procedures whereby one could select the most suitable $\mathbb{H}$-axioms, by studying their mathematically 'optimal' features, in a way which may plausibly recall the *objectivity over objects* account advocated by Hauser, that is by downplaying the role of ontology (in particular, a *universist* ontology). Only, we do not view 'practical consequences' as crucial to this undertaking (although certainly the consequences of maximality principles are worth examining), nor do we subscribe to a holistic view concerning set-theoretic truth: the idea of 'testing' maximality principles to find optimal $\mathbb{H}$-axioms should not be viewed as subservient to the search for *extrinsic* (that is, 'empirical') evidence for new axioms, but rather to the goal of best expressing the maximality of V.

Within scientific procedures, optimality is provided by the fine-tuning of the general statements of a theory through empirically testing its results. Within set theory, it is hard to say what may count as an analogue of this, unless one takes the study of 'consequences' to play the same role as that of confirmation in physics (which is, to say the least, utterly problematic). For our purposes, though, this can hardly be different from the idea of producing *progressive* refinements and strengthenings of higher-order principles.

The idea of progressive refinements of maximality principles adds an interesting 'dialectical' twist to our search for new axioms: the motivating idea is that different principles should be combined to produce syntheses of their features and better candidates as ultimate maximality principles (as illustrated in Fig. 1).

Ideally, then, the study of $\mathbb{H}$-axioms will reach its natural endpoint when optimal maximality principles are found. We believe that the attainment of this might reasonably be described in terms of finding fully *intrinsically justified* new axioms.

At this stage, we cannot say anything definitive, but surely the analysis of the maximality of V conducted within the Hyperuniverse Programme has already led to remarkable findings, that, in conclusion, we may recapitulate as follows:

(1) #-generated V may be viewed as the strongest possible form of reflection construed as 'lengthening' of V
(2) IMH may be viewed as the most natural form of expressing the width maximality of V, insofar as it successfully thrives on a suitable conceptualisation of 'thickenings' (outer models) of V through V-logic
(3) combinations, variants and refinements of these principles construed as quantifying over members of the hyperuniverse ($\mathbb{H}$-axioms) can be shown to have the effect of strongly reducing set-theoretic incompleteness, in such a way as to make it at least plausible to assume that they could be seen as new axioms.

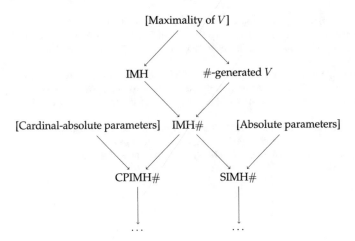

Fig. 1 Maximality principles

References

1. C. Antos, N. Barton, S.-D. Friedman, Universism and Extensions of V. preprint
2. C. Antos, S.-D. Friedman, R. Honzik, C. Ternullo, Multiverse conceptions in set theory. Synthese **192**(8), 2463–2488 (2015)
3. T. Arrigoni, S. Friedman, The hyperuniverse program. Bull. Symb. Log. **19**(1), 77–96 (2013)
4. N. Barton, S.-D. Friedman, Maximality and ontology: how axiom content varies across philosophical frameworks. Synthese 1–27 (2017)
5. P. Benacerraf, H. Putnam (eds.), *Philosophy of Mathematics. Selected Readings* (Cambridge University Press, Cambridge, 1983)
6. P. Bernays, Sur le Platonisme dans les Mathématiques. L'Enseignement Mathématique **34**, 52–69 (1935)
7. G. Boolos, The iterative conception of set. J. Philos. **68**(8), 215–231 (1971)
8. W. Ewald (ed.), *From Kant to Hilbert: A Source Book in the Foundations of Mathematics*, volume II (Oxford University Press, Oxford, 1996)
9. S. Friedman, Internal consistency and the inner model hypothesis. Bull. Symb. Log. **12**(4), 591–600 (2006)
10. S. Friedman, Evidence for set-theoretic truth and the hyperuniverse programme. IfCoLog J. Log. Appl. **3**(4), 517–555 (2016)
11. S. Friedman, R. Honzik, On strong forms of reflection in set theory. Math. Log. Q. **62**(1–2), 52–58 (2016)
12. S. Friedman, P. Welch, W.H. Woodin, On the consistency strength of the inner model hypothesis. J. Symb. Log. **73**(2), 391–400 (2008)
13. K. Gödel, What is Cantor's continuum problem? Am. Math. Mon. **54**, 515–525 (1947)
14. K. Gödel, What is Cantor's continuum problem? ed. by P. Benacerraf, H. Putnam, *Philosophy of Mathematics. Selected Readings*, pp. 470–85 (Prentice-Hall, 1964)
15. K. Hauser, Is the continuum problem inherently vague? Philos. Math. **10**, 257–285 (2002)
16. L. Horsten, P. Welch, Reflecting on absolute infinity. J. Philos. **113**(2), 89–111 (2016)
17. L. Incurvati, Maximality principles in set theory. Philos. Math. **25**(2), 159–193 (2017)
18. R. Jensen, Inner models and large cardinals. Bull. Symb. Log. **1**(4), 393–407 (1995)
19. A. Kanamori, Zermelo and set theory. Bull. Symb. Log. **4**, 487–553 (2004)
20. P. Koellner, On reflection principles. Ann. Pure Appl. Logic **157**(2–3), 206–219 (2009)

21. P. Maddy, Believing the axioms, I. Bull. Symb. Log. **53**(2), 481–511 (1988)
22. C. Parsons, What is the Iterative Conception of Sets? ed. by P. Benacerraf, H. Putnam, in *Philosophy of Mathematics. Selected Readings*, pp. 503–529 (Cambridge University Press, Cambridge, 1983)
23. M. Potter, *Set Theory and Its Philosophy* (Oxford University Press, Oxford, 2004)
24. C. Ternullo, Gödel's Cantorianism, ed. by G. Crocco, E.-M. Engelen, in *Kurt Gödel: Philosopher-Scientist*, pp. 413–442 (Presses Universitaires de Provence, Aix-en-Provence, 2015)
25. H. Wang, *From Mathematics to Philosophy* (Routledge & Kegan Paul, London, 1974)
26. H. Wang, *A Logical Journey* (MIT Press, Cambridge (MA), 1996)
27. E. Zermelo, Über Grenzzahlen und Mengenbereiche: neue Untersuchungen über die Grundlagen der Mengenlehre. Fundam. Math. **16**, 29–47 (1930)

Large Cardinals and the Continuum Hypothesis

Radek Honzik

Abstract This is a survey paper which discusses the impact of large cardinals on provability of the Continuum Hypothesis (CH). It was Gödel who first suggested that perhaps "strong axioms of infinity" (large cardinals) could decide interesting set-theoretical statements independent over ZFC, such as CH. This hope proved largely unfounded for CH—one can show that virtually all large cardinals defined so far do not affect the status of CH. It seems to be an inherent feature of large cardinals that they do not determine properties of sets low in the cumulative hierarchy if such properties can be forced to hold or fail by small forcings.

The paper can also be used as an introductory text on large cardinals as it defines all relevant concepts.

1 Introduction

The question regarding the size of the continuum—i.e. the number of the reals—is probably the most famous question in set theory. Its appeal comes from the fact that, apparently, everyone knows what a real number is and so the question concerning their quantity seems easy to understand. While there is much to say about this apparent simplicity, we will not discuss this issue in this paper. We will content ourselves by stating that the usual axioms of set theory (ZFC) do not decide the size of the continuum, except for some rather trivial restrictions.[1] Hence it is consistent, assuming the consistency of ZFC, that the number of reals is the least possible, i.e. the cardinal $\aleph_1$, but it can be something much larger, e.g. $\aleph_{\aleph_1}$.

The statement that the number of reals is the least possible is known as the *Continuum Hypothesis*, CH, for short:

$$\text{CH:} \quad |\mathbb{R}| = 2^{\aleph_0} = \aleph_1.$$

Originally published in R. Honzik, Large cardinals and CH, AUC Philosophica et Historica. Miscellanea Logica IX **2**, 35–52 (2013).

[1] The cofinality of the size of the continuum must be uncountable.

R. Honzik (✉)
Department of Logic, Charles University, Celetná 20, Praha 1, 116 42, Czech Republic
e-mail: radek.honzik@ff.cuni.cz

© Springer International Publishing AG 2018
C. Antos et al. (eds.), *The Hyperuniverse Project and Maximality*,
https://doi.org/10.1007/978-3-319-62935-3_10

CH was made famous by David Hilbert who included this problem as the first one on his list of mathematical problems for the twentieth century (see for instance [5]).

Since ZFC does not decide CH, are there any natural candidates for axioms which do? That is, is there a statement φ without apparent connection to CH which decides CH one way or the other? In fact there are many of these, such as MA or PFA,[2] but we will require φ to be one of a more special kind. In 1946, that is well before the development of forcing, Gödel entertained the idea of so called *stronger axioms of infinity* deciding CH (and other independent statements as well)[3]:

> It is not impossible that [...] some completeness theorem would hold which would say that every proposition expressible in set theory is decidable from the present axioms plus some true assertion about the largeness of the universe of all sets [1].

A natural way to arrive at "true assertions" about largeness of the universe of sets is to take up analogies with natural numbers. When we compare the theory of arithmetics such as PA with the theory of sets such as ZFC, we can show that the only important strengthening of ZFC over PA is the addition of the axiom of infinity. The axiom can be formulated in many ways, but for our purposes we adopt the following definition:

(*) Axiom of Infinity: *There is an ordinal ω which is the domain of a model for the formalization of PA.*

Because of this axiom, ZFC can not only prove some logical arithmetical statements which PA itself cannot prove (unless it is inconsistent), such as $\mathrm{Con}(PA)$, but also some purely number-theoretical statements as well (such as Goodstein's theorem, see for instance [17]). Gödel suggested that perhaps by adding a stronger axiom of infinity to ZFC, this new theory might decide new statements interesting to set theoreticians.[4] Can we find such an axiom, perhaps similar to (**) or (***) below, which will decide CH?

(**) A strong Axiom of Infinity: *There is a regular cardinal κ such that $\langle V_\kappa, \in \rangle$ is a model of the formalization of ZFC.*

or

(***) A still stronger Axiom of Infinity: *There is a regular cardinal κ such that $\langle V_\kappa, \in \rangle$ is a model of the formalization of ZFC + (**).*

Where V_κ is an initial part of the universe of sets (see Definition 2.1) and is the analogue of ω for sets.

---

[2] See footnote 27.

[3] As regards the intuitive "truth" of such axioms, or why they should be preferable to other types of axioms, see a discussion for instance in [2].

[4] Such extensions will always decide new statement, such as $\mathrm{Con}(ZFC)$, but these are considered too "logical" and not properly set-theoretical.

Remark This paper is in a sense a continuation of [6] which contains an introduction to the axioms of set theory, discusses the basic set-theoretical notions and not so briefly reviews basics of forcing. Of course, any of the standard texts such as [7] or [10] contains all the prerequisites to this article. A standard reference book for large cardinals is [8] where an interested reader can find more details.

2   How to Find Large Cardinals

In this section we survey large cardinals which can be considered as candidates for the stronger axioms of infinity. The selection is rather arbitrary, but does attempt to do justice to the most important concepts.

2.1   Inaccessible Cardinals

In the presence of the Axiom of Foundation,[5] the universe V is equal to the union $V = \bigcup_{\alpha \in \text{ORD}} V_\alpha$, where the initial segments V_α are defined by recursion along the ordinal numbers ORD as follows:

Definition 2.1

$$V_0 = \emptyset,$$
$$V_{\alpha+1} = \mathscr{P}(V_\alpha),$$
$$V_\lambda = \bigcup_{\alpha < \lambda} V_\alpha, \text{ for } \lambda \text{ limit ordinal,}$$
$$V = \bigcup_{\alpha \in \text{ORD}} V_\alpha.$$

If κ satisfies (**) above, we say that V_κ is a "natural model" of set theory. To obtain such a model in set theory, we must transgress the power of the plain ZFC theory—this is a consequence of the second Gödel theorem.

What are the properties of a cardinal κ such that V_κ satisfies (**) above? We postulated that it must be regular (we will later see that we cannot avoid this assumption), but what else?

Definition 2.2 We say that a cardinal μ is *strong limit* if for all $\nu < \mu$, $2^\nu < \mu$.

Notice that every strong limit cardinal is also limit (i.e. does not have an immediate cardinal predecessor).

Lemma 2.3 *Assume κ satisfies (**). Then κ is strong limit.*

---

[5]This axioms states that sets are "well-behaved"; for instance sets x such as $x \in x$ are prohibited by this axiom.

Proof Assume $\mu < \kappa$ is given. ZFC proves that there is a cardinal $\nu \geq \mu^+$ and a bijection $f : \nu \to \mathscr{P}(\mu)$. Since $\langle V_\kappa, \in \rangle$ is a model of ZFC, we have

$$\langle V_\kappa, \in \rangle \models \text{"There is a bijection between } \mathscr{P}(\mu) \text{ and some } \nu \geq \mu^+\text{"}.$$

Since V_κ is transitive, and $\mathscr{P}(\mu) = (\mathscr{P}(\mu))^{V_\kappa}$,[6] any such bijection in V_κ is really (in V) a bijection between $\mathscr{P}(\mu)$ and some ordinal ν in V_κ. As $\nu < \kappa$, $2^\mu < \kappa$. $\square$

Notice that for a regular κ, if $\mu < \kappa$, then $\mathscr{P}(\mu) \in V_\kappa$ (see Footnote 6); however, this does not generally imply that κ is strong limit because the existence of a bijection between $\mathscr{P}(\mu)$ and some ν in V_κ depends on the truth of the Replacement schema in V_κ. In fact, we state without a proof that if κ is a regular cardinal, then all axioms of ZFC, except possibly some instances of the Schema of Replacement, are true in $\langle V_\kappa, \in \rangle$.

Lemma 2.3 motivates the following definitions:

Definition 2.4 We say that a cardinal $\kappa > \omega$ is *weakly inaccessible* if it is *regular* and *limit*.

Definition 2.5 We say that a cardinal $\kappa > \omega$ is *strongly inaccessible* if it is *regular* and *strongly limit*.

Theorem 2.6

 (i) *Every cardinal satisfying (**) is strongly inaccessible.*
 (ii) *Every strongly inaccessible cardinal satisfies (**).*

Proof Ad (i) Obvious from the definitions and Lemma 2.3.

Ad (ii) (Sketch). For every regular κ, $\langle V_\kappa, \in \rangle$ is a model of ZFC without Schema of Replacement (this is easy to check). Strong limitness is used to ensure that Replacement holds as well. $\square$

Although it may not be immediately apparent, the weakly inaccessible cardinal is not weaker in terms of consistency strength than the strongly inaccessible cardinal. Let $\exists \kappa \; \psi_w(\kappa)$ denote the sentence "there exists a weakly inaccessible cardinal", and similarly for the strongly inaccessible $\exists \kappa \; \psi_s(\kappa)$.

Lemma 2.7

$$\text{Con}(\text{ZFC} + \exists \kappa \; \psi_w(\kappa)) \leftrightarrow \text{Con}(\text{ZFC} + \exists \kappa \; \psi_s(\kappa)).$$

Proof The more difficult direction is from left to right. Assume κ is weakly inaccessible. Let L be the universe of constructible set, defined by Gödel. We know that L satisfies ZFC and also GCH.[7] It is immediate to see that in L, κ is strongly

---

[6] $(\mathscr{P}(\mu))^{V_\kappa}$ is the powerset of μ in the sense of $\langle V_\kappa, \in \rangle$. Note that for every limit ordinal α, if $\beta < \alpha$, then $(\mathscr{P}(\beta))^{V_\alpha} = \mathscr{P}(\beta)$ because $\mathscr{P}(\beta) \subseteq V_{\beta+1}$, and so $\mathscr{P}(\beta) \in V_{\beta+2} \subseteq V_\alpha$.

[7] The Generalized Continuum Hypothesis which states that for every cardinal μ, $2^\mu = \mu^+$.

inaccessible because being a limit cardinal together with GCH implies the desired property of strong limitness.   □

Therefore by Gödel's theorem and Lemma 2.7 and Theorem 2.6(ii):

Corollary 2.8 *If ZFC is consistent, it does not prove the existence of a weakly inaccessible cardinal.*

One can also show that if ZFC is consistent, so is the theory ZFC + "there is no strongly inaccessible cardinal", and that ZFC does not prove the implication CON(ZFC) → CON(ZFC+ "there is a strongly inaccessible cardinals").

Usually, when we talk about an *inaccessible cardinal*, we mean the strongly inaccessible, and assumption of existence of such a cardinal number is taken to be the first step in defining strong axioms of infinity. Thus we can reformulate:

$(**)_r$: (Strong Axiom of Infinity) There is a (strongly) inaccessible cardinal.

Remark 2.9 One might wonder if we can remove the assumption of regularity in (**) and have an equivalent notion. We cannot: if κ is strongly inaccessible, we can use the standard Löwenheim-Skolem argument to obtain an elementary substructure $\langle V_\alpha, \in \rangle \prec \langle V_\kappa, \in \rangle$ with $\alpha > \omega$, and $\mathrm{cf}(\alpha) = \omega$. Thus $\langle V_\alpha, \in \rangle$ is a model of ZFC, but α is not regular.[8] That is why we need to explicitly postulate the regularity of κ in (**).

What about (***)? Well, it is not difficult to see that if $\kappa < \kappa'$ are two strongly inaccessible cardinals, then $\langle V_{\kappa'}, \in \rangle$ is the desired model for (***). This is the case because

$$\langle V_{\kappa'}, \in \rangle \models \text{"}\kappa \text{ is a strongly inaccessible cardinal"}.$$

Thus we may reformulate:

$(***)_r$: (Still stronger Axiom of Infinity) There are two (strongly) inaccessible cardinals.

We could repeat this argument many times over, obtaining stronger and stronger axioms of infinity, in the hope of deciding more and more sentences. However, there is a limit to this recursion—so called Mahlo cardinals (see the next section).

2.2 Mahlo Cardinals

We include this cardinal only because it is in a sense a limit to the process of arriving to a large cardinal by a process "from below". Recall that above we have considered one, two, three, and so on inaccessible cardinals. What if we consider $\aleph_0$ or $\aleph_1$ many of them? Do we get something yet stronger? We do, but there is a natural limit to

---

[8]But it may be a singular strong limit cardinal.

this type of strengthening of the notion of a strong axiom of infinity. Consider an inaccessible cardinal κ such that κ is the κ-th inaccessible cardinal—clearly, it is a limit of the process of taking more and more inaccessible cardinals as far as their number is concerned. A Mahlo cardinal is even stronger (although it may not be apparent without a more detailed look which we will not provide here):

Definition 2.10 A cardinal κ is a *Mahlo cardinal* if the set of inaccessible cardinals smaller than κ is stationary in κ.[9]

2.3 Analogies with ω

We said above that Mahlo cardinals are a limit to arriving to larger cardinals "from below" by repeating certain continuous processes applied to inaccessible cardinals. But what other options do we have? Mathematicians found out that it is useful to consider the usual properties of ω and try to generalize them in a suitable fashion. In fact, inaccessible cardinals can be regarded in this way—either as a generalization of the concept of a "model" for a given theory (see above in (**) and (***)), or combinatorially—notice that ω itself is *regular* and *strong limit*, i.e. no finite subset of ω is cofinal in ω and $\forall n < \omega\ 2^n < \omega$. We generalize[10] three other properties of ω[11]:

(C) ω is *compact* in the sense of the compactness theorem for the first-order predicate logic.

(M)   There is a two-valued non-trivial *measure* on ω, i.e. a non-principal ultrafilter on ω. This measure is ω-complete: for every finite number of elements in the ultrafilter, their intersection is still in the ultrafilter.

(R)   The Ramsey property holds for ω: $\omega \to (\omega)^r_k$, for $r, k < \omega$.

Compactness (C) The classical predicate calculus satisfies compactness: for every language and for every set of formulas A (of arbitrary size) in that language if every finite subset $B \subseteq A$ has a model, so does A. In order to generalize this property, we consider an extension of the classical logic denoted as $L_{\kappa,\kappa}$, where κ is a regular cardinal, as follows. A language in $L_{\kappa,\kappa}$ can have up to κ many variables and an unlimited number of non-logical symbols (functions, constants, predicates). We also allow conjunctions and disjunctions of length $< \kappa$ and quantifications over $< \kappa$

---

[9]We will not define the notion of a stationary set here; any standard set-theoretical textbook contains this definition. Roughly speaking, a set is stationary in κ if it intersects every continuous enumeration of unboundedly many elements below κ. In particular, every stationary subset of a regular cardinal κ has size κ.

[10]We assume AC, the Axiom of Choice, in formulating these generalizations.

[11]Note that a priori there is no guarantee that we get anything like a large cardinal in this fashion; the generalization may turn out to be mathematically trivial and uninteresting. The fact that we do get large cardinals seems to indicate that these generalizations are mathematically relevant.

many variables.[12] The classical logic can be denoted as $L_{\omega,\omega}$ under this notation. Now we can formulate the generalization of the compactness theorem in two ways:

(wC) $\kappa > \omega$ is called *weakly compact* iff whenever A is any collection of sentences in $L_{\kappa,\kappa}$ with at most κ many non-logical symbols if every $B \subseteq A$ of size $< \kappa$ has a model, so does A.

(sC) $\kappa > \omega$ is called *strongly compact* iff whenever A is any collection of sentences in $L_{\kappa,\kappa}$ if every $B \subseteq A$ of size $< \kappa$ has a model, so does A.

We will discuss the relationship between (wC) and (sC) later in the text.

Measure (M) One can find a non-principal ultrafilter on ω, i.e. a set $U \subseteq \mathscr{P}(\omega)$ such that for all A, B subsets of ω:

(i) If $A \in U$ and $A \subseteq B$, then $B \in U$.
(ii) If $A, B \in U$, then $A \cap B \in U$.
(iii) For no $n < \omega$, $\{n\} \in U$.
(iv) For all A, either $A \in U$ or $\omega \setminus A \in U$.

Note that by induction, (ii) implies that if $A_0, \ldots, A_n$ are sets in U for $n < \omega$, then their intersection is in U—this property can be called ω-*completeness* to emphasize the analogy with κ-*completeness* for a cardinal $\kappa > \omega$ introduced below. U is non-principal because it is not generated by a single number (property (iii)); (iii) together with other properties implies that every set $A \in U$ is infinite.

(M) $\kappa > \omega$ is called *measurable* iff there is a κ-complete non-principal ultrafilter U on κ:

(i) If $A \in U$ and $A \subseteq B$, then $B \in U$.
(ii) If $\mu < \kappa$, and $\{A_\xi \mid \xi < \mu\}$ are sets in U, then $\bigcap_{\xi<\mu} A_\xi$ is in U.
(iii) For no $\xi < \kappa$, $\{\xi\} \in U$.
(iv) For all A, either $A \in U$ or $\kappa \setminus A \in U$.

Such an ultrafilter U is often called a *measure* because it "measures" subsets of κ by a two-valued κ-complete measure: if $A \in U$, then measure of A is 1, if $A \notin U$, then its measure is 0.

Ramsey Partitions (R) Let f be a function from $[\omega]^r$ to k, where $[\omega]^r$ denotes the set of all subsets of ω with exactly r elements, and $k = \{0, \ldots, k-1\}$ is a set of size k ($r \geq 1$ and $k \geq 2$ to avoid trivialities).

Definition 2.11 We say that $A \subseteq \omega$ is *homogeneous* for $f : [\omega]^r \to k$ if

$$|\mathrm{rng}(f \upharpoonright [A]^r)| = 1.$$

Ramsey proved in 1930 that for any such f there is an infinite homogeneous subset, in the arrow notation:

$$\omega \to (\omega)^r_k, \text{ for } r, k < \omega.$$

---

[12]For instance "$\exists_{\beta<\alpha} x_\beta \varphi$", $\alpha < \kappa$, quantifies over α-many variables in φ.

The argument is by induction on r, and the nontrivial step is to show

$$\omega \to (\omega)_2^2.$$

This we read that for any partition of two-element subsets of ω to two sets we can find an infinite homogeneous set. We therefore attempt to generalize:

(wR) A cardinal $\kappa > \omega$ is called *weakly Ramsey* if $\kappa \to (\kappa)_2^2$, i.e. for every partition of two-element subsets of κ to two sets we can find a homogeneous set of size κ.

We later learn that this generalization is not getting us a new concept, so we will need to strengthen it. That is why we call this property (wR) and not (R). See the next section for the now standard definition of the Ramsey cardinal.

2.4 Compact, Measurable, and Ramsey Cardinals

As we mentioned above, there is a priori no guarantee that the cardinals defined above under (wC), (sC), (M), and (wR) are even inaccessible. However, as it turns out, they are not only inaccessible but even Mahlo. By way of illustration, we show that a measurable cardinal κ is inaccessible.

Theorem 2.12 *Every measurable cardinal is inaccessible.*

Proof Let U be a non-principal κ-complete ultrafilter witnessing measurability of κ. First notice that by κ-completeness and non-principality of U, all elements in U have size κ. κ is clearly regular, otherwise if $\{\xi_\alpha \mid \alpha < \mathrm{cf}(\kappa)\}$ is cofinal in κ for $\mathrm{cf}(\kappa) < \kappa$, then while for each $\alpha < \mathrm{cf}(\kappa)$, $\xi_\alpha \notin U$, $\bigcup_{\alpha < \mathrm{cf}(\kappa)} \xi_\alpha = \kappa \in U$, contradicting the κ-completeness of U.[13] As regards strong limitness of κ, assume for contradiction that for some $\lambda < \kappa$, we have $2^\lambda \geq \kappa$, and let $f : \kappa \to \mathscr{P}(\lambda)$ be an injection. For a fixed $\alpha < \lambda$, we can consider two subsets of κ given by f: $X_0^\alpha = \{\xi < \kappa \mid \alpha \in f(\xi)\}$ and $X_1^\alpha = \{\xi < \kappa \mid \alpha \notin f(\xi)\}$. For each $\alpha < \lambda$, exactly one of the two sets X_0^α and X_1^α is in U; let us denote this set as X^α. By κ-completeness of U, $X = \bigcap_{\alpha < \lambda} X^\alpha$ must be in U. However X can have at most one element since f is an injection—if $\xi \neq \zeta$ are in X, then $f(\xi) \neq f(\zeta)$ and hence at some $\alpha < \lambda$, ξ must be in X_0^α and ζ in X_1^α (or conversely). This contradicts the non-principality of U. It follows that κ is strong limit, and hence inaccessible. □

With nice combinatorial arguments, not always trivial ones, one can show that every strongly compact cardinal is measurable, every measurable is weakly compact, and every weakly compact is Mahlo, and every Mahlo is inaccessible.

---

[13]Notice that κ-completeness can be equivalently expressed as follows: whenever $\mu < \kappa$ and $\{X_\alpha \mid \alpha < \mu\}$ are sets not in U, the the union $\bigcup_{\alpha < \mu} X_\alpha$ is not in U, either.

Thus disparate combinatorial notions gave rise to a linearly ordered scale of cardinals.

What about the weakly Ramsey cardinal? With a little work, it can be shown that the definition (wR) is in fact equivalent to (wC). And so the classes of weakly compact cardinals and weakly Ramsey cardinals are the same. However, there is a way how to generalize the Ramsey property and obtain something stronger than a weakly compact cardinal:

(R)   A cardinal $\kappa > \omega$ is called *Ramsey* if $\kappa \rightarrow (\kappa)_2^{<\omega}$, i.e. for every partition of all finite subsets of κ to two sets we can find a homogeneous set A of size κ.[14]

Many questions concerning these cardinals are quite difficult. For instance, it had long been open (from 1930s to 1960s) whether the least inaccessible cardinal can be measurable. By a new method using elementary embeddings and ultrapowers developed by Scott, it was proved in the 1960s that measurable cardinals are quite large—they can never by the least inaccessible, or the least weakly compact cardinal. In fact if κ is measurable, then it is the κ-th weakly inaccessible cardinal, and more. We will touch briefly on the method of elementary embeddings in Sect. 3.4.

Finally, let us note that measurable cardinals were first used—before the introduction of the Cohen's method of forcing—to argue for the consistency of the statement $V \neq L$, i.e. that there exists a non-constructible set. It was Scott [15] who showed in 1961 that if there exists a measurable cardinal, then $V \neq L$. Nowadays large cardinal which imply that $V \neq L$ are called "large" large cardinals, while others are called "small" large cardinals. Inaccessible, Mahlo, and weakly compact cardinals are "small", while Ramsey, measurable and strongly compact are "large".

2.5 *Motivation*

We showed that by a natural attempt to generalize properties which hold for ω, we arrive at interesting notions in set theory which form a linear scale, as regards the strength of the notions. This is often taken as a heuristical point in favour of the naturalness of the definitions. Not least because by the linearity, no two large cardinals are inconsistent together—so far, no large cardinal was found that prohibits the existence of some other large cardinals. The properties which can be generalized range from purely logical (such as the inaccessible cardinal witnessing (**), or (wC) and (sC)), to combinatorial (wR), (R) and measure-theoretic (M).

On the downside, all these notions substantially increase the consistency strength of the relevant theories, thus increasing the risk of introducing a contradiction. It is conceivable, but not considered probable now, that ZFC is consistent, while ZFC + "there is an inaccessible" is not. Or that ZFC + "there is a weakly compact cardinal"

---

[14]For every $n < \omega$, $|\mathrm{rng}(f \restriction [A]^n)| = 1$.

is consistent while ZFC + "there is a measurable cardinals" is not. See Sect. 4 for more discussion on consistency strength.

Such discussion are not of logical interest only. It can be shown for instance that a certain weakening of the GCH, denoted as SCH,[15] is provable in ZFC if ZFC refutes the existence of inaccessible cardinals.[16] However, with some large cardinals around, SCH cannot be proved, and is therefore independent over the theory ZFC + certain large cardinals.[17]

3 Large Cardinals and CH

As we mentioned earlier, Gödel expressed his hopes that perhaps large cardinals could provide a natural extension of ZFC with interesting set-theoretical consequences such as determining the truth or falsity of CH. However, with the development of forcing on the way, Levy and Solovay in 1967 [11] came with arguments which are almost universal and show that truth or falsity of CH is unaffected by large cardinals. In the following sections, we assume some basic understanding of forcing on the reader's part.

3.1 How to Force CH and ¬CH

A standard forcing notion to force CH, which we will denote as $\mathbb{P}_{CH}$, is composed of functions $f : \omega_1 \to 2$ with the domain of f being at most countable. The extension is by reverse inclusion. $\mathbb{P}_{CH}$ adds a new subset of ω_1, and collapses 2^ω to ω_1 in the process.[18] $\mathbb{P}_{CH}$ is called the *Cohen forcing for adding a subset of* ω_1.

To force ¬CH, we will use ω_2 copies of the *Cohen forcing which adds a new subset of* ω. Formally, a condition in $\mathbb{P}_{\neg CH}$ is a function with finite domain from ω_2 to 2. One can show that $\mathbb{P}_{\neg CH}$ preserves cardinals and forces $2^\omega = \omega_2$.

For our purposes notice that $|\mathbb{P}_{CH}| = 2^\omega$ and $|\mathbb{P}_{\neg CH}| = \omega_2$, i.e. both forcings are quite small, certainly smaller than the first inaccessible.

---

[15]GCH, the Generalized Continuum Hypothesis, states that for all cardinals κ, $2^\kappa = \kappa^+$. SCH, the Singular Cardinal Hypothesis, states that for all singular cardinals κ, $2^\kappa = \max(2^{cf(\kappa)}, \kappa^+)$.

[16]This is true for larger cardinals than just inaccessibles.

[17]For instance if ZFC + (sC) is consistent, so is ZFC + ¬SCH.

[18]Notice that for every $X \subseteq \omega$ in V, it is dense in $\mathbb{P}_{CH}$ that there exists some $\alpha < \omega_1$ and p such that p restricted to $[\alpha, \alpha + \omega)$ is a characteristic function of X. The function defined in a generic extension which takes every $\alpha < \omega_1$ to a subset of ω given by the restriction of the generic filter to $[\alpha, \alpha + \omega)$ is therefore onto $(2^\omega)^V$. It follows that 2^ω of V is collapsed to ω_1.

3.2   Inaccessible and Mahlo Cardinals and CH

We have defined above two "small" forcings which can force CH and ¬CH, $\mathbb{P}_{CH}$ and $\mathbb{P}_{\neg CH}$, respectively. As it turns out, for the preservation of large cardinals, it suffices to assume that the forcing in question has size $< \kappa$.

Theorem 3.1 *Let P be a forcing of size $< \kappa$ and let G be a P-generic filter. Assume κ is inaccessible or Mahlo in V, then κ is inaccessible or Mahlo, respectively, in V[G]. In particular, these large cardinals do not decide CH.*

Proof First notice that the theorem really implies that these large cardinals do not decide CH. Suppose for contradiction that one of these cardinals decides CH; for example let us assume that ZFC + "there is an inaccessible" proves CH. Assume there is an inaccessible and force with $\mathbb{P}_{\neg CH}$; we obtain a generic extension where ¬CH holds and there is still an inaccessible. This a contradiction.

Let us now turn to the proof of the rest of the theorem. By standard forcing technique, if $\lambda < \kappa$ is given, then there are just $2^{|P|\lambda}$-many nice names for subsets of λ in $V[G]$. Since $\mu = 2^{|P|\lambda} < \kappa$ by inaccessibility of κ, we have $V[G] \models 2^\lambda \leq \mu < \kappa$, i.e. κ remain inaccessible in $V[G]$.[19]

To argue for preservation of Mahloness, we show as a lemma that forcings with κ-cc preserve stationarity of subsets of κ.

Lemma 3.2 *Assume Q is a forcing notion. If Q is κ-cc, then it preserves stationary subsets of κ.*

Proof Let $V[E]$ be a Q-generic extension and S stationary subset. We wish to show that S is still stationary in $V[E]$: that is, we need to show that if $C \in V[E]$ is closed unbounded, then $S \cap C \neq \emptyset$. Fix a closed unbounded C and let $p \in E$ force this:

$$p \Vdash \dot{C} \text{ is a closed unbounded subset of } \check{\kappa}.$$

Denote

$$D = \{\xi < \kappa \mid p \Vdash \check{\xi} \in \dot{C}\}.$$

Note that $D \subseteq C$ and $D \in V$. We prove that D is a closed unbounded subset of κ. Now the claim follows because $D \in V$, and so $D \cap S \neq \emptyset$. To prove D is closed unbounded, it suffices to argue that it is unbounded (closure is easy). Let $\alpha < \kappa$ be given. By induction construct for each $n < \omega$ a maximal antichain $A_n = \langle q_n^\xi \mid \xi < \alpha_n \rangle$ of elements below p and an increasing sequence of ordinals $\langle \beta_n^\xi \mid \xi < \alpha_n \rangle$, where $\alpha_n < \kappa$ (this is possible by κ-cc), such that:

(a) $\beta_0^0 \geq \alpha$;
(b) for each n, $\langle \beta_n^\xi \mid \xi < \alpha_n \rangle$ is strictly increasing;

---

[19]Note also that all cardinals $\geq |P|^+$, and hence also μ, remain cardinals in $V[G]$.

(c) if $m < n$ then all elements in the β_n-sequence are above the β_m-sequence;
(d) $q_n^\xi \Vdash \check{\beta}_n^\xi \in \dot{C}$.

Since for every $n < \omega$, $\alpha_n < \kappa$, $\bigcup_{n<\omega}\{\beta_n^\xi \mid \xi < \alpha_n\}$ is bounded in κ.

We show that $\delta = \sup\{\beta_n^\xi \mid n < \omega, \xi < \alpha_n\}$ is in D, that is $p \Vdash \check{\delta} \in \dot{C}$. By forcing theorems, it suffices to show that whenever F is a Q-generic and $p \in F$, then $\delta \in \dot{C}^F$. Since each A_n is maximal below p, $F \cap A_n$ is non-empty for each $n < \omega$. It follows that there is a sequence $\langle q_n \mid n < \omega \rangle$ of conditions in F which force that elements of $\dot{C}$ are unbounded below δ. Hence $\delta \in \dot{C}^F$ as required. □

Since our forcing P has size $< \kappa$, it certainly has the κ-cc, and therefore the set of regular cardinals below κ is still stationary in $V[G]$. That is κ is still Mahlo in $V[G]$. □

3.3 Weakly Compact and Measurable Cardinals and CH

By way of example, we show that if P has size $< \kappa$, then κ is still weakly compact or measurable in $V[G]$ if it was weakly compact or measurable, respectively, in V. In Theorems 3.3 and 3.4 we will give direct arguments, while in Sect. 3.4 we will put large cardinals into a more general picture so that we can formulate a uniform approach to preservation of large cardinals.

Theorem 3.3 *Assume κ is weakly compact in V and P has size $< \kappa$, and G is P-generic. Then κ is weakly compact in $V[G]$.*

Proof As a fact we state that κ is weakly compact iff

$$\kappa \rightarrow (\kappa)_\lambda^n, \text{ for every } n < \omega, \lambda < \kappa. \tag{3.1}$$

Let us fix in $V[G]$ a function $f : [\kappa]^2 \rightarrow 2$; it suffices to find in $V[G]$ a homogeneous set $X \subseteq \kappa$ of size κ. By Forcing theorem, there is $p \in P$ such that

$$p \Vdash \dot{f} : [\kappa]^2 \rightarrow 2.$$

Define back in V,

$$h : [\kappa]^2 \rightarrow \mathscr{P}(P \times 2)$$

by

$$h(s) = \{\langle q, i \rangle \mid q \leq p \ \& \ q \Vdash \dot{f}(\check{s}) = i\}.$$

Since $|\mathscr{P}(P \times 2)| < \kappa$, we can apply (3.1) and find a homogeneous set $X \subseteq \kappa$ for the function h. We claim that

$$p \Vdash \check{X} \text{ is homogeneous for } \dot{f},$$

or equivalently

$$X \text{ is homogeneous for } f \text{ in } V[G].$$

The homogeneity of X for h means that for all $s \in [X]^2$, $h(s)$ is equal to some fixed set of the form $A = \{\langle q, i \rangle \mid q \leq p \ \& \ q \Vdash \dot{f}(\check{s}_0) = i\}$, for some $s_0 \in [\kappa]^2$. Notice that because p forces that $\dot{f}$ is a function, there can be no "contradictory pairs" $\langle q, 0 \rangle$ and $\langle q, 1 \rangle$ in A; that is for each $q \leq p$ occurring at the first coordinate of a pair in A there is unique $i(q)$ such that $\langle q, i(q) \rangle$ is in A. Assume F is any P-generic with $p \in F$. For each $s \in [X]^2$, there is some $q(s) \in F$ such that $\langle q(s), i(q(s)) \rangle$ is in A. If s_1, s_2 are in $[X]^2$, then $q(s_1)$ and $q(s_2)$ are compatible in F by some r which thus decides both $\dot{f}(\check{s}_1)$ and $\dot{f}(\check{s}_2)$; furthermore, there is a unique $i(r)$ such that $\langle r, i(r) \rangle$ is in A and so $i(q(s_1)) = i(q(s_2)) = i(r)$. This proves that p forces that X is homogeneous for $\dot{f}$. □

Theorem 3.4 *Assume κ is measurable in V and P has size $< \kappa$. Then κ is measurable in V^P.*

Proof Let G be a P-generic filter, and let U be a κ-complete non-principal ultrafilter on κ in V. We will show that

$$W = \{A \subseteq \kappa \mid \exists B \in U \ B \subseteq A\}$$

is a κ-complete non-principal ultrafilter in $V[G]$; we say that W is generated by U. It is easy to show that W is non-principal, closed upwards, and κ-complete—that is that is a κ-complete non-principal filter:

 (i) Non-principality. Since U is non-principal and every element of W is above an element of U, the argument follows.
(ii) κ-completeness. Fix in $V[G]$ a sequence $\langle A_\xi \mid \xi < \lambda \rangle$, $\lambda < \kappa$ of sets in W. By definition of W, there is $p \in G$ such that

$$p \Vdash \text{"There exists a sequence } \langle \dot{B}_\xi \mid \xi < \lambda \rangle \text{ of sets in } U \text{ such that}$$

$$\text{for every } \xi < \lambda, \dot{B}_\xi \subseteq \dot{A}_\xi." \quad (3.2)$$

By $|P| < \kappa$, there is for each ξ and $\dot{B}_\xi$ a family $\mathscr{B}_\xi$ of size $< \kappa$ of sets in U such that

$$p \Vdash \dot{B}_\xi \in \check{\mathscr{B}}_\xi.$$

By κ-completeness of U, for every ξ, $b_\xi = \bigcap \mathscr{B}_\xi$ is in U. The sequence $\langle b_\xi \mid \xi < \lambda \rangle$ exists in V, and therefore by κ-completeness of U in V, $\bigcap_{\xi < \lambda} b_\xi$ is in U. It follows

$$p \Vdash \check{b}_\xi \subseteq \dot{A}_\xi \text{ for every } \xi < \lambda \text{ and } p \Vdash \bigcap_{\xi < \lambda} \check{b}_\xi \subseteq \bigcap_{\xi < \lambda} \dot{A}_\xi,$$

and hence $\bigcap_{\xi < \lambda} A_\xi$ is in W.

It remains to show that W is an ultrafilter. Let $\dot{X}$ be a name for a subset of κ. For each $p \in P$, let

$$A_p = \{\alpha < \kappa \mid p \text{ decides whether } \check{\alpha} \in \dot{X}\}.$$

Notice that

$$D = \{p \in P \mid A_p \in U\} \text{ is dense in } P.$$

This is because for each $q \in P$,

$$\bigcup_{p \leq q} A_p = \kappa$$

and by κ-completeness of U and the fact that $|P| < \kappa$, there must be some $p \leq q$ such that $A_p \in U$. Let r be in $D \cap G$—then $A_r \in U$ where A_r can be written as a disjoint union of $A_0 = \{\alpha < \kappa \mid r \Vdash \check{\alpha} \in \dot{X}\}$ and $A_1 = \{\alpha < \kappa \mid r \Vdash \check{\alpha} \notin \dot{X}\}$. If $A_0 \in U$, then $\dot{X}^G \in W$, and if $A_1 \in U$, then $\kappa \setminus \dot{X}^G \in W$. $\square$

3.4 A Uniform Approach

So far we have argued that inaccessible, Mahlo, weakly compact, and measurable cardinals do not decide CH. This, *per se*, is not an argument that other large cardinals cannot behave differently in this respect—after all, every argument we gave was unique to a given large cardinal concept, and not directly generalizable to other large cardinals. As it turns out, however, many large cardinals can be formulated in terms of elementary embeddings, and there is a uniform approach which shows that such cardinals do not affect CH. Among the cardinals with definitions through elementary embeddings are weakly compact cardinals, measurable cardinals, strongly compact cardinals, supercompact cardinals and many others.

Definition 3.5 Let M and N be two transitive classes. We say that $j : M \rightarrow N$ is an elementary embedding if for every formula and every n-tuple $m_0, \ldots, m_n$ of elements in M, if $\varphi^M(m_0, \ldots, m_n)$, then $\varphi^N(j(m_0), \ldots, j(m_n))$.

The notation φ^M is defined recursively and subsists in replacing every occurrence of an unbounded quantifier Qx with $Qx \in M$. Note that M, N and j may be proper classes.[20]

We say that κ is a critical point of $j : M \to N$ if for all $\alpha < \kappa$, $j(\alpha) = \alpha$, and $j(\kappa) > \kappa$. One can show that if j is not the identity it has a critical point which is always a regular uncountable cardinal in M.

Theorem 3.6 *The following are equivalent for a cardinal $\kappa > \omega$:*

(i) κ is measurable.
(ii) There is an elementary embedding $j : V \to M$ with critical point κ, where M is some transitive class.

Proof Ad (i)→(ii). (Sketch) A generalization due to Scott [15] of the ultrapower construction can be used to form the ultrapower of the whole universe V via a κ-complete ultrafilter U witnessing the measurability of κ. One can show that this construction is well defined and yields a proper class ultrapower model, denoted as $\text{Ult}_U(V)$. Since the U is ω_1-complete, one can further show that that the ultrapower is well-founded and can therefore be collapsed using the Mostowski collapsing function. Thus there is an elementary embedding

$$j : V \to \text{Ult}_U(V) \cong M$$

to a transitive isomorphic image of $\text{Ult}_U(V)$. κ-completeness of U is invoked to prove that j is the identity below κ, and $j(\kappa) > \kappa^+$.

Ad (ii)→(i). Let $j : V \to M$ be elementary with critical point κ. Let us define

$$U = \{X \subseteq \kappa \mid \kappa \in j(X)\}.$$

We will show that U is a κ-complete non-principal ultrafilter. It is non-principal because for every $\alpha < \kappa$, $j(\{\alpha\}) = \{\alpha\}$ and therefore $\{\alpha\} \notin U$. κ-completeness follows by the following argument: if $\{A_\xi \mid \xi < \mu\}$ are sets in U for $\mu < \kappa$, then

$$j(\{A_\xi \mid \xi < \mu\}) = \{j(A_\xi) \mid \xi < \mu\}$$

---

[20]There are some logical issues here because ZFC does not formalize satisfaction for proper classes, and hence one should be careful in saying that some φ holds in M, or that j is elementary. The relativation φ^M solves the issue to a certain extent, but it is not entirely optimal (for instance the property "j is elementary" is a schema of infinitely many sentences). Luckily, as always with issues like these, there are ways to make these concepts completely correct from the formal point of view. See for instance [4] for a nice discussion of approaches to formalizing large cardinal concepts which refer to elementary embeddings.

because $j(\mu) = \mu$ and therefore the j-image of the system $\{A_\xi \mid \xi < \mu\}$ is just the system of the j-images of the individual sets. Therefore

$$\kappa \in \bigcap_{\xi < \mu} j(A_\xi)$$

and hence

$$\bigcap_{\xi < \mu} A_\xi \in U.$$

Thus U is a κ-complete non-principal filter. It remains to show that U is an ultrafilter—but this is easy: if $X \subseteq \kappa$ is given, then $\kappa = X \cup (\kappa \setminus X)$, and so

$$\kappa \in j(\kappa) = j(X) \cup j(\kappa \setminus X)$$

by elementarity. Hence $\kappa \in j(X)$ or $\kappa \in j(\kappa \setminus X)$.

Notice that U is generated by a single element—κ. But it is not principal because κ is not in the range of j. If ξ is in the range of j, then any attempt to define an ultrafilter as we did ends up with a principal ultrafilter because the singleton of $j^{-1}(\xi)$ would be in the filter. Conversely, if we defined our U with any other ξ in the interval $[\kappa, j(\kappa))$, we would get a non-principal κ-complete ultrafilter by an identical argument.

The importance of U, as generated by κ, is that U is *normal*, but this goes beyond the scope of this paper. □

The above theorem provides a new tool to show that a measurable cardinal is preserved by forcing. It suffices to show that in a generic extension $V[G]$, there exists an elementary embedding with critical point κ. The following lemma is very useful for finding elementary embeddings in the generic extensions.

Lemma 3.7 (Silver) *Assume* $j : M \to N$ *is an elementary embedding between transitive classes* M, N. *Let* $P \in M$ *be a forcing notion and let* G *be* P-generic over M.[21] *Assume further that* H *is* $j(P)$-generic over N such that

$$\{j(p) \mid p \in G\} \subseteq H. \tag{3.3}$$

Then there exists elementary embedding $j^* : M[G] \to N[H]$ *such that:*

(i) j^* *restricted to* M *is equal to* j,
(ii) $j^*(G) = H$.

We call j^* *a* lifting of j to $M[G]$.

---

[21]This means that G meets every dense open set which is an element of M.

Proof We first show how to define j^*. Let x be an element of $M[G]$ and let $\dot{x}$ be a name for x so that $\dot{x}^G = x$. We set

$$j^*(\dot{x}^G) = (j(\dot{x}))^H.$$

This definition is correct because by elementarity $j(\dot{x})$ is a $j(P)$-name; further if $\dot{y}$ is another name for x, $\dot{y}^G = \dot{x}^G = x$, then there is some $p \in G$ such that $p \Vdash \dot{y} = \dot{x}$. By elementarity,

$$j(p) \Vdash j(\dot{x}) = j(\dot{y}).$$

By (3.3), $j(p) \in H$ and therefore $(j(\dot{x}))^H = (j(\dot{y}))^H$.
   j^* is elementary by the following implications:

$$\varphi^{M[G]}(x, \ldots) \to \exists p \in G \; p \Vdash \varphi(\dot{x}, \ldots) \to \exists p \in G \; j(p) \Vdash \varphi(j(\dot{x}), \ldots) \to$$
$$\varphi^{N[H]}(j^*(x), \ldots), \qquad (3.4)$$

where the last implication follows by (3.3).
   Ad (i). For $x \in M$, $j^*(x) = (j(\check{x}))^H = j(x)$, by the properties of the canonical name $\check{x}$.
   Ad (ii). Let $\dot{g}$ be a canonical name for the generic filter, i.e. a name which always interprets by the generic filter. Then $\dot{g}^G = G$, and $j^*(G) = (j(\dot{g}))^H = H$. $\square$
   Silver's "lifting lemma" allows us to reprove Theorem 3.4 in a more straightforward way:

Theorem 3.8 *Assume κ is measurable in V, P has size $< \kappa$ and G is P-generic. Then κ is measurable in $V[G]$.*

Proof By Theorem 3.6, there is an embedding $j : V \to M$ with critical point κ (this embedding exists, i.e. is definable, in V). Since j is the identity below κ, one can easily show that $V_{\kappa+1} = (V_{\kappa+1})^M$ and $j(x) = x$ for every $x \in V_\kappa$. In particular

$$j(P) = P$$

because $|P| < \kappa$ implies that we can assume $P \in V_\kappa$.[22] It follows by Silver's lemma, when we substitute G for H, that there exists a lifting

$$j^* : V[G] \to M[G],$$

where $j^*(G) = G$. Since j^* is definable in $V[G]$, it shows by Theorem 3.6 that κ is still measurable in $V[G]$. $\square$

---

[22]If $|P| < \kappa$, then there is an isomorphic copy of P which is in V_κ.

3.5 Other Large Cardinals

Many large cardinals can be formulated in terms of elementary embeddings and hence the proof in Theorem 3.8 can be straightforwardly generalized to argue that these cardinals do not decide CH. Here is a list of some more known large cardinals defined using elementary embeddings satisfying certain properties, where $\kappa > \omega$:

- κ is *weakly compact* iff κ is inaccessible and for every transitive model M of ZF without the powerset axiom such that $\kappa \in M$, M is closed under $< \kappa$-sequences and $|M| = \kappa$, there is an elementary embedding $j : M \to N$, N transitive, with critical point κ.
- κ is *strongly compact* iff for every $\gamma > \kappa$ there is an elementary embedding $j : V \to M$ with critical point κ, $j(\kappa) > \gamma$, and for any $X \subseteq M$ with $|X| \leq \gamma$, there is a $Y \in M$ such that $Y \supseteq X$ and $(|Y| < j(\kappa))^M$.
- κ is *supercompact* iff for every $\gamma > \kappa$ there is an elementary embedding $j : V \to M$ with critical point κ, $j(\kappa) > \gamma$, and ${}^{\gamma}M \subseteq M$.[23]
- κ is *strong* iff for every $\gamma > \kappa$ there is an elementary embedding $j : V \to M$ with critical point κ, $j(\kappa) > \gamma$, and $V_{\gamma} \subseteq M$.

Even Ramsey cardinals can be formulated in terms of elementary embeddings, see for instance [12]. All the cardinals considered so far are linearly ordered in terms of strength: for instance every supercompact is strongly compact, and every strongly compact is strong.

Note that by a celebrated result by Kunen [9], there can be, in ZFC, no cardinal κ such that there exists an elementary embedding $j : V \to V$ with critial point κ. This sets an upper bound on the large cardinal concept which we can consider.[24]

4 On the Consistency Strength

Large cardinals are interesting set-theoretical objects with beautiful combinatorics and surprising connections among themselves; for instance many of these can be defined in apparently disparate ways—using elementary embedding, satisfaction in various structures, or by partition properties. However, this does not fully explain the willingness with which large cardinals are almost universally accepted by the set theoreticians. To explicate the wider role of large cardinals we need to introduce the notion of a consistency strength over ZFC.

---

[23]${}^{\gamma}M \subseteq M$ is true if for every sequence of length γ of elements in M, the whole sequence is in M. This a non-trivial requirement because the sequence itself is in general only in V, and not in M.

[24]Rather surprisingly, it is still open whether this limiting result holds in ZF.

Definition 4.1 A statement σ in the language of set theory is *stronger in terms of consistency* then another statement σ' if

$$\mathrm{CON(ZFC} + \sigma) \to \mathrm{CON(ZFC} + \sigma').$$

We denote here this relation by

$$\sigma' \leq_c \sigma.$$

Statements are called *equiconsistent* if

$$\mathrm{CON(ZFC} + \sigma) \leftrightarrow \mathrm{CON(ZFC} + \sigma').$$

For instance, GCH $\equiv_c \neg$CH $\equiv_c V = L \equiv_c V \neq L \equiv_c \Diamond \equiv_c$ "There are no ω_1-Souslin trees".[25] Moreover we have

$$\mathrm{CON(ZF} - \text{ Axiom of Foundation }) \to \mathrm{CON(ZFC} + \sigma) \qquad (4.5)$$

for any σ from the class $[\mathrm{GCH}]_{\equiv_c}$.[26]

Note that the relation of equiconsistency $\equiv_c$ is an equivalence relation, and the relation $\leq_c$ is an ordering on the equivalence classes given by $\equiv_c$. What is the structure of this ordering? In principle, it might be highly non-linear. However, large cardinal concepts can be used to show that it is in fact mostly linear: for many combinatorial statements σ and σ' considered in practice, we either have $\sigma \leq_c \sigma'$ or $\sigma' \leq_c \sigma$. The key here is that large cardinal concepts themselves are linearly ordered under $\leq_c$, and very often one can show that a statement σ is equiconsistent with a certain large cardinal axiom.

By way of example, considered the following three statements (see [7] for the definitions of the concepts mentioned):

(A) (Over ZF) All sets of reals are Lebesgue measurable.
(B) (Over ZFC) Every ω_2-tree has a cofinal branch.
(C) (Over ZFC) SCH fails.

A priori, they might be incomparable under $\leq_c$; however, one can prove:

Theorem 4.2 (Solovay [18], Shelah [16])

$$(A) \equiv_c \text{ "there exists an inaccessible cardinal"}.$$

---

[25]This can be shown using Gödel's class of constructible sets L, or by forcing.
[26]Notice that by (4.5), $[\mathrm{GCH}]_{\equiv_c}$ is equal to $[\nu]_{\equiv_c}$ for any ν such that ZFC $\vdash \nu$.

Theorem 4.3 (Mitchell [13])

$$(B) \equiv_c \text{ "there exists a weakly compact cardinal".}$$

Theorem 4.4 (Mitchell [14], Gitik [3])

$$(C) \equiv_c \text{ "there exists a measurable cardinal of Mitchell order } o(\kappa) = \kappa^{++} \text{ ".}$$

Corollary 4.5 $GCH <_c (A) <_c (B) <_c (C)$.

The above theorems are proved using two complementary methods: (i) forcing over a model with the given large cardinal, and (ii) technique of inner models to find a large cardinal (in some model of set theory) from the given combinatorial statement. For instance Theorem 4.3 is proved by iterating a certain forcing notion (such as the Sacks forcing at ω) along κ, where κ is weakly compact: this gives

$$(B) \leq_c \text{ "there exists a weakly compact cardinal".}$$

Conversely, one can show that if (B) holds, then ω_2 of V is a weakly compact cardinal in L, and hence there is a model with weakly compact cardinal. This gives:

$$\text{"there exists a weakly compact cardinal" } \leq_c (B).$$

Problems arise when the large cardinal in question is inconsistent with L (such as a measurable cardinal), then to obtain the consistency of the large cardinals, a generalization of L must be defined which allows large cardinals. This is the field of *inner model theory*. So far, inner models were devised for infinitely many Woodin cardinals (Woodin cardinals are much stronger than measurable cardinals in terms of consistency strength), but not—crucially—for strongly compact or supercompact cardinals. This inability to find suitable inner models for such large cardinals is one of the most pressing problems in current set theory. Because of this, the following is still open for a certain important combinatorial statement denoted as PFA (Proper Forcing Axiom)[27]:

Open Question We know: PFA $\leq_c$ "there exists a supercompact cardinal". Does the converse hold as well, i.e. is PFA equiconsistent with a supercompact cardinal?

There is a general agreement that this is the case, but we cannot prove it.[28]

The following is also long open, probably for the similar reason as the case of PFA:

---

[27] PFA, a strengthening of MA—the Martin's Axiom-, implies $2^\omega = \omega_2$ and thus decides CH. However, PFA itself is not a large cardinal axiom in the strict sense. Also, PFA trivially implies $2^\omega > \omega_1$ the way it is set up, so what is surprising is that it also implies $2^\omega \leq \omega_2$, and not that it implies failure of CH. MA, on the other hand, is consistent with any reasonable value of $2^\omega > \omega_1$.

[28] We do know that PFA implies consistency of many Woodin cardinals, and so PFA is sandwiched between "many Woodins" and "supercompact". But this gap is quite substantial.

Open Question By definition, every supercompact cardinal is strongly compact. We also know that κ can be measurable + strongly compact but not supercompact. However, we do not know, but consider probable: Are strongly compact and supercompact cardinals equiconsistent?

5 Conclusion

Large cardinals considered in this article do not decide CH one way or another. In fact no commonly considered large cardinals decide CH, which can be shown by similar methods.[29] However, notice that we cannot prove a statement such as "no large cardinal decides CH" because in this statement we quantify over a vague domain of "large cardinals" and hence such a statement is not in the language of set theory. It may be, but it is not considered probable, that a new large cardinal will be devised which will be more susceptible to effects of small forcings. At present, no such cardinal is known.

Acknowledgements The author acknowledges the generous support of JTF grant Laboratory of the Infinite ID35216.

References

1. S. Feferman, J. Dawson, S. Kleene, G. Moore, J. Van Heijenoort, (eds.), *Kurt Gödel. Collected Works. Volume II* (Oxford University Press, Oxford, 1990)
2. Sy-D. Friedman, T. Arrigoni, Foundational implications of the inner model hypothesis. Ann. Pure Appl. Logic **163**, 1360–1366 (2012)
3. M. Gitik, The negation of singular cardinal hypothesis from $o(\kappa) = \kappa^{++}$. Ann. Pure Appl. Logic **43**, 209–234 (1989)
4. J.D. Hamkins, G. Kirmayer, N.L. Perlmutter, Generalizations of the Kunen inconsistency. Ann. Pure Appl. Logic **163**, 1872–1890 (2012)
5. D. Hilbert, Mathematical problems. Bull. Am. Math. Soc. **8**, 437–479 (1902)
6. R. Honzik, Quick guide to independence results in set theory. Miscellanea Logica **7**, 89–131 (2009)
7. T. Jech, *Set Theory* (Springer, New York, 2003)
8. A. Kanamori, *The Higher Infinite* (Springer, New York, 2003)
9. K. Kunen, Elementary embeddings and infinitary combinatorics. J. Symb. Log. **407–413**, 21–46 (1971)
10. K. Kunen, *Set Theory: An Introduction to Independence Proofs.* (North-Holland, Amsterdam, 1980)
11. A. Levy, R.M. Solovay, Measurable cardinals and the continuum hypothesis. Isr. J. Math. **5**, 234–248 (1967)
12. W.J. Mitchell, Ramsey cardinals and constructibility. J. Symb. Log. **44**(2), 260–266 (1979)

---

[29]We consider κ to be large when it is at least inaccessible. If we drop this requirement, the situation is more complex.

13. W.J. Mitchell, Aronszajn trees and the independence of the transfer property. Ann. Math.
    Logic **5**(1), 21–46 (1972/1973)
14. W.J. Mitchell, The core model for sequences of measures. I. Math. Proc. Camb. Phil. Soc. **95**,
    229–260 (1984)
15. D.S. Scott, Measurable cardinals and constructible sets. Bull. de l'Académie Pol. Sci. **9**,
    521–524 (1961)
16. S. Shelah, Can you take Solovay's inaccessible away? Isr. J. Math. **48**, 1–47 (1984)
17. A. Sochor, *Klasická matematická logika* (Karolinum, 2001)
18. R.M. Solovay, A model of set theory in which every set of reals is Lebesgue measurable. Ann.
    Math. **92**, 1–56 (1970)

Gödel's Cantorianism

Claudio Ternullo

Abstract Gödel's philosophical conceptions bear striking similarities to Cantor's. Although there is no conclusive evidence that Gödel deliberately used or adhered to Cantor's views, one can successfully reconstruct and see his "Cantorianism" at work in many parts of his thought. In this paper, I aim to describe the most prominent conceptual intersections between Cantor's and Gödel's thought, particularly on such matters as the nature and existence of mathematical entities (sets), concepts, Platonism, the Absolute Infinite, the progress and inexhaustibility of mathematics.

"My theory is rationalistic, idealistic, optimistic, and theological."
Gödel, in [48, p. 8]

"Numeros integros simili modo atque totum quoddam legibus et relationibus compositum efficere"
Cantor, *De transformatione formarum ternariarum quadraticarum*, thesis III, in [5, p. 62]

1 Introductory Remarks

There is no conclusive evidence, either in his published or his unpublished work, that Gödel had read, meditated upon or drawn inspiration from Cantor's philosophical doctrines. We know about his philosophical "training", and that, since his youth, he had shown interest in the work of such philosophers as Kant, Leibniz and Plato. It is also widely known that, from a certain point onwards in his life, he started reading and absorbing Husserl's thought and that phenomenology proved to be one of the most fundamental influences he was to subject himself to in the course of

Originally published in C. Ternullo, Goedel's cantorianism, in *Kurt Gödel: Philosopher-Scientist*, ed. by G. Crocco, E.-M. Engelen (Presses Universitaires de Provence, Aix en Provence, 2015), pp. 413–442.

C. Ternullo (✉)
KGRC, University of Vienna, Vienna, Austria
e-mail: claudio.ternullo@univie.ac.at

the development of his ideas.[1] But we do not know about the influence of Cantor's thought.

In Wang's book containing reports of the philosophical conversations the author had with Gödel, one can find only a few remarks by Gödel concerning Cantor's philosophical conceptions. Not much material do we get from the secondary literature either. For instance, if one browses through the indexes of Dawson's fundamental biography of Gödel [9], or those of Wang's three ponderous volumes [46–48], one finds that all mentions of Cantor in those works either refer to specific points of Cantorian set theory, as discussed by the authors of these books, or, more specifically, to Gödel's paper on Cantor's continuum problem,[2] wherein references to Cantor, once again, are not directed at the examination of the latter's philosophical work and conceptions.[3]

As a consequence, we do not know whether Gödel had direct or indirect acquaintance with Cantor's thought and how he judged it.[4] Indeed, Gödel's ostensible lack of interest in Cantor's philosophy in the first place might be one of the reasons why the varied and multi-faceted connections between his views and Cantor's have, with a few exceptions, gone altogether unnoticed.[5] To give you a taste of what such connections look like, let me briefly anticipate some of the material, which I will discuss in greater depth in the next sections.

There is hardly need, I believe, to emphasise both Cantor's and Gödel's commitment to a peculiar form of belief in the existence of the *actual infinite*. Cantor saw transfinite numbers as a *natural* generalisation of the natural numbers, while Gödel expressed the thought that the unbounded continuation of the process of formation of such numbers would have a deep impact on most fundamental mathematical issues.

At least from a certain point on in their lives, both embraced a thoroughgoing and unabated form of *realism*. Cantor acknowledged and reconstructed Plato's

---

[1] Wang [48, p. 164].

[2] Gödel [16], revised and extended version, [20].

[3] Further information on biographical and philosophical aspects of Gödel's life can be found in Feferman's introduction to [21, pp. 1–34]. A precise and exhaustive reconstruction of the development of Gödel's thought is also carried out by van Atten and Kennedy in [45]. None of these works mentions specific connections between Cantor and Gödel.

[4] I have only found two passages in Wang [48], where Gödel says something directly about Cantor. The first, on p. 175, concerns the philosophy of physics: "5.4.16 The heuristics of Einstein and Bohr are stated in their correspondence. Cantor might also be classified together with Einstein and me. Heisenberg and Bohr are on the other side. Bohr [even] drew metaphysical conclusions from the uncertainty principle." The second, on p. 276, is about the distinction between *set* and *class* (for whose relevance see Sect. 6): "8.6.13 Since concepts can sometimes apply to themselves, their extensions (their corresponding classes) can belong to themselves; that is, a class can belong to itself. Frege did not distinguish sets from proper classes, but Cantor did this first." Both remarks show at least some familiarity with Cantor's writings.

[5] Among these few exceptions should be counted some remarks by Wang and van Atten, in, respectively, [48] and [44], concerning the so-called Cantor-von Neumann axiom, for which also see Sect. 6.

characterisation of *ideas* as the ontological basis of his own transfinite numbers.[6] But while casting transfinite numbers as purely 'ideal entities', he also vested them with a *trans-subjective* meaning, as being *(meta)physically* instantiated. Thus, he could legitimately claim that his conception encompassed 'idealist' and 'realist' features.

Similarly, Gödel strove for a theoretical synthesis between the idealist and realist position and eventually found it in Husserl's conceptions. In a sense, as we shall see, Cantor's philosophical doctrines about concepts may be viewed as reaching their theoretical completion in Gödel's *conceptual realism*.

In general, although the scope of their philosophical sources is wide and varied, they have a clear preference for authors belonging to the *rationalist* (as opposed to *empiricist*) tradition. In Cantor's works, one can find references to or quotations from works of Plato, Spinoza, Leibniz, Augustine, Origen, Euclid, Nicomachus of Gerasa, Boethius and others, and most of these references and quotations are generally accompanied by extolling comments and used in support of Cantor's own theses. On the other hand, references to Aquinas, Aristotle, Locke, Hume and Kant are mostly made by Cantor with the purpose of refuting or discarding these thinkers' views.

Analogously, Gödel, at least half a century later, fosters Plato's, Leibniz's and Husserl's conceptions, that he classifies as "objectivistic", often contrasting them with what he takes to be the opposite point of view, that is philosophical "subjectivism" or conventionalism, which is represented by such authors as Kant (but only to a certain extent), Carnap and Wittgenstein. His characterisation of philosophical conceptions as dividing into "left-wing" and "right-wing" is in line with such presuppositions. It is worth quoting in full the crucial passage where such classification and its underlying rationale are introduced:

> I believe that the most fruitful principle for gaining an overall view of the possible world-views will be to divide them up according to the degree and manner of their affinity to or, respectively, turning away from metaphysics (or religion). In this way we immediately obtain a division into two groups: skepticism, materialism and positivism stand on one side, spiritualism, idealism and theology on the other. ([19], in [23, p. 375])[7]

Both authors lived in an age of disillusionment with, if not outright refusal of, the *metaphysical* tradition. The strong pressure exerted by *positivism* on all undue "metaphysical pre-conceptions" may have been the main reason why Cantor was so philosophically meticulous in presenting his work. In a similar vein, but many years later, against all conventionalist and formalist reductions of mathematics, Gödel felt the urge to explain carefully why set-theoretic problems such as the Continuum Problem retain a meaning, if mathematics is construed as referring to an independently existing realm of objects. Cantor and Gödel never abandoned their fundamentally metaphysical outlook on mathematics, and, more generally, on the world. This attitude they pursued so coherently, that they sometimes seem to fall

---

[6]See Sect. 5.

[7]Throughout this article, all quotations from Gödel's published and unpublished works reproduce the established text in the II and III volume of his *Collected Works* [22, 23].

prey to what some commentators believe to be rationally untenable, even "bizarre", beliefs.

An integral part of this general attitude, not unsurprisingly, is their frequent appeal to and increasing fascination with *theology*, displaying itself more overtly in Cantor's work, less so in Gödel's.[8] Theology, with all its traditional theoretical artifices, gave both authors wide scope for speculations about the nature of the infinite, the set-theoretic hierarchy and its connections to the *phenomenal* world. But what is, perhaps, most interesting is their peculiar construal of theology (and theological arguments), as connected to, if not quite part of, the theory of sets. For instance, one can view the emergence of Cantor's theological *Infinitum Absolutum* in connection with the emergence of the notion of the *absolute infinity* of the *universe of sets*, on which Gödel had subsequently much to say.

Gödel's Cantorianism is transparent throughout Gödel's philosophical work, although, as I said, it has been largely overlooked by the scholarly literature. Its examination meets several purposes. First of all, I believe that its description may help us put into focus more accurately Gödel's philosophy, its developmental stages and history. As any other philosophical "transformative" conception, it may also shed new light on other important connections, such as those that bind Husserl, Frege and Gödel together. Finally, it might also help us discuss the peculiar form of realism that Gödel advocated, and that still pays an influential role in the contemporary debate on the foundations of mathematics.

A few methodological remarks. I quote a lot of text from the primary sources, as is necessary for a study of this sort, and I will also indicate, when necessary, the relevant secondary literature. Although I will essentially be concerned with proposing connections between the two, I will also pay attention to, and comment on, the substance of Cantor's and Gödel's philosophical arguments.

2   Intra-Subjective ("Immanent") Existence

One of the most characteristic and widely discussed traits of Gödel's thought is his *conceptual realism*. Given Gödel's own theoretical oscillations, it is perhaps somewhat problematic to say crisply what this position consists in.[9]

What seems to be fairly certain is that it includes, at least, the two following claims:

1. Concepts exist, in a way, which is similar, although not reducible to, the existence of physical objects.
2. Mathematical truths express relations of concepts.

As we shall see, both claims are in accordance, and could even be entirely derived from Cantor's conceptions.

---

[8]However, Gödel's interest in theology is noticeable in the Max-Phil Notebooks.

[9]See Parsons [38], Martin [35] and Crocco [7] for a careful examination of the issues related to this position.

Of Gödel's early adherence to conceptual realism, we are informed by Wang. In a letter sent to him in 1975, Gödel says: "0.1.3 I have been a conceptual and mathematical realist since about 1925."[10] Explicit references to the *existence* of concepts are contained, respectively, in two important passages of [15] and in [18]. The first one presents Gödel's argument that logical paradoxes do not affect set theory, insofar as the formation of all purely *mathematical* sets does not involve such paradoxical notions as that of a "set of all sets not belonging to themselves".[11] Sets formed through the standard iterative procedures have proved to be free from the kind of contradiction involved in the logical paradoxes.

Gödel says:

> Classes and concepts may, however, be also conceived as real objects, namely classes as "pluralities of things" or as structures consisting of a plurality of things and concepts as the properties and relations of things existing independently of our definitions and constructions. [...] They are in the same sense necessary to obtain a satisfactory system of mathematics as physical bodies are necessary for a satisfactory theory of our sense perceptions and in both cases it is impossible to interpret the propositions one wants to assert about these entities as propositions about the "data", i.e., in the latter case the actually occurring sense perceptions. ([15], in [22, p. 128])

Incidentally, in the second part of this quotation, Gödel seems to be adumbrating his conception that mathematical concepts are "grasped" (certainly "understood", but probably also "perceived") by some special faculty in the same way as physical objects are perceived by the senses.[12] In this latter as in the former case, however, our "perceptual data" would not be the result of the mere interaction between the objects and the corresponding perceiving faculty. I will return to this point later.

The second important passage can be found in Gödel's famous Gibbs lecture, wherein he states that:

> ...it is correct that a mathematical proposition says nothing about the physical or psychical reality existing in space and time, because it is true already owing to the meaning of the terms occurring in it, irrespectively of the world of the real things. What is wrong, however, is that the meaning of the terms (that is, the concepts they denote) is asserted to be something man-made and consisting merely in semantical conventions. The truth, I believe, is that

---

[10]However, as late as 1933, Gödel stated ([14], in [23, p. 50]): "The result of the preceding discussion is that our axioms, if interpreted as meaningful statements, necessarily presuppose a form of Platonism, which cannot satisfy any critical mind and which does not even produce the conviction that they are consistent." See Feferman's comments on this in his Introduction to [14]. But, apart from that, it seems very plausible that Gödel embraced Platonism, in at least some of its forms, at a very early stage in his career.

[11]Purely mathematical sets, in ZFC, or in alternative systems, with or without *urelements*, are sets formed through the iteration of the power-set operation at successor-stages and the union of all previous stages at limit-stages, starting from ∅ or ur-elements.

[12]However, it is not wholly uncontroversial what Gödel thought to be the objects of "perception", whether mathematical objects or concepts or both. For instance, in Wang [48, p. 253], Gödel is reported to have said: "7.3.12 Sets are objects but concepts are not objects. We perceive objects and understand concepts. Understanding is a different kind of perception: it is a step in the direction of reduction to the last cause." I thank Eva-Maria Engelen for pointing me to this quotation and to the subtle difference between these two forms of "perception" in Gödel's thought.

these concepts form an objective reality of their own, which we cannot create or change, but only perceive and describe. ([18], in [23, p. 320])

As far as claim (2) is concerned, that mathematical truths express relations of concepts, one must turn one's attention to the following key passage in the same work:

Therefore a mathematical proposition, although it does not say anything about space-time reality, still may have a very sound objective content, insofar as it says something about relations of concepts. The existence of non-"tautological" relations between the concepts of mathematics appears above all in the circumstance that for the primitive terms of mathematics, axioms must be assumed, which are by no means tautologies (in the sense of being in any way reducible to $a = a$), but still do follow from the meaning of the primitive terms under consideration. ([18], in [23, pp. 320–321])

This passage requires an extended commentary. In his Gibbs lecture, Gödel challenges two positions concerning the nature of mathematical truth. The first assumes that mathematical truths are *tautologies*, that is, are *analytic*, in the sense that they can be reduced to basic logical laws such as the *identity law*.[13] One main reason provided by Gödel for countering such conception is the following: the axioms of sets are *non-tautological*, insofar as they refer to irreducible primitive concepts (such as the very concept of "set" or "plurality").

The second conception he wishes to oppose is the "conventionalist" one (due to Poincaré and Carnap), whereby the axioms, and the theorems derivable from them, only have a "conventional" character, and do not express an objective mathematical content. Against the Carnapian, Gödel claims that, although mathematical truths do not refer to any spatio-temporal property of reality, they still refer to something, namely, the *objectively given* realm of concepts itself and, furthermore, express relations among concepts.

Well, as I said at the beginning of this section both, this position and claim (1) above had already been expressed by Cantor. In a crucial passage concerning the existence of mathematical objects, Cantor says:

We can speak of the actuality of the integers, finite as well as infinite, in *two* senses; but strictly speaking they are the same two relationships in which in general the reality of any concepts and ideas can be considered. First, we may regard the integers as actual insofar as, on the basis of definitions, they occupy an entirely determinate place in our understanding, are well distinguished from all other parts of our thought, and stand to them in determinate relationships, and thus modify the substance of our mind in a determinate way; let us call this kind of reality of our numbers their *intrasubjective* or *immanent reality*. ([1], in [10, pp. 895–896])[14]

---

[13]No doubt, this conception has a Leibnizian ancestry, but Gödel may have also deliberately wanted to refer to the logicist standpoint. For instance, see Frege [13, p. 85]. "Thus, arithmetic becomes only a further developed logic, every arithmetical proposition a logical law, albeit a derivative one." However, Frege never affirmed that arithmetical truths are *tautologies*. In any case, as we have seen, in the passage quoted, Gödel fosters a different notion of "analytic", meaning: "owing to the meaning of the terms occurring in it".

[14]The English translation of all Cantor's quotations from [1] comes from Ewald [10, pp. 878–920]. In reproducing it, I have also kept Ewald's annotations in square brackets.

In a footnote, Cantor gives a characterisation of concepts and ideas in terms of Spinozian *ideae verae*, which, in turn, can be assimilated to Plato's *ideas*:

> What I here call the "immanent" or "intrasubjective" reality of concepts or ideas ought to agree with the adjective "adequate" in the sense in which Spinoza uses this word when he says (*Ethica*, part II, def. IV): 'Per ideam adequatam intelligo ideam, quae, quatenus in se sine relatione ad objectum consideratur, omnes verae ideae proprietates sive denominationes intrinsecas habet [By adequate idea I mean an idea which, as far as it is considered as not having a relationship with an object, enjoys all intrinsic properties and designations of a real idea (my translation)].' ([1], in [10, p. 918])

The notion of an *idea vera* can be glossed in the following way: an idea which does not lead to contradictions and, which is, in addition, self-subsistent. Self-subsistent concepts, in Cantor's view, are those concepts that truly *exist*. As Hallett carefully explains in his [24], there is no doubt that the kind of existence that Cantor is referring to here is genuine *platonic* existence.[15]

As we will see in the next section, Cantor also explains that concepts, although self-subsistent, should also be viewed as nodes in a logical network, where the new ones are connected to the older already found to be existing. Furthermore, concepts are connected to each other in a well-determined, non-arbitrary way. That is, new concepts must have *determinate* properties, which distinguish them from, but, at the same time, connect them to older concepts.

3 Concepts as Objective Constructs

Both Gödel's *conceptual realism* and Cantor's *immanence conception* are committed to the view that concepts have an objective status, namely, that they are independent, to a certain extent, from our mental faculties. Yet, using our mental faculties, we can "perceive" their objectivity, through a process of logical refinement and sharpening of the properties entering their definitions.

Both Gödel and Cantor investigated this process, but neither of them ever gave a systematic account of it. We could say that Husserl's phenomenology plays a "linking" role in the transmission of Cantor's doctrines to Gödel. But Husserl was not the only one who held such *objectivistic* views about concepts at the time. Frege was one further major proponent of objectivism, although the influence of Frege's conceptions upon Gödel's thought might have been considerably weaker than Husserl's.[16]

---

[15] See Hallett [24, p. 17].

[16] Frege's "objectivistic" views about concepts can be found, in particular, in Frege [12]. It should be noticed that Frege thought that the main value of his work had consisted, among other things, precisely in the clarification of the essence of *concepts* (see the letter quoted by Ricketts in [40, p. 149]).

For the time being, I want to focus my attention on Cantor's theory. In a footnote of the *Grundlagen*, we find a remarkable passage concerning the crucial point of the correct procedure to generate new concepts.

> The procedure in the correct formation of concepts is in my opinion everywhere the same. One posits [setzt] a thing with properties that at the outset is nothing other than a name or a sign *A*, and then in an orderly fashion gives it different, or even infinitely many, intelligible predicates whose meaning is known on the basis of ideas that are already at hand, and which may not contradict one another. In this way one determines the connection of *A* to the concepts that are already at hand, in particular to related concepts. If one has reached the end of this process, then one has met all the preconditions for awakening the concept *A* which slumbered inside us, and it comes into being accompanied by the intrasubjective reality which is all that can be demanded of a concept; to determine its transient meaning is then a matter for metaphysics. ([1], in [10, pp. 918–919])

This passage requires some detailed interpretative work. The tacit assumption we have to keep in mind preliminarily is that concepts should be viewed as "existents" in the way indicated in the preceding section, that is, insofar as they have such a high degree of *determinacy* as to be distinguished from other existing concepts, but, at the same time, be *consistent* with them. Now, the passage under consideration tells us how the formation of new concepts conforms to such requirements.

The procedure envisaged by Cantor has three parts. One first starts with the elaboration of "signs", which may have (infinitely) many properties (in other terms, satisfy (infinitely) many predicates), all of which should not be *inconsistent* with each other. If such process is carried out successfully, then one can proceed to the next stage, wherein one declares the "birth" of a new concept. In this second stage, a new concept can be successfully declared to be born if and only if the sign "created" is correlated to something "slumbering" within us, that is, if the sign is perceived as a "reminisced" concept. Only through that can one secure the grasp of the concept itself, and proceed to determine all of its properties and connections with the already available concepts. I will talk about the third stage of the process in Sect. 7.

In view of what I have said in Sect. 2 and of the procedure described above, Cantor's doctrines on concepts can be summarized in the following way:

1. Concepts are mental constructs that have a *meaning*, which consists of all the *properties* that can be attributed to them. Meaning is independent from the structure of the associated mental construct.
2. The meaning of a concept exists independently of our minds.
3. The formation of new concepts consists in the assignment or *clarification* of the *meaning* of new "signs", and in the *study* of their *relationships* to already existing concepts.

Let me now turn back to the Husserl point. It may not be accidental that all the doctrines in the bullet points above are in line with Husserl's phenomenological doctrines. The relationship between Cantor and Husserl has been recently investigated by Claire Ortiz Hill. The upshot of Hill's examination is that Husserl may have been influenced by Cantor's Platonism and by his emphasis on *objectivism* in his early years (especially at the time of composition of the *Philosophy of Arithmetic*), and maybe also in the subsequent years, when he created his phenomenological method.[17] Regardless of this, we may still acknowledge Husserl's phenomenology as a major *trait d'union* between Cantor's and Gödel's conceptions.

Now, I do not want to present well-charted facts about Gödel's adherence to phenomenology and his use of Husserl's philosophy here.[18] I am more interested in how he saw phenomenology in connection with his and Cantor's strive for conceptual objectivity.

In [19], phenomenology figures prominently among the philosophical conceptions Gödel surveys.

> Now in fact, there exists today the beginning of a science which claims to possess a systematic method for such a clarification of meaning, and that is the phenomenology founded by Husserl. [...] But one must keep clearly in mind that this phenomenology is not a science in the same sense as the other sciences. Rather it is [or in any case should be] a procedure or technique that should produce in us a new state of consciousness in which we describe in detail the basic concepts we use in our thought, or grasp other basic concepts hitherto unknown to us. ([19], in [23, p. 383])

The reader will have noticed the strong Cantorian overtones of this statement. Phenomenology is said to be a technique for generating a "new state of consciousness", which allows us to describe concepts hitherto unknown to us, something which seems to echo the "awakening" of concepts Cantor was referring to in the passage quoted above.

As already said, Gödel absorbed phenomenology especially in the late stage of his thought and found in it an answer to the problem of whether we have a method to establish the "foundedness" of concepts.

In essence, Gödel was looking for a "science" of concepts, which may have realised the Leibnizian ideals of a *characteristica universalis* and *calculus ratiocinator*. Such a science, producing a clarification of the *meaning* of concepts, he found in Husserl's phenomenology. But with this choice, it seems to me, he was also completely and determinately fulfilling Cantor's ideal, as described and encapsulated in the conception we have reviewed above.

It should be noticed that Gödel's project is always subservient to his programme of investigating new axioms of set theory. Therefore, his reliance upon the idea of finding a systematic method for analysing concepts should be seen in connection

---

[17]The bulk of Hill's careful work on the relationships between Cantor and Husserl can be found in [36]. See also [37]. In [36], Hill identifies three stages of influence of Cantor's thought on Husserl. But she clearly acknowledges, although only conjecturally, that there might be a further, fourth stage, that she does not examine, which "...would consist of the assimilation of certain of Cantor's ideas into Husserl's phenomenology and extends far beyond the compass of this study. Here it would be a matter of studying the relationship between Cantor's theories and, for example, Husserl's *Mannigfaltigkeitslehre*, his theories about eidetic intuition, the phenomenological reductions, noemata, horizons, infinity, whole and part..." (p. 166).

[18]On this point, see, in particular, Wang's mentioned books, Kennedy and van Atten [45], Tieszen [41–43], Hauser [25] and Crocco [7]. Føllesdal's introduction to [19, pp. 364–373] also provides interesting insights.

with what he says in his Cantor paper about the role of set-theoretic axioms and concepts underlying them:

Similarly also the concept "property of set" (the second of the primitive terms of set theory) can constantly be enlarged, and furthermore concepts of "property of property of set" etc. be introduced whereby new axioms are obtained, which, however, as to their consequences for propositions referring to limited domains of sets (such as the continuum hypothesis) are contained in the axioms depending on the concept of set. ([16], in [22, fn. 17, p. 181])

Further, in the same work, he suggests:

But probably there exist others based on hitherto unknown principles; also there may exist, besides the ordinary axioms, the axioms of infinity and the axioms mentioned in footnote 17, other (hitherto unknown) axioms of set theory which a more profound understanding of the concepts underlying logic and mathematics would enable us to recognize as implied by these concepts. ([16], in [22, p. 182])

Conceptual objectivity, in turn, obtained through progressive logical refinements and clarifications, lies at the roots of *mathematical evidence*, the ultimate ideal that Gödel was pursuing. The belief in mathematical evidence is what makes him conjecture that undecidable statements such as CH (Continuum Hypothesis)[19] might be settled in the future. He seems to draw such faith in mathematical evidence from phenomenology, but it is unclear whether there is any direct connection between Husserl's ideas and Gödel's belief in the solvability of all set-theoretic problems.[20]

In any case, the following statement can be read in the light of such belief:

The mere psychological fact of the existence of an intuition which is sufficiently clear to produce the axioms of set theory and an open series of extensions of them suffices to give meaning to the question of the truth or falsity of Cantor's continuum hypothesis. ([20], in [22, p. 268])

At a certain stage, this faith in the objectivity of concepts led Gödel to proceed well beyond the Cantorian (and maybe Husserlian) ideal of a clarification of their meaning, to set up a general theory of concepts, which would work a sort of axiomatised metaphysics.

In his conversations with Wang, Gödel says, among other things:

8.6.20 Even though we do not have a developed theory of concepts, we know enough about concepts to know that we can have also something like a hierarchy of concepts (or also of classes) which resembles the hierarchy of sets and contains it as a segment. But such a hierarchy is derivative from and peripheral to the theory of concepts; it also occupies a quite different position; for example, it cannot satisfy the condition of including the concept of *concept* which applies to itself or the universe of all classes that belong to themselves.

---

[19]The Continuum Problem is the problem of determining the cardinality of $\mathbb{R}$ (denoted c). The Continuum Hypothesis (CH) is Cantor's conjecture that $c = \aleph_1$. See footnote 27 below.

[20]The problem with Gödel's claim that set-theoretic statements might be shown to have a determinate and unique truth-value as a result of conceptual refinements is explained very neatly by Hauser in [25, pp. 539–540]: "On this view, the meaning of the continuum problem is tied to an unfolding of concepts through successive refinements of mathematical intuition. One difficulty is why it should lead to a unique resolution of CH, for our intuitions could conceivably evolve into different directions inducing us to formulate axioms with opposite outcomes of CH."

To take such a hierarchy as the theory of concepts is an example of trying to eliminate the intensional paradoxes in an arbitrary manner. [48, p. 278]

In the passage above, Gödel hints at connections between objects and concepts, as being reflected by the connection between sets and concepts. We will see later on how this project was further substantiated by further philosophical conceptions, which were also discussed by Cantor.

4 Anti-subjectivism

The frequent target of Cantor's philosophical invective is Kant. In Cantor's view, Kant was held to be the main person responsible for introducing and advocating *subjectivism* in philosophy. A subjectivistic conception can only give us knowledge of "appearances", not of "stable", unchanging forms, as Plato wanted. Cantor says:

> Only since the growth of modern empiricism, sensualism, and scepticism, as well as of the Kantian criticism that grows out of them, have people believed that the source of knowledge and certainty is to be found in the senses or in the so-called pure form of intuition of the world of appearances, and that they must confine themselves to them. ([1], in [10, p. 918])

In that footnote, Cantor also contrasts Kant's conception with Plato's, Spinoza's and Leibniz's. The association of *empiricism*, *sensualism* and *scepticism*, on the one hand, and of Plato, Leibniz and Spinoza, on the other, seems, to say the least, too quick. Such schematisations may have had an echo in that presented by Gödel in [19], which grouped philosophical conceptions into left-wing and right-wing. But maybe we should not ascribe too much value to such quick distinctions. Cantor contrasts platonic "objectivism" to Kantian "subjectivism" for his own purposes, that is, defending the conceptual legitimacy of the *actual infinite*, something that Kant would have certainly found preposterous. Strongly related to such "objectivistic" attitudes is also Cantor's inclination to *arithmetical purism*, which he must have subscribed to at a very early stage in his career. As a pupil of Weierstrass', Cantor had witnessed and, in a sense, had been involved in the programme of *arithmetisation* of analysis. In his work, the programme is, in a sense, further reflected by the creation of the *transfinite*, and by his general belief that numbers, both finite and transfinite, are the building blocks of reality.

One further instance of this attitude can be seen at work in Cantor's conception of the *continuum*. This latter, he thought, could be understood only through a process of logical simplification, which could provide us with solid conceptual knowledge of its internal components. This process he thought he had carried out with his theory of derived point-sets and related cardinal powers. But Kant's notion of continuity, like that of Aristotle and of the Scholastic philosophers, he suggests, is dependent upon the a priori intuition of *space* and *time*, which, in turn, cannot provide us with any conceptual knowledge of the phenomenal world.

> ...the continuum is thought to be an unanalysable concept or, as others express themselves, a pure a priori intuition which is scarcely susceptible to a determination through concepts.

Every arithmetical attempt at determination of this *mysterium* is looked on as a forbidden encroachment and repulsed with due vigour. ([1], in [10, p. 903])

Time and space, he continues, are only *syncategorematic*, that is, *relational* concepts, and they have failed to produce, via Kant's conception, any tangible progress in our knowledge. He continues:

Such a thing as *objective* or *absolute time* never occurs in nature, and therefore *time* cannot be regarded as the measure of *motion*; far rather motion as the measure of *time* - were it not that *time*, even in the modest role of a *subjective necessary a priori* form of intuition [Anschauungsform] has not been able to produce any fruitful, incontestable success, although since Kant the time for this has not been lacking. ([1], in [10, p. 904])

Gödel expressed the same dissatisfaction with Kant's "subjectivism". For a Platonist, and a conceptual realist like him, this is hardly surprising. However, his relationship with Kant's doctrines is more articulated. The main point of friction with Kant is Gödel's notion of *intuition*.

While intuition in Kant's system is related to the work of the intellect, which "elaborates" the representations of our senses, Gödel's intuition is something stronger and deeper: it should give us knowledge of *objects* (and *concepts*) which are seemingly formed by us, but, in fact, already *exist* within us. Intuition [*Anschauung*], in Kant's view, serves the purpose of "constructing" concepts; in Gödel's, that of "seeing" or "perceiving" objects (and concepts thereof).[21] However, this latter process also needs some form of elaboration. Gödel crucially explains:

It should be noted that mathematical intuition need not be conceived as a faculty giving an *immediate* knowledge of the objects concerned. Rather it seems that, as in the case of physical experience, we *form* our ideas also of those objects on the basis of something else which *is* immediately given. [...] Evidently the "given" underlying mathematics is closely related to the abstract elements contained in our empirical ideas. It by no means follows, however, that the data of this second kind, because they cannot be associated with actions of certain things upon our sense organs, are something purely subjective, as Kant asserted. ([20], in [22, p. 268])

Gödel seems to envisage a role for intuition which is analogous to that envisaged by Kant in his account, that is, that of a faculty which operates on something *given* in order to "derive" something else, but the two conceptions differ to a substantial extent. While, for Kant, intuition [*Anschauung*] acts on sensory data to derive conceptual information, Gödel's intuition acts on some given conceptual contents, to produce other forms of conceptual contents (such as *mathematical objects*). But, for Gödel, such an intellectual operation has the character of objectivity, insofar as it provides us with knowledge of *ideal forms*. Such move is seen by most commentators as connected to Gödel's adhering to Husserl's *transcendentalism*. Husserl's transcendental intuition, unlike Kant's, has a fully objective character, as it is directed at "things themselves", and operates through the process called *eidetic*

---

[21] See footnote 12 above.

reduction.[22] Thus, Gödel seems to re-use Kant's original conception to produce something like a more powerful version of it.[23]

As already said, Gödel's attitude to Kant is more articulated and varied than Cantor's. His interpretation of Kant's notion of "phenomenon" is, in this respect, revealing. In his [17], Gödel makes the somewhat baffling claim that Einstein's relativity theory, by showing the deceitfulness of the notion of *temporal simultaneity*, has confirmed

> [...] the view of those philosophers who, like Parmenides, Kant, and the modern idealists, deny the objectivity of change and consider change as an illusion or an appearance due to our special mode of perception. ([17], in [22, p. 202])

In this passage, Kant is associated with such philosophers as Parmenides, who had explicitly denied the reality of motion. In Gödel's perspective Kant is more the philosopher who has revealed the deceitfulness of the phenomenal world, rather than, as in Cantor, the strong advocate of subjectivism. However, we have seen that, elsewhere, Gödel, not unlike Cantor, had judged Kant's subjectivism a major shortcoming.

5 Set-Theoretic Platonism

Gödel's ontological conceptions gradually evolved towards a form of thoroughgoing Platonism, more specifically, Platonism about *sets*. In published work, his platonistic leanings are declared first in his Russell paper (1944) and, afterwards, re-asserted in the Cantor paper (1947). In 1944, he wrote:

> It seems to me that the assumption of such objects is quite as legitimate as the assumption of physical bodies and there is quite as much reason to believe in their existence. ([15], in [22, p. 128])

In the Cantor paper, the *existence* of a set-theoretic reality is even more unequivocally and firmly asserted. Gödel's argument, expounded there, aims to show that the Continuum Hypothesis (CH) necessarily has a determinate truth-value, if one believes that set theory describes a *well-determined reality*. As a

---

[22]The bulk of Husserl's phenomenological ideas can be found in the three volumes of the *Ideen*, [26, 27] and [29]. See also [28]. A quick review of the main concepts of phenomenology can be found in one of the articles/books I mentioned above, footnote 16, or, for instance, in Christian Beyer's entry "Husserl" in the *Stanford Philosophical Encyclopedia*, 2013, which also includes an up-to-date bibliography.

[23]See, in particular, Parsons [38, pp. 56–70], concerning the difficulties with Gödel's notion of intuition. Parsons' interpretation, especially of Gödel's quotations from [20], seems inclined to explain away the presence of phenomenological elements in Gödel's thought. For instance, with regard to the notion of "immediately given", he says: "The picture resembles Kant's, for whom knowledge of objects has as "components" a priori intuition and concepts. It is un-Kantian to think of pure concepts as given, immediately or otherwise. But Gödel's picture seems clearly to be that our conceptions of physical objects have to be constructed from elements, call them primitives, that are given, and that some of them (whether or not they are much like Kant's categories) must be abstract and conceptual." (p. 68).

consequence, it makes sense to search for its truth-value, even after one has shown
that such truth-value cannot be determined by the ZFC axioms.

> It is to be noted, however, that on the basis of the point of view here adopted, a proof
> of the undecidability of Cantor's conjecture from the accepted axioms of set theory (in
> contradistinction, e.g., to the proof of the transcendency π of) would by no means solve the
> problem. For if the meanings of the primitive terms of set theory [. . .] are accepted as sound,
> it follows that the set-theoretical concepts and theorems describe some well-determined
> reality, in which Cantor's conjecture must be either true or false. Hence its undecidability
> from the axioms being assumed today can only mean that these axioms do not contain a
> complete description of that reality. ([20], in [22, p. 260])

Later, in his unpublished Gibbs lecture [18], Gödel sketched some further
arguments in favour of Platonism. I will not deal with those arguments here. Rather,
I wish to examine whether and to what extent Gödel's set-theoretic Platonism is
indebted to Cantor's set-theoretic Platonism.

In Sect. 2, I have briefly discussed the notion of *intra-subjective* existence in
Cantor's *Grundlagen* and shown how it is connected to a peculiar form of conceptual
realism. In the same footnote in which he gives an account of the notion of *idea
vera*, as constituting the historical and conceptual ground for his notion of *immanent
existence*, Cantor also claims that his conception of "set" fits perfectly into Plato's
conception of a *third gender* of being which is defined, in the *Philebus*, as μικτόν
[*miktón*].

> In general, by a "manifold" or "set" I understand every multiplicity [jedes Viele] which can
> be thought of as one, i.e. every aggregate [Inbegriff] of determinate elements which can be
> united into a whole by some law. I believe that I am defining something akin to the Platonic
> εἶδος or ἰδέα as well as to that which Plato called μικτόν in his dialogue "Philebus or the
> Supreme Good". He contrasts this to the ἄπειρον (i.e. the unbounded, undetermined, which
> I call the improper infinite) as well as to the πέρας i.e the boundary; and he explains it as
> an ordered "jumble" of both. Plato himself indicates that these concepts are of Pythagorean
> origin. ([1], in [10, p. 916])

This scanty remark affords us three important pieces of information about
Cantor's conception of sets:

1. Sets arise from putting together elements of a *multitude*, by using a specified
   'uniting' law.
2. The notion of "set" can be successfully compared to Plato's notion of ἰδέα, that
   is, "intelligible form". Such is the ontological status also of the μικτόν, that is,
   mixed entity, that Cantor is referring to here as corresponding to his notion of set.
3. Via Plato's conceptualisation, the notion of "set" is, in turn, related to some
   Pythagorean conception of "set".

The ideas underlying (1) and, partly (3), have been extensively explored by
Hallett in his [24], and here I will use his interpretation of Cantor's set-theoretic
conception as based on a theory of "ones".[24]

---

[24]See, in particular, Hallett [24, pp. 128–142].

Sets would be constituted of *irreducible unities*, which are, afterwards, transformed into *new unities*, which are the sets themselves. Such conception is used by Cantor to define numbers. Hence, it is an essential ingredient of Cantor's strategy to reduce all numbers, finite and transfinite, to *sets*. In other passages, cardinal numbers are also defined as being *abstracted* from certain particular collections (which are, in turn, collections of *ordinals*). Cantor's set-theoretic ontology thus collapses to well-ordered collections of given number-sets consisting of "unities", the *ordinals*, from which *cardinal numbers* are subsequently abstracted. On this picture, sets are already (ideal) numbers. This is clearly shown by the reference to platonic ἰδέαι[*idéai*] in (2).

As we have seen, Cantor's definition of number-sets uses a complicated theory appearing in Plato's *Philebus*, whereby the generation of numbers is seen as the outcome of a dialectical process involving the interaction of the ἄπειρον [*ápeiron*], the "Unlimited" with the πέρας [*péras*] the "Limit".

Although we are not very sure what the nature of the ἄπειρον to which Plato refers in the *Philebus* is, it is most likely that the concept can be interpreted as corresponding to the *potential infinite*. The πέρας, thus, operates on (merges with) the potential infinite in a *process of determination*, which yields (ideal) numerical entities (this also explains why such entities are called μικτά [*miktá*], "mixed": because they participate of both concepts).[25]

It should be noted that Cantor needs to construe the platonic process in a different way: μικτά have now become determinations of the *actual*, not the *potential* infinite. But determinations of the actual infinite are nothing other than transfinite numbers. It is no surprise, then, that in another passage from [2], Cantor describes *transfinite ordinals* as being ἀριθμοὶ νοητοί [*arithmoì noetoí*] or εἰδητικοί [*eidetikoí*], using a terminology that is reminiscent of that used by Plato.[26]

In published work, Gödel never gave any further details about his set-theoretic Platonism. However, in conversations with Wang, we find quite a few remarks about the nature of "sets", which seem to align his position with, or even clearly echo, Cantor's conceptions. For instance, he says:

> 8.2.1 [...] It is a primitive idea of our thinking to think of many objects as one object. We have such ones in our mind and combine them to form new ones.

---

[25]In [4], reprinted in [5, p. 380], Cantor also uses the Greek word μονάς [*monás*] to refer to number-sets, a term which is borrowed from a definition in Euclid's *Elements* he mentions in that work. He says (my translation): "Cardinal numbers as well as order-types are *simple* conceptual formations; each of them is a *true Unity* (μονάς), as in them a plurality and multiplicity of *Ones* is unitarily bound together [Die Kardinalzahlen sowohl, wie die Ordinungstypen sind *einfache* Begriffsbildungen; jede von ihnen ist eine *wahre Einheit* (μονάς), weil in ihr eine Vielheit und Mannigfaltigkeit von *Einsen einheitlich* verbunden ist]." He also reports instances of the notion of μονάς as can be found in Nicomachus' *Institutio Arithmetica* and Leibniz. Nicomachus' neo-Pythagorean view about numbers also implied that they are συστήματα μονάδων [*systémata monádon*], that is, *aggregations of unities (monads)*. In Cantor's quoted passage from Leibniz's *De arte combinatoria*, Leibniz says: "Abstractum autem ab uno est unitas, ipsumque totum abstractum ex unitatibus, seu totalitas, dicitur numerus".

[26]Cantor [2], reprinted in [5, p. 372].

8.2.2 [. . .] Sets are multitudes which are also unities.

8.2.3 This [fact]—that sets exist—is the main objective fact of mathematics which we have not made in some sense: it is only the evolution of mathematics which has led us to see this important fact. [. . .] there must be something objective in the forming of unities. [48, all p. 254]

Gödel uses a language, which overlaps Cantor's: he talks of "unities", "multitudes", "many objects as one". Such language aims to convey the idea, very similar to Cantor's, that sets are *new unities*, arising out of multitudes. This is particularly relevant, insofar as, as we shall see in Sect. 7, Gödel may have thought to embed Cantor's conception of sets, qua μονάδες [*monádes*], into Leibniz's conception of *monads*.

In what he says above, a second fundamental point is Gödel's idea that the process of "uniting" objects into such "unities" is proof of the objectivity of mathematics. One should be wary of seeing any "constructive" overtone in evoking a process of "unification". If the process itself is possible, it is only because such unifications can be carried out successfully a priori.

I will describe one more, fundamental feature of this form of set-theoretic Platonism in the next section.

6   The Absolute Infinite and the Universe of Sets

In his Cantor paper, Gödel famously proposes the extension of the system of axioms of set theory to settle open problems in set theory and mathematics. His argument views *strong* axioms of infinity, that is, axioms positing the existence of *large cardinals*, as the most suitable axiom candidates for extending ZFC (Zermelo–Fraenkel set theory with the Axiom of Choice). At that time, Gödel thought that they might, in particular, have a significant impact on such problems as the Continuum Problem.[27]

Gödel's argument in favour of the acceptance of such large cardinal hypotheses is an argument from "intrinsic necessity", that is, it is based on considerations related to the features of the *iterative concept of set*. The case for the extension of ZFC is introduced in the following way:

First of all the axioms of set theory by no means form a system closed in itself, but, quite on the contrary, the very concept of set on which they are based suggests their extension by new axioms which assert the existence of still further iterations of the operation "set of". ([20], in [22, p. 260])

---

[27]At the time of composition of [16], and of its revision in [20], it was not known that large cardinal axioms do not fix the power of the continuum, as Gödel had conjectured that they might. This was first shown by Solovay and Lévy in [32] using measurables, but the result generalises to all known large cardinals. An analogous result for smaller large cardinals had already been proved by Cohen in [6]. See Kanamori [31, p. 126].

In footnote 18, he explains:

> Similarly the concept "property of set" (the second of the primitive terms of set theory) suggests continued extensions of the axioms referring to it. Furthermore, concepts of "property of property of set" etc. can be introduced. ([20], in [22, p. 260])

In his conversations with Wang, we find further details about how new axioms should reflect the iterative concept of set. Some remarks contain mention of such properties of the universe of sets as *reflection, uniformity, closure*, etc., all of which could orient the selection of new axioms.[28]

The principle I am mostly interested in here is *reflection*. This is how Gödel comments on it:

> 8.7.3 *Reflection principle*. The universe of sets is structurally indefinable. One possible way to make this statement precise is the following: The universe of sets cannot be uniquely characterized (i.e., distinguished from all its initial segments) by any internal structural property of the membership relation [. . .]. This principle may be considered a generalization of the closure property. Further generalizations and refinements are in the making in recent literature. The totality of all sets is, in some sense, indescribable. [48, pp. 280–281]

In this account of the principle, the notion of the *indescribability* of the universe of sets plays a crucial role. But where did Gödel get it? And why did he see it as essential for his purposes? In order to answer these questions, we have to do some careful interpretive work.

In another revealing passage reported by Wang, Gödel talks about the so-called von Neumann axiom, and describes it in the following way:

> 8.3.7 As has been shown by von Neumann, a multitude is a set if and only if it is smaller than the universe of all sets. This is understandable from the objective viewpoint, since one object in the whole universe must be small compared with the universe and small multitudes of objects should form unities because being small is an intrinsic property of such multitudes. [48, p. 262]

As Wang explains, the "axiom" Gödel attributes to von Neumann, is, in fact, due to Cantor,[29] and made his first appearance in a letter that this latter sent to Dedekind in 1899.[30] In that letter, responding to Dedekind's concerns that Cantor's notion of "sethood" might be unclear and paradox-laden, Cantor draws a distinction between *consistent* and *inconsistent* multiplicities. The latter are what would become to be described subsequently by von Neumann as "classes", whereas the former are "sets" in the proper sense. Classes are collections that are too "big" to be considered sets.[31]

This conception lies at the roots of the so-called *limitation of size* doctrine. It implies, in particular, that classes do not have a transfinite size. To be more precise,

---

[28]Wang discusses them in both [46] and [48], but this latter contains a more detailed account.

[29]Wang [48, p. 261]: "[..] Cantor called multitudes "like" V *inconsistent multitudes*, and introduced a general principle to distinguish them from sets."

[30]English translation in [10, pp. 931–935].

[31]However, as we have seen (footnote 4), Gödel was fully aware of the fact that Cantor, not von Neumann, had first introduced the distinction between sets and classes. On this point, see also Van Atten [44].

the size of inconsistent multiplicities is that of the *absolute infinite*. In turn, the absolute infinite is the infinite of God.[32]

In [2], a summary of the different forms of *actual infinite* provides the following definition of the absolute infinite:

> One can question the Actual Infinite in three main forms: first, insofar as it is in *Deo extramundano aeterno omnipotenti sive natura naturante*, that is the *Absolute*, second, insofar as it can be found *in concreto seu in natura naturata*, and I call it *Transfinitum*, and third, it can be questioned *in abstracto*, that is, insofar as it can be understood by human beings in the form of *actual infinite* or, as I have called them, of *transfinite numbers*, or in the more general form of *transfinite order-types* (ἀριθμοί or εἰδητικοί) [Man kann nämlich das A.-U. in *drei Hauptbeziehungen* in Frage stellen: *erstens, sofern es in Deo extramundano aeterno omnipotenti sive natura naturante*, wo es das *Absolute* heißt, *zweitens* sofern es *in concreto seu in natura naturata* vorkommt, wo ich es *Transfinitum* nenne und *drittens* kann das A.-U. *in abstracto* in Frage gezogen werden, d.h. sofern es von der menschlichen Erkenntnis in Form von *aktual-unendlichen*, oder wie ich sie gennant habe, von *Transfiniten Zahlen* in der noch allgemeineren Form der *Transfiniten Ordnungstypen* (ἀριθμοί oder εἰδητικοί) aufgefaßt werden könne] ([2], in [5, p. 372] my translation).

Some years before, in the *Grundlagen*, after expounding the point of view of various authors, he had referred to the absolute infinite in the following way:

> However different the theories of these writers may be, in their judgement of the finite and infinite they essentially agree that finiteness is part of the concept of number and that the true infinite or Absolute, which is in God, permits no determination whatsoever. As to the latter point I fully agree, and cannot do otherwise; the proposition: "omnis determinatio est negatio" is for me entirely beyond question. ([1], in [10, pp. 890–891]).

In footnote, he adds:

> The absolute can only be acknowledged [anerkannt] but never known [erkannt] – and not even approximately known. [...] As Albrecht von Haller says of eternity: 'I attain to the enormous number, but you, O eternity, lie always ahead of me.' ([1], in [10, p. 916])

In that work, Cantor uses theological tones, whereas in his letter to Dedekind, the *absolute infinite* is viewed as an eminently mathematical phenomenon, relating to collections too "big" to be measured (such as the class of all ordinals or the class of all cardinals). Such oscillation (and maybe tension) has been acknowledged and examined by the secondary literature.[33] It seems to me that, in view of all these interpretive efforts, one can, at least, say that Cantor's absolute infinite plays a dual

---

[32]See [24], and particularly, p. 164–194. With regard to von Neumann's re-statement of Cantor's principle (what Gödel calls "von Neumann axiom"), see, in particular, pp. 286–298.

[33]See, in particular, Hallett [24, pp. 41–48 and 165–176]. Wang discusses Cantor's conception in connection with Gödel's criteria for introducing new axioms especially in Wang [46, pp. 188–190]. Jané addresses Cantor's conception in full, in his [30]. Jané lays strong emphasis on the tension between the idea that the *Absolute* cannot be measured (and determined), and the fact that it can still be seen as a sort of "quantitative maximum" for the actual infinite, a tension which was perceived and addressed by Cantor in different ways over his career. Jané thinks that, in the end, God's *absoluteness* and *mathematical absoluteness* fell apart, as Cantor was forced to accept, mathematically, that the *absolute infinite* is not a form of the *actual infinite*.

role in his work:

1. To provide a general justification for his *limitation of size* doctrine, whereby one should distinguish between *consistent* and *inconsistent* multiplicities.
2. To rebuke successfully the objection that measuring the infinite might lead to some kind of rational *pantheism*. God's infinity is still clearly and determinately distinguished from that of the transfinite, through attributing to him an absolute infinity.

As we have seen, the absolute infinite characterises itself for being, essentially, indescribable, indefinable. However, in a sense, although indefinable, it can still be seen as endowed with some properties, and be thought of as a sort of *aggregate*. However, these properties cannot be ascribed to it directly (in view of its *indefinability*), but to initial segments of it, which would, thus, in a sense, reflect it. This form of *reflection* Gödel sees at work in one further set-theoretic principle, that he defines as the "basic axiom of set theory", Ackermann's axiom. This axiom does not mention the reflection principle directly, but can be seen as a consequence of it.

(A) *Ackermann's axiom.* Let y and z be in V and $F(x, y, z)$ be an open sentence not containing V, such that, for all x, if $F(x, y, z)$, then x is in V. There is then some u in V, such that, for all x, $F(x, y, z)$ if and only if x belongs to u. (in [48, p. 282])

Ackermann's axiom can be glossed in the following way: if there exists a set-theoretic property, whose formulation does not mention V, then there must be some set *in* V which satisfies such property. It is crucial to mention that the set under consideration is in V, that is, it is crucial to "reflect" V onto a set, which has that property.

It is Gödel himself who acknowledges the connection between Ackermann's axiom and Cantor's absolute. He says in 8.7.9: "All the principles for setting up the axioms of set theory should be reducible to a form of Ackermann's principle: The Absolute is unknowable." [48, p. 283]

So, we have finally made the following picture available to ourselves. The doctrine of the Absolute, first expounded by Cantor, became a basic principle of his *limitation of size* doctrine. This latter, in turn, was made into an axiom by von Neumann. Such an axiom has consequences on the *definability* of the universe of sets, and, thus, encourages the discovery of such properties as reflection. The reflection principle, in the version given by Gödel, is, in turn, encapsulated by one single set-theoretic principle: Ackermann's axiom, that Gödel believed should be considered the most basic axiom of set theory. Thus, Gödel's doctrine concerning the extension of ZFC, insofar as it is essentially based on the *reflection principle*, is largely indebted to one single conception, Cantor's conception of the *absolute infinite*.[34]

---

[34]However, this form of reflection principle does not justify *very large* large cardinal hypotheses. Gödel was maybe already aware of this fact, when, in a footnote added in 1966 to his [20], referring to the axiom asserting the existence of a *measurable cardinal*, he stated (pp. 260–261): "That these axioms are implied by the general concept of set in the same sense as Mahlo's has not been made

The use of Cantor's conception might also be viewed as instrumental for Gödel's own parallel belief of the *inexhaustibility of mathematics*. Gödel addresses the notion of *inexhaustibility* in his Gibbs lecture, but there is no connection there between this latter and Cantor's absolute infinite. However, the connection can be reconstructed indirectly. In the text we possess, Gödel says that the collection of all mathematical *truths* represents "objective" mathematics, as opposed to "subjective" or "mechanised" mathematics, which consists of all *demonstrable* propositions. The first incompleteness theorem shows that the collection of mathematical truths is *larger* than that of the demonstrable truths.[35]

Thus, Gödel construes his incompleteness theorem as proof that the realm of "objective mathematics" is *larger* than that of "mechanical" (or "subjective") mathematics.[36] The gap between the two cannot be filled, because our grasp of objective mathematics is incomplete. This fact is seen, Gödel continues, in relation to the goal of providing something like a "definitive" axiomatisation of set theory. The task is impossible, for the following reasons:

> ... if one attacks this problem, the result is quite different from what one would have expected. Instead of ending up with a finite number of axioms, as in geometry, one is faced with an infinite series of axioms, which can be extended further and further, without any end being visible and, apparently, without any possibility of comprising all these axioms in a finite rule producing them. ([18], in [23, p. 306])

In turn, the process of formation of ever new axioms never ceases, since, as Gödel explains, further ordinals can always be formed. This leads him to conclude that:

> You will realize, I think, that we are still not at an end, nor can there ever be an end to *this* procedure of forming the axioms, because the very formulation of the axioms up to a certain stage gives rise to the next axiom ([18], in [23, p. 307])

To summarise, Cantor's conception of the absolute infinite has two bearings on Gödel's set-theoretic Platonism. One can see it as connected, essentially, to two doctrines, which Gödel, at some point, held:

1. The universe of sets is *indescribable*, yet is, in a sense, characterisable. Since it is not describable, though, all of its characterising properties reflect onto initial segments (*reflection principle*).
2. There is no endpoint in the series of axioms expressing ways to generate sets. This is also the reason for the *incompletability* of mathematics.

---

clear yet [...]", whereas he is aware of the fact that small large cardinals such as Mahlo's can be connected to *reflection* successfully: "Mahlo's axioms have been derived from a general principle about the totality of sets which was first introduced by Levy (1960). It gives rise to a hierarchy of precise formulations". For details about this hierarchy, see Kanamori [31, pp. 57–59].

[35]This claim can be made precise by saying that Gödel's First Incompleteness Theorem shows that any theory of arithmetic of the same strength as PA (Peano Arithmetic) is *incomplete*, namely, that it does not prove all *arithmetical truths*.

[36]However, it could actually turn out that "subjective" mathematics is larger than "mechanical" mathematics, should human minds prove to be *stronger* than machines, something Gödel had already cast as a conjecture in the Gibbs lecture and which is again reported by Wang in [48, p. 186]. I am indebted to Gabriella Crocco for pointing me to the subtle difference between "subjective" and "mechanical" in Gödel's formulations.

7   Trans-Subjective ("Transient") Existence

Alongside the "immanent" one, Cantor also mentions one further form of "existence" of mathematical objects:

> But then, reality can also be ascribed to numbers to the extent that they must be taken as an expression or copy of the events and relationships in the external world which confronts the intellect, or to the extent that, for instance, the various number-classes (I), (II), (III), etc. are representatives of powers that actually occur in physical and mental nature. I call this second kind of reality the *transsubjective* or the *transient reality* of the integers. ([1], in [10, pp. 895–896])

Cantor's mention of a *trans-subjective* form of reality of mathematical entities and, in particular, his idea that number-classes (that is, the $\aleph$s) are, somehow, instantiated in the physical reality have baffled many commentators.

In order to make sense of such statements and see their connections with Gödel's analogous claims, we have to expound some further Cantorian conceptions relating to the nature of physical reality.

As I have pointed out many times, Cantor's views are essentially metaphysical. For instance, concerning the status of *analytic mechanics* and *mathematical physics*, he says:

> These disciplines are, in my opinion, in their foundations as well as in their aims *metaphysical*; if they seek to make themselves free from metaphysics, as has been recently proposed by a celebrated physicist [Gustav Kirchhoff], they degenerate into a "description of nature" in which the fresh breeze of free mathematical thought - as well as the power of *explaining* and *justifying* natural phenomena - must be absent. ([1], in [10, p. 897])

Always in the *Grundlagen*, he encourages physicists' efforts to develop a different form of physics, which may overcome the limitations of a purely *mechanical explanation* of natural forces, and, instead, be committed to some form of *organicism*, which he saw in essential accordance with the systems of Leibniz and Spinoza. This leads him to assert that:

> For, alongside of (or in place of) the mechanical explanation [Erklärung] of nature (which inside its proper domain has all the aids and advantages of mathematical analysis at its disposal, but whose one-sidedness and insufficiency have been strikingly exposed by Kant) there has until now not been even the start of an *organic* explanation of nature that would attempt to go further and that would be armed with the same mathematical rigour; this organic explanation can, I believe, be initiated only by taking up afresh and continuing the works and endeavours of those thinkers [i.e., Leibniz and Spinoza, *my note*]. ([1], in [10, p. 892])

But, in just what sense would an organic explanation of the physical world be in accordance with Leibniz's and Spinoza's intuitions? Cantor's use of Spinozian *ideae verae* and *adaequatae* has already been discussed. Now, in Spinoza's conception, ideas are features of reality belonging to both the *material* and the *mental* ontological realms. Cantor's views seem to follow suit. *Transfinite* concepts, insofar as they exist immanently, should have some physical correlates as any other concept.

In particular, this justifies Cantor's claim that the $\aleph$s (cardinal powers) will *occur in* or *represent* aspects of the real world.

Some further details of this conception are tied more prominently to Leibniz's doctrines. Such phenomenal aspects of the real world as the *continuum, matter* and *corporeal forms* should be reduced to logically simpler entities, that Cantor calls *atoms* in the *Grundlagen*. These atoms should not be construed as the atoms of the *atomistic* tradition, but rather as something comparable to Leibniz's *monads*, infinitely small points without parts. In the *Grundlagen*, he mentions their presence in the physical reality incidentally, in the following passage about Bolzano's view on the actual infinite:

> The proper-infinite, as we find it in, for example, well-defined point-sets or in the construction of bodies from point-atoms [punktuellen Atomen] (I thus do not mean here the chemical-physical (Democritean) atoms, because I cannot hold them for existent, either in thought or reality, although much that is useful has been achieved up to a certain limit by this fiction) has found its most determined defender etc. ([1], in [10, p. 894])

However, in [3], Cantor's Leibnizian characterisation of "atoms" is more explicit. In the following passage, he also clarifies the connection between monads and transfinite cardinal numbers. He says:

> Following Leibniz, I call the *simple* elements of nature, from whose composition *matter* is, in some sense, constituted, *Monads* or *Unities*. [...] it is many years that I have formulated the *hypothesis* that the *power* of *corporeal matter* is what I have called, in my investigations, the *first power* and that, on the contrary, the power of *ethereal matter* is the second class [Ich nenne in Anschluss an Leibniz die *einfachen* Elemente der Natur, aus deren Zusammensetzung in gewißem Sinne die *Materie* hervorgeht, *Monaden* oder *Einheiten* [...] ...in dieser Beziehung habe ich mir schon vor Jahren die *Hypothese* gebildet, daß die Mächtigkeit der *Körpermaterie* diejenige ist, welche ich in meinem Untersuchungen die *erste* Mächtigkeit nenne, daß dagegen die Mächtigkeit der *Äthermaterie* die *zweite* ist]. ([3], in [5, pp. 275 and 276], my translation)[37]

Mention of monads, of point-atoms, of an *organic*, as opposed to *materialist*, explanation of the physical world, gives us a taste of the broadness of Cantor's conception of the *trans-subjective*.[38]

Is there an analogous mention of a trans-subjective form of existence in Gödel's thought? Although we cannot find any direct connection to Cantor's concept, we find, in some of Gödel's incidental observations, reference to a well-structured metaphysical ontology. Especially on the grounds of what Wang reports in his [48], it seems reasonable to assert that also Gödel took *monads* to be the essential constituents of the world. Like Cantor, he also thought that *monadology* might

---

[37]Using the subsequent $\aleph$-notation, the *powers* of the first and the second class are, respectively, $\aleph_0$ and $\aleph_1$. Under CH, the power of ethereal matter is c. For further details on this claim, and its connection to Cantor's set-theoretic work, see Dauben [8, p. 126] and Ferreirós [11], in particular, pp. 75–77.

[38]An articulated review of extra-mathematical themes in Cantor's thought is the aforementioned Ferreirós [11]. Cantor's ideas related to *organicism* were not altogether foreign to the scientific debate, as Ferreirós shows (see p. 77). On this point, see also Purkert-Ilgauds [39, pp. 67–68].

provide a different, alternative description of the physical world. My claims are substantiated by what he says in the following remarks:

> 9.1.20 We should *describe* the world by applying these fundamental ideas: the world as consisting of monads, the properties (activities) of the monads, the laws governing them, and the representations (of the world in the monads).
> 9.1.21 The simplest substances of the world are the monads.
> 9.1.22 Nature is broader than the physical world, which is inanimate. It also contains animal feelings, as well as human beings and consciousness. [48, all p. 295]

However, things are not so simple as they appear. As explained by Tieszen in [43], Gödel may have thought of Leibniz's monads in terms of *transcendental egos*, in the way indicated by Husserl.[39]

But there are other passages in Wang's book, which have been extensively examined by van Atten in [44], wherein Gödel would seem to hold a more "standard" view concerning monads. In particular, he would seem to have been inclined to identify *sets* with *monads*.[40]

In connection, again, with Ackermann's axiom, which I have discussed in the preceding section, Gödel also says:

> 8.7.14 There is also a theological approach, according to which V corresponds to the whole physical world, and the closeness aspect to what lies within the monad and in between the monads. According to the principles of rationality, sufficient reason, and pre-established harmony, the property $P(V, x)$ of a monad x is equivalent to some intrinsic property of x, in which the world does not occur. In other words, when we move from monads to sets, there is some set y to which x bears intrinsically the same relation as it does to V. Hence, there is a property $Q(x)$, not involving V, which is equivalent to $P(V, x)$. According to medieval ideas, properties containing V or the world would not be in the essence of any set or monad. [48, p. 284]

And in one further remark, he says:

> 9.1.27 *Objects.* Monads are objects. Sets (of objects) are objects. A set is a unity (or whole) of which the elements are constituents. [...] Sets are the limiting case of spatiotemporal objects and also of wholes. [48, p. 296]

In published work, Gödel was more wary of proposing connections between ideal entities and physical reality. In the Cantor paper, he advocates a sharp distinction between the epistemic status of Euclid's fifth postulate and CH. The former is based on our interpretation of the physical reality, whereas the latter can only be settled by purely mathematical considerations. However, Gödel's remarks on this point leave

---

[39]Cf., in particular, Tieszen [43, p. 38]. According to Husserl, *"Monads are transcendental egos in their full concreteness.* Transcendental egos in their full concreteness are not "mere poles of identity", but are rather egos with all the predicates that attach to these poles of identity, so that each monad is distinct from every other monad. We know that Leibniz has a range of different kinds of monads, but Husserl's focus is much narrower. It is on the kind of 'monads' that we are". Tieszen also observes that we do not know to what extent Gödel wanted to use Leibniz's monadological conception in a way which would conform to Husserl's.

[40]Van Atten notices (p. 4), that the sole fact that "Leibniz denies the existence of infinite wholes of any kind" would doom Gödel's attempt to failure.

some room for the possibility that, in the future, we could find a trans-subjective meaning for transfinite set theory as well:

> On the other hand, the objects of transfinite set theory, [...], clearly do not belong to the physical world, and even their indirect connection with physical experience is very loose (owing primarily to the fact that set-theoretical concepts play a minor role in the physical theories of today). ([20], in [22, p. 267])

To sum up, I believe that Cantor's and Gödel's conceptions of the trans-subjective can be successfully compared. At bottom, what they seem to share is a parallel attitude to apply to the physical reality the same attitude they applied to mathematics, which I have defined in Sect. 4 "arithmetical purism", namely a tendency to reduce complex phenomena to simpler elements. This attitude seems to be aptly reflected by their thick speculations on the existence of atomic (monadic) constituents of the reality, whatever these latter might be, either simple substances or vitalistic soul-like principles.

8 Connection of Immanent and Transient

Cantor distinguishes two forms of "existence". However, he also commits himself to the belief in their fundamental identity. This belief is expressed in the following passage:

> Because of the thoroughly realistic but, at the same time, no less idealistic foundation of my point of view, I have no doubt that these two sorts of reality always occur together in the sense that a concept designated in the first respect as existent always also possesses in certain, even infinitely many, ways a transient reality. ([1], in [10, p. 896])

Further, in the subsequent paragraph, he adds: "This linking of both realities has its true foundation in the *unity* of the all to which we ourselves belong." [1, p. 896].

In footnote, he provides us with some further details about the nature and the origin of his conception. In particular, he assures that his position is in accordance, as ever, with Plato's, Spinoza's and Leibniz's philosophical systems, and quotes one famous proposition from Spinoza's *Ethics*: "As for Spinoza, I need only mention his statement in *Ethica*, part II, prop. VII: '*ordo et connexio idearum idem est ac ordo et connexio rerum*' [The order and connection of ideas is the same as the order and connection of things (my translation)]" ([1], in [10, p. 918]).

However, the emphasis on Spinozian "monism", in Cantor's philosophy, is tied to a specific goal, that of emphasising, once more, the objectivity of mathematical *concepts*. In a sense, such monism only serves as a strengthening of conceptual objectivity. It can be paraphrased in the following way: the conceptual constructs we seem to build up (but, in fact, *reminisce*) are related to equally objective constructs, which are given, trans-subjectively, in the physical reality. The emphasis on the connection between the two realms works, thus, essentially as a striking epistemological metaphor: finding new mathematical concepts is equivalent to detecting new "forms" in the physical reality.

This leads Cantor to conclude that:

> The mention of this linking has here only one purpose: that of enabling one to derive from it a result which seems to me of very great importance for mathematics, namely, that mathematics, in the development of its ideas, has *only* to take account of the *immanent* reality of its concepts and has *absolutely no* obligation to examine their *transient* reality. ([1], in [10, p. 896])

It seems reasonable to assume that also Gödel might have conceived of a sort of connection between concepts and objects. An instance of such connection could be the one we have described in the preceding section, that between *sets* (concepts) and *monads* (objects).

In Wang's book, we find one observation he makes concerning axioms and models, which resumes the aforementioned Spinozian doctrine that appealed so much to Cantor, that of the connection between *ordo idearum* and *ordo rerum*. He says:

> 4.3.9 The axioms correspond to the concepts, and the models which satisfy them correspond to the objects. The representations give the relation between concepts and objects. For Spinoza the connection of things are connections of ideas. [...] We have here a general proportionality of the membership relation (the concept) and the sets (the objects). The original difference is that concepts are abstract and objects are concrete. In the case of set theory, both the membership relation and the sets are abstract, but sets are more concrete. [48, p. 141]

However, whether Gödel here refers to objects as *trans-subjective* constructs is unclear. In any case, even so, we do not know what relevance the reported Spinozian point of view would globally have in Gödel's conceptions. As a matter of fact, we know that Gödel rejected Spinoza's *pantheistic* views and held a *theist* conception.[41] But he could have been inclined to accept some form of *monism* regardless of any commitment to Spinoza's doctrines. An alternative way could be suggested, once more, by monadology. After all, monads have a dual aspect: on the one hand, they are *simple substances* which also interact with the physical reality and, on the other, they are soul-like forms, or, as Husserl would put it, *transcendental egos*. In the first aspect, they bear on the structure of reality, including the physical reality, in the other, they belong to the realm of the "intelligible", the "immaterial", the "conceptual". Thus, in a sense, the two aspects of reality, to which Cantor was referring, would be successfully *unified* in them.

This interpretation might help us to understand the following very famous, but also cryptic, Gödelian statement in the Cantor paper. As we know, Gödel assimilates the "data" resulting from *conscious* (*introspective*) *elaboration* to "data" resulting from *sensory elaboration*. Such data, he says, need not, as Kant thought, be conceived as an "expression" of human subjectivity.

---

[41]Cf. Wang [48, p. 112]: "Gödel gave his own religion as "baptized Lutheran" (though not a member of any religious congregation) and noted that his belief was *theistic*, not pantheistic, following Leibniz, rather than Spinoza. [...] Gödel was not satisfied with Spinoza's impersonal God."

Rather they, too, may represent an aspect of objective reality, but, as opposed to sensations, their presence in us may be due to another kind of relationship between ourselves and reality. ([20], in [22, p. 268])

The relationship Gödel is referring to here might be the one we would naturally expect to find in a monadological system: that between conceptual, mental, intellectual aspects of monads and their physical instantiation in reality. In this sense, we could say that the connection between "immanent" and "trans-subjective", posited by Cantor, could have also appealed to Gödel.

9   The Development of Mathematics

In one of her most widely known papers, [33], Maddy brings to the fore and stresses the importance of a fundamental aspect of Gödel's philosophical production, that is, his defence of "extrinsicness" as legitimate evidence in favour of the acceptance of the axioms. As we will see, also Cantor fostered the importance of some sort of "extrinsicness" in his *Grundlagen*. In his Cantor paper, Gödel had stated that:

> ...even disregarding the intrinsic necessity of some new axiom, and even in case it has no intrinsic necessity at all, a probable decision about its truth is possible also in another way, namely, inductively, by studying its "success". Success here means fruitfulness in consequences, in particular, in "verifiable" consequences, i.e., in consequences demonstrable without the new axiom [...]. There might exist axioms so abundant in their verifiable consequences, shedding so much light upon a whole field, and yielding such powerful methods for solving problems (and even solving them constructively, as far as that is possible) that, no matter whether or not they are intrinsically necessary, they would have to be accepted at least in the same sense as any well-established physical theory. ([20], in [22, p. 261])

In Maddy's view, this statement is proof that Gödel's "ontological" realism is mitigated by more pragmatic, "naturalistic" concerns, which would coalesce into the claim that a decision concerning the "acceptability" of an axiom should also be tied to considerations stemming from *intra-mathematical* practice. Overall, what comes out of Maddy's reconstruction is a different picture of Gödel's realism, as laying as much emphasis on "intra-mathematical practice" as on "intrinsic (conceptual) necessity" of new set-theoretic axioms. However, she says:

> I don't claim that this second picture of Gödel's views is completely accurate, any more than the first one was, but I do think it provides a useful perspective. Gödel's views are often presented in connection with the first of the two aforementioned attractions of realism: its faithfulness to mathematical experience. This alternative reading highlights the second attraction of realism - an account of the meaningfulness of the independent questions - but its working parts bypass realism altogether; they argue directly for the meaningfulness of those questions on purely mathematical grounds, not via philosophical realism. [33, p. 498]

Maddy has further elaborated upon her interpretation of Gödel's naturalism in [34]. If her interpretation is correct, then there must necessarily be some tension between these two sets of criteria for the acceptance of the axioms, "intrinsic" and "extrinsic". I will say something on this point later, but for the time being I will try to show that Gödelian naturalism seems to square well with some Cantorian remarks in the *Grundlagen* I am going to quote. While formulating the *generating principles* for the ordinals,[42] Cantor warns that:

> It is not necessary, I believe, to fear, as many do, that these principles present any danger to science. For in the first place the designated conditions, under which alone the freedom to form numbers can be practised, are of such a kind as to allow only the narrowest scope for discretion. Moreover, every mathematical concept carries within itself the necessary corrective: if it is fruitless or unsuited to its purpose, then that appears very soon through its uselessness, and it will be abandoned for lack of success. ([1], in [10, p. 896])

The first part of this quotation introduces the Cantorian theme of mathematicians' freedom, and argues that such freedom does not imply any form of *arbitrariness*. In a sense, freedom already contains some sort of *necessity* in it, insofar as one is only free to *follow* the "designated conditions" for the regulated *awakening* of a logically transparent concept. The second part introduces what Maddy would call "naturalistic" concerns, which seem to clash with the "intrinsic" ones. What Cantor says seems to imply that concepts, even those consistent with older ones, that is, *coherent* with a previously established conceptual network, might nonetheless be abandoned as *useless*, for lack of success.

As a consequence, there seems to be also some tension in Cantor between a realistic ontology and this kind of "pragmatism" about the introduction of new concepts. It should be noticed that this tension is not eased by the fact that "extrinsic evidence" only counts after one has fully carried out the process of ascertaining whether a concept has "intrinsic necessity". As a matter of fact, what Cantor says above does not seem to exclude that a concept is consistent, *sufficiently determinate* and distinguished from the other ones, in other terms that it complies with the "intrinsic" requirements, but that, at the same time, it is abandoned as *useless*. The importance of such considerations, however, should not be overestimated. We have to keep in mind that Cantor, in the *Grundlagen*, is also concerned with responding successfully to his critics, who had argued that using the actual infinite might turn out to run the risk of bringing inconsistencies into mathematical thought.

Cantor's counterargument has two parts: the first, by emphasising the existence of conceptual constraints within mathematicians' thought, has the effect of dismantling the objection that the introduction of "new numbers" is arbitrary. The second is directed at denying the risk that science might be harmed by novelties, by emphasising what, in his eyes, was natural and even obvious: mathematicians are able to detect the potential inconsistency or unsuitability of a concept and, in that case, they can always retreat on their steps. This second part does not need to be seen as a naturalistic doctrine, although it certainly contains an element of pragmatism.

The case of Gödel's naturalism is, maybe, slightly different, as his "pragmatism" seems more structured. He seems to take "inductive" evidence quite seriously,

---

[42]Generating principles are the three principles Cantor uses in the *Grundlagen* to "construct" the whole series of transfinite ordinals (see [1], in [10, pp. 907–909]), *viz.*, the *successor*, the *limit* and the *restriction* principle, whereby one can build, respectively, successor-ordinals ($\omega+1, \omega+2, \ldots$), limit-ordinals ($\omega, \omega + \omega, \ldots$) and initial ordinals ($\omega_0, \omega_1, \ldots$).

and the analogy between physics and mathematics is certainly striking. However, he sometimes uses the analogy between physics and mathematics in a different way, namely to prove that mathematical knowledge is dependent upon the internal elaboration of the "perceiving" subject in the same way as physical knowledge is (i.e., to claim that the "immediately given" of both does not merely consist of "data"). Which of the two analogies is, then, more faithful to his conceptions?

Elsewhere, Gödel seems to bring forward some sort of *indispensability argument*, particularly when he says that the assumption of the existence of mathematical objects can be compared to the assumption of the existence of bodies, both assumptions being *indispensable* for obtaining a *satisfactory* theory of mathematics and physics, respectively.

Cantor seems to have used the same argument when, in presenting the transfinite ordinals, he asserted:

> I am so dependent on this extension of the number concept that without it I should be unable to take the smallest step forward in the theory of sets [Mengen]; this circumstance is the justification (or, if need be, the apology) for the fact that I introduce seemingly exotic ideas into my work. ([1], in [10, p. 882])

To conclude, in both Gödel's and Cantor's conceptions, there is some emphasis on extrinsic evidence. However, the full import of their arguments considered globally, within their respective conceptions, is difficult to judge. What seems to be certain is that Gödel could, again, see Cantor's thought as an antecedent of his views concerning the importance of "extrinsicness".

10 Concluding Remarks

It is time to make some final considerations. I have been suggesting, in the introduction, that Gödel's conceptions are strikingly similar to Cantor's and that he may have been consciously or unconsciously influenced by them, and also that he may have used them in some particular circumstances. Now that a more detailed picture is available, I want to reconsider my claim more systematically and add some further remarks.

As said at the beginning, we do not know whether and to what extent Gödel knew Cantor's work, but the comparison of many passages from both authors' works has shown that there is a high number of textual overlapping, conceptual resemblances, even similarities in the use of linguistic expressions. Consequently, one could reasonably conjecture that Gödel drew upon Cantor's thought, or that, at least, he was fully aware that his philosophical positions were in line with Cantor's. However, I believe that the point may not be so crucial. It would be worth exploring such similarities regardless of whether they have been produced intentionally or not. Philosophically, these similarities involve, very frequently, the use of the same sources. Among them, in particular, Plato, Leibniz, Kant, Spinoza. Their philosophical choices reveal a substantial identity of tastes: objectivistic, metaphysical, systematic, theological, "right-wing" conceptions are preferred to subjectivist, positivist, "left-wing" ones.

Another crucial point of contact is that, although both are essentially concerned with mathematics, their philosophical projects seem, at times, more ambitious and, in fact, all-encompassing, spanning material and ideal objects, the finite and the infinite, numbers, sets, V, God. As a consequence of these similarities, in comparing Gödel's and Cantor's thought, one can sometimes get a more precise picture of what Gödel is aiming at. For instance, this is what happened when I struggled to understand Gödel's reference to a "different relationship between ourselves and reality". The knowledge of Cantor provided some grounds to assert that those words referred to a connection between *conceptual* and *non-conceptual* aspects of reality. Similarly, Gödel's remarks on the nature of the Absolute and his mention of the von Neumann axiom strike the right note in anyone acquainted with Cantor's work and, in turn, become more intelligible because of that.

Sometimes, Gödel's views are just a *transformation* (in a certain sense, a continuation) of Cantor's. Take the example of the notion of "intuition". Both are committed to believing that there exists a mathematical intuition that allows us to "perceive" mathematical objects/concepts. In Gödel's view, such an intuition should, in principle, also allow us to solve all mathematical problems *uniquely* and *determinately*.[43] Cantor was only initiating such conception, which perhaps had some influence on the young Husserl, who, in turn, provided Gödel with the theory Cantor was looking for. In this sense, Gödel was continuing investigations or developing intuitions fully in the wake of Cantor's work. Gödel's conceptions can even be seen, sometimes, as improvements on Cantor's.

Some other time, Gödel's thoughts make apparent what is concealed or follows from Cantor's conceptions. Take the idea of inexhaustibility of mathematics. In Gödel's conception, we never cease to produce ever new axioms, and this is, in a sense, a consequence of the existence of an inexhaustible infinite (Cantor's *absolute infinite*). Cantor used this concept to forestall the paradoxes, and as the logical basis of the distinction between classes and sets, whereas Gödel successfully connected it to its incompleteness theorems. Gödel also understood that the *indescribability* of V could be turned into a positive phenomenon, that is, into a justification of the reflection principle.

The work I am presenting here may be subject to further generalisation. For instance, a natural corollary would be to imagine a new brand of realism, Cantor-Gödel Platonism, based on the conceptual intersections between Gödel's and Cantor's thought I have described, and that I would like to resume briefly:

1. Mathematical entities (essentially, *sets*) exist both as *conceptual* and as *trans-subjective* objects. Their relationships and properties, both at the conceptual and at the trans-subjective level, exist independently of our mind.

---

[43] In the Cantor paper, Gödel's optimism is very robust, but elsewhere (the Gibbs lecture and some other remarks in Wang [48]), it is mitigated by the observation that the intuition of mathematical objects may be fallible and, what is more important, *incomplete*, thus leaving it open whether we are able to find solutions to all set-theoretic problems.

2. There is essentially one correct method to develop mathematics, which consists in using "intuition" to grasp concepts and their relations. Such intuition is strong enough to decide, in principle, all problems of set theory (incidentally, this is also the correct procedure to determine whether a concept is consistent).

3. Although there is only one correct method to develop mathematics, in certain circumstances it is useful for mathematicians to look at extrinsic criteria for deciding whether a concept (or an axiom) are legitimate.

4. Set theory and logic are inexhaustible, as the actual infinite is (given the existence of the absolute infinite). A parallel phenomenon is the *indescribability*, in a sense, of the universe of sets, which encourages the re-iteration of processes of formation of ordinals (and the postulation of axioms positing new ordinals).

5. God is absolutely infinite.

I believe that we are not very far from being able to delineate something like a Cantor-Gödel Platonism as a determinate and distinguished form of realism. Of course, one may legitimately ask how the identification of such a conception could benefit us. I may suggest one way in which this could happen. The issue of realism in the contemporary philosophy of mathematics is still thriving and widely debated. Gödel was instrumental in revitalising it, and his connection to Cantor should be paid the right attention and given proper emphasis, precisely because of our purpose to evaluate the tenability of realism.

However, if and how Cantor-Gödel Platonism might be thought of as a successful or adequate reconstruction of mathematical thought and of set theory as a whole or whether it could be used for present mathematical purposes is, of course, far from being clear.

Acknowledgements The writing of this article has been supported by the JTF Grant ID35216 (research project "The Hyperuniverse. Laboratory of the Infinite"). A preliminary version was presented and discussed at the conference *Kurt Gödel Philosopher: from Logic to Cosmology* held in Aix-en-Provence, July, 9–11, 2013. I wish to thank Richard Tieszen for insightful comments on my presentation and Mark van Atten for pointing me to bibliographical material which has proved of fundamental importance for the subsequent writing of this article. Gabriella Crocco and Eva-Maria Engelen read earlier drafts of this work, providing me with extremely helpful comments and suggestions. Finally, I owe a debt of gratitude to Mary Leng, whose advice and help in the final stage of revision have been invaluable.

References

1. G. Cantor, *Grundlagen einer allgemeinen Mannigfaltigkeitslehre. Ein mathematisch-philosophischer Versuch in der Lehre des Unendlichen* (B. G. Teubner, Leipzig, 1883)
2. G. Cantor, Über die verschiedenen Standpunkte auf das aktuelle Unendlichen. Zeitschrift für Philosophie und philosophische Kritik **88**, 224–233 (1885)
3. G. Cantor, Über verschiedene Theoreme aus der Theorie der Punktmengen in einem n-fach ausgedehnten stetigen Raume G_n (Zweite Mitteilung). Acta Math. **7**, 105–124 (1885)
4. G. Cantor, Mitteilungen zur Lehre vom Transfiniten, I-II Zeitschrift für Philosophie und philosophische Kritik, I **91**, 81–125 (1887); II **92**, 240–265 (1888)

5. G. Cantor, *Gesammelte Abhandlungen mathematischen und philosophischen Inhalts* (Springer, Berlin, 1932)
6. P.J. Cohen, The independence of the continuum hypothesis. Proc. Natl. Acad. Sci. **5**, 105–110 (1964)
7. G. Crocco, Gödel on concepts. Hist. Philos. Log. **27**(2), 171–191 (2006)
8. W. Dauben, *Georg Cantor. His Mathematics and Philosophy of the Infinite* (Harvard University Press, Harvard, 1979)
9. J.W. Dawson, *Logical Dilemmas: The Life and Work of Kurt Gödel* (A K Peters, Wellesley, MA, 1997)
10. W. Ewald (ed.), *From Kant to Hilbert: A Source Book in the Foundations of Mathematics, Volume II* (Oxford University Press, Oxford, 1996)
11. J. Ferreirós, The Motives behind Cantor's set theory. Sci. Context **12/1–2**, 49–83 (2004)
12. G. Frege, Der Gedanke. Beiträge zur Philosophie der deutschen Idealismus **1**, 58–77 (1918)
13. G. Frege, *The Foundations of Arithmetic* (Pearson-Longman, New York, 2007)
14. K. Gödel, The present situation in the foundations of mathematics, in *Collected Works*, vol. III, ed. by K. Gödel (Oxford University Press, Oxford, 1995), pp. 45–53
15. K. Gödel, Russell's mathematical logic, in *The Philosophy of Bertrand Russell*, ed. by P.A. Schilpp (Open Court Publishing, Chicago, 1944), pp. 125–153
16. K. Gödel, What is Cantor's continuum problem? Am. Math. Mon. **54**, 515–525 (1947)
17. K. Gödel, A remark on the relationships between relativity theory and idealistic philosophy, in *Albert Einstein: Philosopher-Scientist*, ed. by P.A. Schilpp (Open Court Publishing, La Salle, IL, 1949), pp. 555–562
18. K. Gödel, Some basic theorems on the foundations of mathematics and their implications, in *Collected Works*, vol. III, ed. by K. Gödel (Oxford University Press, Oxford, 1995), pp. 304–323
19. K. Gödel, The modern development of the foundations of mathematics, in *Collected Works*, vol. III, ed. by K. Gödel (Oxford University Press, Oxford, 1995), pp. 375–387
20. K. Gödel, What is Cantor's continuum problem?, in *Philosophy of Mathematics. Selected Readings*, ed. by H. Putnam, P. Benacerraf (Prentice-Hall, New York, 1964), pp. 470–485
21. K. Gödel, *Collected Works, I: Publications 1929–1936* (Oxford University Press, Oxford, 1986)
22. K. Gödel, *Collected Works, II: Publications 1938–1974* (Oxford University Press, Oxford, 1990)
23. K. Gödel, *Collected Works, III: Unpublished Essays and Lectures* (Oxford University Press, Oxford, 1995)
24. M. Hallett, *Cantorian Set Theory and Limitation of Size* (Clarendon Press, Oxford, 1984)
25. K. Hauser, Gödel's Program revisited. Part I: the turn to phenomenology. Bull. Symb. Log. **12/4**, 529–590 (2006)
26. E. Husserl, *Ideas Pertaining to a Pure Phenomenology and to a Phenomenological Philosophy. Book 3: Phenomenology and the Foundation of Sciences* (Kluwer, Dordrecht, 1980)
27. E. Husserl, *Ideas Pertaining to a Pure Phenomenology and to a Phenomenological Philosophy. Book 1: General Introduction to a Pure Phenomenology* (Nijhoff, The Hague, 1982)
28. E. Husserl, *Cartesian Meditations* (Kluwer, Dordrecht, 1988)
29. E. Husserl, *Ideas Pertaining to a Pure Phenomenology and to a Phenomenological Philosophy. Book 2: Studies in the Phenomenology of Constitution* (Kluwer, Dordrecht, 1989)
30. I. Jané, The role of the absolute infinite in Cantor's conception of set. Erkenntnis **42**, 375–402 (1995)
31. A. Kanamori, *The Higher Infinite* (Springer, Berlin, 2003)
32. A. Lévy, R. Solovay, Measurable cardinals and the continuum hypothesis. Isr. J. Math. **5**, 234–248 (1967)
33. P. Maddy, Set-theoretic naturalism. Bull. Symb. Log. **61/2**, 490–514 (1996)
34. P. Maddy, *Naturalism in Mathematics* (Oxford University Press, Oxford, 1997)
35. D.A. Martin, Gödel's conceptual realism. Bull. Symb. Log. **11/2**, 207–224 (2005)
36. C. Ortiz Hill, Did Georg Cantor influence Edmund Husserl? Synthèse **113**, 145–170 (1997)

37. C. Ortiz Hill, Abstraction and idealization in Edmund Husserl and Georg Cantor, *Poznan Studies in the Philosophy of the Sciences and in the Humanities*, vol. 82/1 (2004), pp. 217–244

38. C. Parsons, Platonism and mathematical intuition in Kurt Gödel's thought. Bull. Symb. Log. **1/1**, 44–74 (1995)

39. W. Purkert, H.J. Ilgauds, *Georg Cantor* (Teubner, Leipzig, 1985)

40. T. Ricketts, M. Potter (eds.), *The Cambridge Companion to Frege* (Cambridge University Press, Cambridge, 2010)

41. R. Tieszen, *Phenomenology, Logic and the Philosophy of Mathematics* (Cambridge University Press, Cambridge, 2005)

42. R. Tieszen, *After Gödel. Platonism and Rationalism in Mathematics* (Oxford University Press, Oxford, 2011)

43. R. Tieszen, Monads and mathematics: Gödel and Husserl. Axiomathes **22**, 31–52 (2012)

44. M. Van Atten, Monads and sets. On Gödel, Leibniz, and the reflection principle, in *Judgment and Knowledge. Papers in Honour of G.B. Sundholm* (King's College Publications, London, 2009), pp. 3–33

45. M. Van Atten, J. Kennedy, On the philosophical development of Kurt Gödel. Bull. Symb. Log. **9/4**, 425–476 (2003)

46. H. Wang, *From Mathematics to Philosophy* (Routledge & Kegan Paul, London, 1974)

47. H. Wang, *Reflections on Kurt Gödel* (MIT Press, Cambridge, MA, 1987)

48. H. Wang, *A Logical Journey* (MIT Press, Cambridge, MA, 1996)

Remarks on Buzaglo's Concept Expansion and Cantor's Transfinite

Claudio Ternullo

Abstract Historically, mathematics has often dealt with the 'expansion' of previously accepted concepts and notions. In recent years, Buzaglo (The logic of concept expansion, 2002) has provided a formalisation of concept expansion based on *forcing*. In this paper, I briefly review Buzaglo's logic of concept expansion and I apply it to Cantor's 'creation' of the transfinite. I argue that, while Buzaglo's epistemological considerations fit well into Cantor's conceptions, Buzaglo's logic of concept expansion might be unsuitable to justify the creation of the transfinite in terms of a logically rigorous derivation of concepts.

1 Preliminaries

Concept expansion is a widespread phenomenon in mathematics. Historically, there is plenty of examples that show how mathematical functions were 'modified' or new mathematical entities were 'created' as a result of the expansion of previously established concepts. Consider, for instance, the simple *exponential function*, a^x. In order to make sense of the previously meaningless expression a^0, mathematicians had to somehow 'stretch' the definition of *exponential*, and, in general, if they had not 'forced' certain other mathematical concepts or families of numbers beyond their definitional limits, now mathematics would not deal with i, ω, $log(-1)$, π, e, the concept of *irrational* number.

Now, assuming concept expansion is an entirely rational process, is there any *logic* governing it? If yes, what would one gain by describing such a logic?

A possible answer to the latter question would be that a logic of concept expansion would show us how to derive concepts from concepts in a *non-arbitrary* way. It is also imaginable that, in some special cases, such a logic would allow

Originally published in C. Ternullo, *Remarks on Buzaglo's Concept Expansion and Cantor's Transfinite*, preprint.

C. Ternullo (✉)
Kurt Gödel Research Center for Mathematical Logic, University of Vienna, Vienna, Austria
e-mail: ternulc7@univie.ac.at

© Springer International Publishing AG 2018
C. Antos et al. (eds.), *The Hyperuniverse Project and Maximality*,
https://doi.org/10.1007/978-3-319-62935-3_12

us to demonstrate that *only* certain concepts can be derived from other concepts. Such a logic, therefore, might embody the ideal of what Leibniz called *calculus ratiocinator*, essentially, a method to formally derive concepts from other concepts.[1]

Another fundamental purpose of such a logic would be descriptive, in the sense that it would help us understand how mathematicians concretely do mathematics, giving up any pretence to prescribe a general method of expanding concepts. In particular, studying the development of concepts would allow us to make sense of the use of, e.g., 'metaphors' or 'analogies' in mathematical contexts, an area of study historically neglected by professional philosophers (perhaps with the notable exception of Wittgenstein).

Buzaglo, in his [1], pursues both these goals. While he is interested in giving an account of concept expansion to describe the way concepts develop, he also thinks that his logic of concept expansion may work as a formal template. This is how he characterises his project:

> I claim that there can be a logic that includes non-arbitrary expansions, and that there are convincing reasons to believe that a certain type of expansion expresses human rationality. Therefore, instead of allowing some principles to place this phenomenon outside logic, the principles must be changed so as to include this process. [1, p. 3]

As far as the former goal is concerned, in short, Buzaglo describes a type of thoughts that he calls *inchoate*. Sentences containing inchoate thoughts are neither true nor false. For instance, if I do not know what -2 means, then the sentence $-2 < 0$ is not true, it is *indeterminate*. Concepts which 'evolve' go through inchoateness, in the sense that they become inchoate thoughts which have not yet reached the status of concepts. -2, in the example above, can initially be thought of only as a meaningless combination of two concepts, '$-$' and '2'. It is only through some logical concept forcing that we become convinced that -2 instantiates a 'new' concept, that of negative integer, and then proceed to describe the laws governing its use. Only then we become aware of the fact that the whole $\mathbb{Z}$ can be derived from the whole $\mathbb{N}$ through a non-arbitrary expansion of the latter. Now, what has been crucial for the process to be initiated is the perceived 'analogy' between $\mathbb{N}$ and $\mathbb{Z}$, which must have been, subsequently, made sense of through purely logical means (concept forcing).

The second goal is partly spelt out by concept forcing, and will be described in the next section.

---

[1]In more recent times, this ideal seems to have been resurrected by none other than Gödel. Concerning the nature of *logic*, Gödel says: 'Logic is the theory of the formal. It consists of set theory and the theory of concepts [...] Set is a formal concept. If we replace the concept of set by the concept of concept, we get logic. The concept of concept is certainly formal and, therefore, a logical concept.' Wang comments thus: 'It is clear that Gödel saw concept theory as the central part of logic and set theory as a part of logic. It is unclear whether he saw set theory as belonging to logic only because it is, as he believed, part of concept theory, which is yet to be developed.' Wang [9, p. 247].

2 Concept Forcing

2.1 Preliminaries

The crucial point in defining concept forcing is to assimilate concepts to *unsaturated functions*. This move had already been carried out by Frege (see, in particular, [6]), but Buzaglo also significantly departs from Frege's theory, insofar as he admits of functions which are *undefined* at certain *points*, that is, at certain elements of their domain.[2]

For instance, take, again, an exponential function, such as 2^x, where $x \in \mathbb{N}$. There are certain values of $\mathbb{N}$, where 2^x is, *prima facie*, undefined: in particular, it is not clear what 2^0 means. Now, the way mathematicians solve this sort of problems is, in Buzaglo's view, through *forcing* concepts, that is, through identifying ways to force functions to yield a value on points where they were previously undefined. $2^0 = 1$ would, thus, be the result of a concept forcing over 2^x. If concept forcing goes through, then one can justify the 'expansion' of the concept 2^x defined at $\mathbb{N} - \{0\}$ to the concept 2^x defined at all $\mathbb{N}$, and say that the concept 2^x has been 'expanded'. But there are, also, 'external' expansions, that is, expansions which yield 'new' objects. For instance, the expansion of $\mathbb{Z}$ from $\mathbb{N}$ is an external expansion.

Now, Buzaglo's aim is to set up a *general* logic of concept forcing, that is, a formal theory of concept expansions, both internal and external.

The idea of concept forcing is clearly indebted to the method of forcing introduced by Cohen in the '60s. Forcing is a very sophisticated technique, whereby one may produce models of (fragments of) the axioms of set theory (e.g., ZF or ZFC) with certain specified properties. Very roughly, Cohen's construction yields an extension of a ground model M to a model $M[G]$, where G is a *generic filter* over M. The 'expansion', as it were, from M to $M[G]$, through accurate choice of G, *forces* $M[G]$ to have or contradict a certain property.

For instance, given a countable transitive model M and a partial order $\mathbb{P}$ in M, Cohen defined a $\mathbb{P}$-generic real r which is provably not in M. By adding transfinitely many *generic* r's to M, one gets, for instance, that $M[G] \models \mathfrak{c} \neq \aleph_1$, thus contradicting CH.

The crux of forcing is that, given what are called *forcing conditions*, certain facts necessarily follow. For instance, in Cohen's forcing, forcing conditions are $p \in \mathbb{P}$, where $\mathbb{P}$ is a partial order in M. In the case above, $p \in \mathbb{P}$ may be thought of as finite sequences of 0 and 1, and the appropriate choice of $p \in \mathbb{P}$ forces $M[G]$ to have or

---

[2]However, Buzaglo contends that his theory of *partially defined* functions does not counter Frege's point of view, but rather expands on it (for discussion of this, see [1], pp. 24–30 and 59–63). Frege opposed concept expansion, as he thought that concepts were rigid constructs, instantiated by a fixed domain of objects. Buzaglo's theory does not deny this, while, at the same time, conjecturing that concepts *qua* functions may have undefined values, which can be subsequently somehow 'filled out' to *produce* new concepts.

contradict a certain sentence ϕ, which means: given a suitable choice of $p \in \mathbb{P}$, then $M[G] \models \phi$. We say, then, that p forces ϕ ($p \Vdash \phi$).

In a similar vein, Buzaglo's formalisation of the expansion of concepts aims at forcing the values of functions at certain points at which those functions were previously undefined. Forcing conditions are given by accurately chosen properties ('laws') holding in a theory T, such as formal arithmetic. For instance, let us resume the example of the exponential 2^x. How do mathematicians get $2^0 = 1$? Because, as is known, $\frac{2^x}{2^y} = 1$, when $x = y$. But we also know that

$$\frac{2^x}{2^y} = 2^{x-y} \tag{1}$$

Now, 2^{x-y} is undefined, when $x = y$, but, thanks to (1), we know that it must be equal to 1. Therefore, we can say that: $\frac{2^x}{2^y} = 2^{x-y} \Vdash 2^0 = 1$.

2.2 Formalisation of Concept Expansion

Following Buzaglo [1], pp. 40–56, I shall now give a proper definition of concept forcing. Buzaglo's definition is based on the standard Tarskian definition of satisfaction in first-order logic for a theory T in a language $\mathcal{L}$, with a notable difference: he introduces one more value beyond 'true' and 'false', which is the 'undefined' ($\underline{X}$).

For instance, before concept forcing, the function 2^x is defined as follows:

$$2^x = \begin{cases} \underbrace{2 \cdot 2 \cdot \ldots}_{x \text{ times}} & \text{iff } x > 0 \\ \underline{X} & \text{iff } x = 0 \end{cases}$$

but, after forcing, we have that:

$$2^x = \begin{cases} \underbrace{2 \cdot 2 \cdot \ldots}_{x \text{ times}} & \text{iff } x > 0 \\ 1 & \text{iff } x = 0 \end{cases}$$

In order to formalise processes of expansion, Buzaglo uses the notion of *embedding*. An embedding of two structures $j : \mathcal{M} \to \mathcal{N}$ is the isomorphism between $\mathcal{M}$ and a submodel $\mathcal{M}' \subset \mathcal{N}$. This may be sufficient, as far as external expansions are concerned, in the sense that one can see the expansion of a domain of objects as the embedding of the original domain into a larger domain. But this is not the case when the expansion is merely *internal*, that is, when it does not give rise to new objects, but only forces a function to yield a certain value at a point where it was previously undefined (as happens with 2^x above). The key definition, here, is, instead, that of:

Definition 1 (Forced Internal Expansion) Given two models M and N of a first-order theory T in a language $\mathcal{L}$, and a set of forcing conditions $S \in T$, such that $M \models S$, we say that N is an internal expansion of M (in symbols, $N \gg M$) if and only if:

(1) $N \models S$.
(2) $dom(M) = dom(N)$.
(3) $(\forall f) f \in M \leftrightarrow f \in N$.
(4) $(\exists x, a \in M, N) \, M \models f(x) = \underline{X} \wedge S \Vdash N \models f(x) = a$.
(5) Given all K such that $K \gg M$ as above, then K must agree with N on the value of the functions they share with N.

Now, a forced internal expansion is particularly interesting when it is *invariant* (*absolute*) in a collection of models. This invites the following strengthening of the internal forced expansion:

Definition 2 (Absolutely Forced Internal Expansion) Given a model M of a first-order theory T in a language $\mathcal{L}$, we say that S absolutely forces $N \gg M$, if the expansion occurs in all $N \gg M$, including all N which are external to M (that is, which contain elements which are not in M).

Absolute forcing is particularly relevant for Buzaglo's purposes, insofar as he thinks that it may be compared to a form of *deduction*: given a choice of forcing conditions S, and any model M which satisfies S, if $S \Vdash \phi$ *absolutely*, then ϕ is valid in all extensions of M and, thus, the choice of M is, in a sense, irrelevant. Therefore, using 'absolute concept forcing', we may be able to assert that ϕ is deducible from S. Absolute concept forcing is thus the closest Buzaglo gets to a 'concept calculus', insofar, through it, one may substantiate the idea of inevitability which is inherent in concept derivation.

3 Cantor on Concept Expansion

Cantor's early works in set theory contain several references to 'construction', 'generation', 'expansion' of concepts and objects.[3]

It is, *prima facie*, hard to make sense of such expressions, particularly as they clash with Cantor's subsequent tendency to attribute to the transfinite a rather firm *mind-* and *construction*-independent status. I argue that a way to understand this is through conjecturing that, at the time of the introduction of the transfinite machinery,

---

[3]Cf. Hallett, in [7], p. xi: '[…] mixed in with Cantor's prevailing realism are splashes of what could well be called constructivism, and this applies particularly to two crucial elements of his theory, the notion of well-ordering and the set concept itself.'

Cantor was thinking of something along the lines of a 'regulated concept expansion' similar to that sketched by Buzaglo.

Let us proceed in an orderly fashion. First, let us see what Cantor says concerning concepts and their extensions. The first and foremost passage to be mentioned may be found in his most fundamental work on the transfinite, that is, the *Grundlagen einer allgemeinen Mannigfaltifgkeitslehre* [2]. There, infinite numbers are presented as a *necessary expansion* of the notion of natural number:

> The foregoing account of my investigations in the theory of manifolds has reached a point where further progress depends on *extending* [my italics] the concept of real integer beyond the previous boundaries; this extension lies in a direction, which, so far as I know, nobody has yet attempted to explore. ([2], in Ewald [5], p. 883)

Later in the text, he adds:

> I am so dependent on this *extension* [my italics] of the number concept that without it I should be unable to take the smallest step forward in the theory of sets [Mengen]; this circumstance is the justification (or, if need be, the apology) for the fact that I introduce seemingly exotic ideas into my work. For what is at stake is the *extension* or *continuation* [my italics] of the sequence of integers into the infinite; and daring though this step may seem, I can nevertheless express, not only the hope, but the firm conviction that with time this extension will have to be regarded as thoroughly simple, proper, and natural. (*ibid.*, p. 883)

Although later in the same work concepts are referred to as 'ideas' in a robustly platonistic sense, throughout the *Grundlagen*, Cantor's justificatory strategy is very pragmatic, and, at times, he even advocates a line of defence which would now be viewed as 'naturalistic'[4]: it might be that concepts are objective constructs, but this does not mean that they should be accepted without testing their fruitfulness. Therefore, Cantor can conclude that:

> ...every mathematical concept carries within itself the necessary corrective: if it is fruitless or unsuited to the purpose, then that appears very soon through its *usefulness* [my italics], and it will be abandoned for lack of success. [2, p. 896]

However, the introduction of new concepts, it is explained in other passages, is an essentially regulated process, which might, therefore, be described in a fully logical way. How this may happen is described more accurately in the following passage, where Cantor says that:

> In particular, in the introduction of the new numbers it is only obligated to give definitions of them which will bestow such a determinacy and, in certain circumstances, such a relationship to the older numbers that they can, in any given instance, be precisely distinguished. As soon as a number satisfies all these conditions it can and must be regarded in mathematics as existent and real. [2, p. 896]

Let us pause a moment to comment on the passage. There are two main criteria, which Cantor is specifically mentioning here. One is 'determinacy': new concepts should be introduced in such a way as to guarantee their distinguishability

---

[4]For this, see my [8], pp. 440–443.

from the older concepts. The second feature is 'inter-dependence': new concepts should somehow be related to existing concepts. As far as the latter is concerned, Cantor devised a procedure which comes in stages. There is one revealing passage concerning this:

> The procedure in the correct formation of concepts is in my opinion everywhere the same. One posits a thing with properties that at the outset is nothing other than a name or a *sign* [my italics] A, and then in an orderly fashion gives it different, or even infinitely many, intelligible predicates whose meaning is known on the basis of ideas that are already at hand, and which may not contradict one another. In this way one determines the connection of A to the concepts that are already at hand, in particular to related concepts. [2, p. 918–9]

Clearly this is a procedure which defines a 'regulated concept expansion'. The procedure has three stages, which may be summarised as follows:

(1) Create a new sign (A).
(2) Assign properties to A (i.e., 'conceptualise' the sign), through concept forcing, that is, through 'ideas already at hand'.
(3) Explain, in full, how the new concept works.

An immediate case study for the procedure sketched above is the creation of the least transfinite ordinal, ω and of the whole series of ordinals. In the early 1880s, Cantor had found a general way to study point sets P by defining their derivative, P', the collection of the accumulation points of P. Now, for some P's, it may happen that, for all $n \in \mathbb{N}$, $P^n \neq P$. Cantor was, then, able to introduce the notion of transfinite derivative:

$$P^\infty = \bigcap_n^\infty P^n$$

but then he saw that the process could be continued by defining $P^{\infty+1} = (P^\infty)'$. The symbol ∞ was soon to be replaced by ω, and with that, the whole sequence of transfinite ordinals $\omega, \omega + 1, \omega + 2, \ldots, \omega_\omega, \ldots$ was born. Each of them was a new sign, at the beginning, but then Cantor moved on to step 2: he conceptualised ω as an *order-type* of a *well-ordered* set. What prompted him to do that? Possibly, the realisation that all $n < \omega$ are well-ordered, and that their well-ordering may be very naturally extended also to all ordinals *past* all finite ordinals.

Once the three steps in the process above have been carried out, and provided one has checked that the new concept is sufficiently determinate, then one may declare that the concept should be added to our network of concepts.

Thus far, Cantor's creation (and justification) of the transfinite seems to be in line with Buzaglo's characterisation of expansions and, in particular, the three-stage procedure described above will strike anyone as being very similar to Buzaglo's. Furthermore, very often, Cantor himself points to revealing analogies. A very well-known and oft-quoted analogy is that between irrationals and transfinite ordinals. He says:

> The transfinite numbers are in a certain sense *new irrationalities*, and in my view the best method of defining the *finite* irrational numbers is quite similar to, and I might even say in principle the same as, my method of introducing transfinite numbers. One can say unconditionally: the transfinite numbers stand or fall with the finite irrational numbers: they are alike in their innermost nature, since both kinds are definitely delimited forms or modifications of the actual infinite. [3, pp. 395–96]

The passage is also interesting for another reason, that is, because the identified structural analogy between $\mathbb{R}$ and *Ord* presumably also played a major role in prompting Cantor's conviction that $\mathfrak{c} = \aleph_1$.[5]

But, if we are to follow Buzaglo's theory to the end, just what properties of previously established concepts did Cantor take as forcing conditions for his new concepts? We shall explore this topic in the next section.

4   The Creation of the Transfinite

Cantor's theory of sets grew out of mathematical work arising in analysis and topology. It is hardly surprising, therefore, that the expansion of concepts already used in those areas played a major role in his shaping of the theory of the transfinite.

However, while the metaphors and analogies used by Cantor at a pre-theoretic level certainly helped him develop the theory of the transfinite to a sufficient extent, the way he forced concepts does not necessarily fit into Buzaglo's machinery. In particular, it is not clear that Cantor took some set S of mathematical laws as the starting point to force his new concepts to take specific properties.

To check whether this is the case, we may, again, turn our attention to Cantor's works. In the mentioned *Grundlagen*, Cantor goes on to explain that his 'new' numbers are generated using three principles:

(1) The *successor principle*. Given α, generate $\alpha + 1$.
(2) The *least upper bound principle*: Given an infinite increasing sequence of numbers $\{a_n\}$, generate $sup\{a_n\}$.
(3) The *'limitation' principle*: Given a set of ordinals α, generate $\beta \notin \alpha$ which has a cardinality greater than that of any α.[6]

It is easy to show that the principles above are sufficient to produce all ordinals. Using (1), one can generate: $\omega + 1, \omega + 2, \ldots$. Using (2), one can generate $\omega, \omega + \omega, \ldots, \omega^2, \omega^\omega, \ldots$. Using (3), one generates: $\omega_1, \omega_2, \ldots, \omega_\omega, \ldots$. These last numbers (initial ordinals) are the $\aleph$s.

---

[5]This argument is briefly introduced and assessed by Hallett in [7], pp. 74–81.

[6]Cantor did not use specific names for the first and the second principle, whereas the third principle is explicitly called *Hemmungsprinzip* (limitation principle) or *Beschränkungsprinzip* (restriction principle). See Cantor [2], in Ewald [5], p. 883.

Now, how did Cantor motivate the principles above? Very trivially, all of these occur in mathematics before being used in set theory. The successor principle is the standard successor of arithmetic: $s(n) = n + 1$. The least upper bound principle is less trivial. The principle is moulded on Cantor's very definition of irrationals as infinite sequences of rationals. Always in [2], Cantor introduces his theory of irrationals, through the use of *fundamental sequences*. Fundamental sequences are converging sequences of rationals $\{a_n\}_{n \in \mathbb{N}}$ such that, given any two consecutive terms a_k and a_{k+1}, of the sequence it happens that, for all $M > 0$, and some $k > 0$, $|a_{k+1} - a_k| < M$. We have that the irrational associated to the sequence, say, b, is

$$\lim_{n \to \infty} a_n = b$$

Analogously, one can define:

$$\lim_{n \to \infty} n = \omega$$

and this procedure extends to all other limit ordinals. The third principle is entirely new, as it cannot be derived from any other previously defined domain of numbers. Ordinal and cardinal notions in the natural numbers are equivalent. But certainly there are ways to make sense of the principle also in the finite. For instance, if one defines numbers using von Neumann ordinals, then the ordinal $\overline{4} = \{\overline{0}, \overline{1}, \overline{2}, \overline{3}\}$. Moreover, one could say that $\overline{4}$ has cardinality 4, and, as is clear, $4 \notin \{0, 1, 2, 3\}$.

These are all trivial facts, which could go entirely unnoticed. However, if we were to use Buzaglo's methodologies and conceptualisation, then we could say that what Cantor was aiming at in defining the generation principles were forced expansions of basic functions arising in the finite. Of course, Cantor's expansions are also 'external' expansions, insofar as new entities are created along with new functions.

For instance, one could say that the successor function for the ordinals, let us call it s_Ω, is *forced* by the standard successor function $s_\mathbb{N}$ for the natural numbers. But the forcing conditions are not only represented by the behaviour of $s_\mathbb{N}$, but, presumably, also by the other axioms of arithmetic, say PA (Peano Arithmetic). Conceptually, then, one could say that $PA \Vdash s_\Omega$, that is, the axioms of arithmetic force the successor function $s_\Omega(\alpha)$ to take the value $\alpha + 1$.

Analogously, the least upper bound principle is something Cantor might have derived from the axioms of the reals (AR). In particular, the *completeness axiom* guarantees that there exist a $sup\{a_n\} = b$ for converging sequences of rationals a_n. Limit ordinals may, therefore, be conceived of analogously, as least upper bounds of sequences of ordinals $\{\alpha_n\}$. Accordingly, one could say that AR force the least upper bound principle in the infinite, that is, that $AR \Vdash \lim_{n \to \infty} n$ to take the value ω.

Now, are all of these new concepts *absolutely* forced, according to Buzaglo's definitions? Recall that absolute forcing implies that the internal expansion holds in all models N which extend M in the way indicated by the relevant definitions.

However, strictly speaking, here we do not have internal expansions of concepts: the functions holding in the finite which forced the values of the analogous functions holding in the infinite are not undefined at points at which their transfinite analogues are defined, simply because those points were not in the domain of those functions! Cantor's expansion is, therefore, quite remarkable: it is an external expansion which also carries with it an entirely new set of 'expanded' functions.

Moreover, it should be noted that there are some properties holding in the infinite which are not forced by any mentioned set of axioms. For instance, while in the finite $n+1 = 1+n$, in the infinite $\alpha+1 \neq 1+\alpha$, where α is a transfinite ordinal. Therefore, it might be true that the successor function is somehow forced by PA, but the same does not apply to other *functions* (such as '+') which take, as arguments, transfinite ordinals. This shows that concept forcing, in the way described by Buzaglo, may not be able to cover all aspects (some of which of great consequence) of the *expansion* of concepts.

One further issue is represented by the fact that essentially 'non-Cantorian' expansions of concepts and properties arising in arithmetic and analysis may be equally well accounted for using concept forcing. For instance, there may be alternative definitions of 'infinite cardinality' which contradict Cantor's definition. Cantor had set forth the well-known principle:

Principle 1 (Cantor's Cardinality Principle) Given two sets A and B, we say that A has the same cardinality as B if and only if there is a bijection $f : A \to B$.

The principle guarantees that there are different levels of infinity and, to begin with, two, that of $\mathbb{N}$, the countable, and that of $\mathbb{R}$, the uncountable. Now, Buzaglo himself shows in his book that one may introduce a different definition of cardinality:

Principle 2 (Buzaglo's Cardinality Principle) Given two sets A and B, $|A| = |B|$ if and only if $A \not< B$ and $B \not< A$.
where $A \not< B$ is defined in the following way:

Definition 3 $A \not< B$ if and only if, whenever there is a function $f : A \to B$ such that, for some $b \in B$ and for all $a \in A$, $b \neq f(a)$, we can always find a function g which is greater than f, that is, such that $ran(f) \subset ran(g)$.

Now, it is easy to verify that, by Buzaglo's definition of cardinality, all infinite sets have the *same* cardinality, a result which clearly runs counter to the whole of Cantor's theory of the transfinite.

As one further example, it is, at least, imaginable to force such expressions as $1/\omega$ from AR. In fact, as is known, $\mathbb{R}$ has a non-standard extension, $\mathbb{R}^*$, which contains numbers (*infinitesimals*) which may be indicated by $1/\omega$. Infinitesimals were fiercely rejected by Cantor, as he was adamantly against the violation of the Axiom of Archimedes.[7] But non-Archimedeanness is as easily forceable from AR as is Archimedeanness! Cantor himself seemed to acknowledge that the main problem

---

[7]In fact, he was fiercely against a conception of 'linear numbers' from which a violation of the Axiom of Archimedes could be inferred. For this, cf. the thorough discussion in Dauben [4],

with non-Archimedean *quantities* was not that they could not be consistently conceived of and defined, but rather that they could not be viewed as 'real' objects, that is, that they did not have a sufficiently high degree of ontological *sharpness*.[8]

But, whatever the matter be, it is clear that it is hard to view concept forcing, in all such cases, as a sort of 'concept calculus'. The only thing that Buzaglo's concept expansion is able to tell us in all such cases is that alternative expansions are all equally justified on the grounds of the acceptance or rejection of certain (collections of) previously accepted axioms, such as Cardinality Principles or the Axiom of Archimedes.[9] But this seems to me tantamount to saying that from those principles follow certain consequences, something which does not enrich significantly our knowledge of how concepts may be derived in a logically rigorous fashion.

5   Concluding Remarks

Buzaglo's logic of concept expansion is epistemologically significant, as there is no doubt that concept expansion captures one of the most important aspects of mathematical creativity, that based on 'stretching' concepts as far as possible, in order to derive further concepts.

In the case of the birth of Cantor's transfinite, concept expansion was robustly at work in many ways, and it was Cantor himself who acknowledged this in his works. In particular, he thought that hadn't he taken expansion of concepts seriously, he could not have continued his investigations in the areas of *analysis* and *topology*. It happened, then, that while establishing mathematical results in those areas, he discovered an entirely new area of research, *set theory* (in fact, only the theory of the transfinite). Cantor's expansion of concepts was accompanied by lots of 'metaphorical' and 'analogical' remarks, which clearly fit into Buzaglo's observations. However, as we have seen, when one applies Buzaglo's concept forcing to the way Cantor created the transfinite, what we may infer at most, are very trivial expansions of concepts used in analysis and arithmetic.

Moreover, if we take Buzaglo's attempts to be equivalent to something like a 'concept calculus', then we might just be going the wrong way. Very incompatible theories of sets may all be accommodated by some form of concept forcing. But

---

pp. 33–36, of Cantor's correspondence with Veronese, Vivanti and Peano concerning the concept of *infinitesimal*.

[8]For instance, he says: 'But all attempts to force this infinitely small into a proper infinite must finally be given up as pointless. If *proper* infinitely-small quantities exist at all, that is, are definable, then they certainly stand in no direct relationship to the familiar quantities which *become* infinitely small.' ([2], in Ewald [5], p. 888).

[9]However, Buzaglo introduces a strengthening of the 'forced internal expansion', that is, a 'strongly forced internal expansion', whereby, given *any* set of forcing conditions $S \in T$, a specific expansion should take place in a *unique* way. It is clear, however, that none of the cases examined meets this notion.

even if Buzaglo's methodology was really be describable as something akin to a concept calculus, it would not provide us with much more information than standard mathematical proofs do.

One further interesting question is whether an account of concept expansion may also include an account of *maximal expansions*. Set theory, in particular, seems to be concerned, in many ways, with maximality. For instance: is there a maximal expansion of L which could be said to be 'conceptually forced' by the other set-theoretic, or even non-set-theoretic, axioms or principles? Are maximality principles such as S. Friedman's IMH similarly conceptually forced? Is a maximal expansion of a non-trivial elementary embedding $j : V \to M$, with $V = M$, consistent with ZF, conceptually forced?

All of these issues are momentous for contemporary set theory, and one could only hope that Buzaglo's logic of concept expansion would provide a way to address them. However, given the perplexities expounded above, it is hard to see how.

Acknowledgements The writing of this paper has been supported by the John Templeton Foundation grant ID35216 "The Hyperuniverse: Laboratory of the Infinite".

Reference

1. M. Buzaglo, *The Logic of Concept Expansion* (Cambridge University Press, Cambridge, 2002)
2. G. Cantor, *Grundlagen einer allgemeinen Mannigfaltigkeitslehre. Ein mathematisch-philosophischer Versuch in der Lehre des Unendlichen* (B. G. Teubner, Lepzig, 1883)
3. G. Cantor, Mitteilungen zur Lehre vom Transfiniten, I-II. *Zeitschrift für Philosophie und Philosophische Kritik* **91**, 81–125; **92**, 240–65, I (1887); II (1888)
4. W. Dauben, *Georg Cantor. His Mathematics and Philosophy of the Infinite* (Harvard University Press, Harvard, 1979)
5. W. Ewald (ed.), *From Kant to Hilbert: A Source Book in the Foundations of Mathematics*, vol. II (Oxford University Press, Oxford, 1996)
6. G. Frege, *Grundgesetze der Arithmetik, Begriffsschriftlich abgeleitet*, vol. I-II (Pohle, Jena, 1893–1903)
7. M. Hallett, *Cantorian Set Theory and Limitation of Size* (Clarendon Press, Oxford, 1984)
8. C. Ternullo, Gödel's Cantorianism, in *Kurt Gödel: Philosopher-Scientist*, ed. by G. Crocco, E.-M. Engelen (Presses Universitaires de Provence, Aix-en-Provence, 2015), pp. 413–442
9. H. Wang, *A Logical Journey* (MIT Press, Cambridge, MA, 1996)

Printed in the United States
By Bookmasters